Artificial Intelligence and Large Language Models

With the latest developments in generative artificial intelligence (AI) and large language model (LLM) applications, like ChatGPT and Gemini, millions of people have started using them in their personal and professional lives. Interest in them continues to grow, and they are becoming the go-to applications for students, academics, engineers, lawyers, designers, writers, media experts, programmers, politicians, artists, and private citizens across the globe. However, understanding these applications can be technical with precious little advice out there. This book cuts through the jargon to explain the "what," "why," and "how" of generative AI and LLM applications.

This book summarizes experiences from a variety of applications relevant to our daily lives. It brings together scientifically grounded experience reports from different perspectives into a one-stop title. With an international outlook and featuring global case studies, it offers insights into the beneficial use of LLMs and generative AI including future trends for how the journey of using such amazing tools will continue in the future. This book details the history of AI, reviews LLM literature, and covers neural computing and the technical foundation of LLMs. Furthermore, the use and control of LLMs in education are covered. Through reading this book, the reader will develop a full understanding of AI and LLM applications, including knowledge of the key issues of trust, legality, security, and technical foundations.

Artificial Intelligence and Large Language Models: A Scientific Perspective will appeal to students and academics, as well as people interested in the use of LLMs and generative AI for professional or everyday tasks, including those in Human Factors, Human–Computer Interaction, Computer Science, Software Engineering, and Industrial Engineering.

The Human Element in Smart and Intelligent Systems

Series Editor: Abbas Moallem

This series aims to cover the role of the human element in all aspects of smart and intelligent systems. It will include a broad range of reference works, textbooks, and handbooks. The series will look for single-authored works and edited collections that include, but not limited to: automated driving, smart networks and devices, cybersecurity, data visualization and analysis, social networking, smart cities, smart manufacturing, trust and privacy, artificial intelligence, cognitive intelligence, pattern recognition, computational intelligence, and robotics. Both introductory and advanced material for students and professionals will be included.

Smart and Intelligent Systems: The Human Elements in Artificial Intelligence, Robotics, and Cybersecurity
Abbas Moallem

Understanding Cybersecurity Technologies: A Guide to Select Right Cybersecurity Tools
Abbas Moallem

Human Factors Analysis of 23 Cyberattacks
Abbas Moallem

Handbook of Human Factors in Cybersecurity Systems: A Human-Centered Approach
Edited by Crystal Fausett, Joseph R. Keebler, David Schuster and Elizabeth H. Lazzara

For more information on this series, please visit: please visit: https://www.routledge.com/The-Human-Element-in-Smart-and-Intelligent-Systems/book-series/CRCHESIS

Artificial Intelligence and Large Language Models
A Scientific Perspective

Edited by
Abbas Moallem, Helmut Degen, and
Stavroula Ntoa

CRC Press
Taylor & Francis Group
Boca Raton London New York

CRC Press is an imprint of the
Taylor & Francis Group, an **informa** business

Designed cover image: iStock.

First edition published
by CRC Press
2385 NW Executive Center Drive, Suite 320, Boca Raton FL 33431

and by CRC Press
4 Park Square, Milton Park, Abingdon, Oxon, OX14 4RN

CRC Press is an imprint of Taylor & Francis Group, LLC

ISBN: 9781032775128 (hbk)
ISBN: 9781032794754 (pbk)
ISBN: 9781003492252 (ebk)

DOI: 10.1201/9781003492252

Typeset in Times
by codeMantra

Contents

About the Editors

Abbas Moallem is a Consultant and Adjunct Professor in the Department of Industrial & Systems Engineering, San José State University, California, USA, where he teaches courses in human–computer interaction, cybersecurity, information visualization, and human factors. He serves as the Program Chair of HCI-CPT: The International Conference on HCI for Cybersecurity, Privacy, and Trust, and is the Chair of the International Conference on Human Factors in Cybersecurity. He has worked with, and consulted for, numerous organizations, including PeopleSoft, Oracle Corporation, Tumbleweed, Axway, Applied Materials, Lam Research, and ASM International. His consulting expertise spans a variety of industries across Europe, Canada, and the United States. He is the editor of *Human-Computer Interaction and Cybersecurity Handbook* (2024) and *Smart and Intelligent Systems* (2024), and the author of *Cybersecurity Awareness Among College Students and Faculty* (2019), *Understanding Cybersecurity Technologies* (2024), and *Human Factors Analysis of 23 Cyberattacks* (2025), all published by CRC Press. Additionally, he is the editor of a book series from CRC Press titled *The Human Element in Smart and Intelligent Systems.*

Helmut Degen is a Senior Key Expert for User Experience at Siemens Corporation in Princeton, NJ, USA. His research focuses on the intersection of trust, value, and efficiency in human–computer interaction, with a particular emphasis on explainable artificial intelligence (XAI). At Siemens, he leads initiatives that explore how AI systems can be made trustworthy and user-centric. He also serves as co-chair of the annual international conference "AI in HCI," affiliated with the HCI International Conference. He holds a Master of Science degree from the Karlsruhe Institute of Technology, Germany. He earned his Ph.D. in Information Science from the Freie Universität Berlin, Germany.

Stavroula Ntoa is a Computer Scientist specializing in Design for All, software accessibility, usability engineering, and User Experience (UX) research and design. She is a principal researcher at the HCI Laboratory of the Institute of Computer Science of the Foundation for Research and Technology—Hellas (ICS-FORTH), Greece, leading the accessible UX research and design activities of the lab. Her areas of research interests focus on design for all and universal access of modern interactive technologies, adaptive and intelligent interfaces, as well as inclusiveness and user experience research in intelligent and artificial intelligence environments. She serves as co-chair of the annual International Conference "AI in HCI," affiliated with the HCI International Conference.

Contributors

Waleed Akhtar
School of Marketing and Communication
University of Vaasa
Vaasa, Finland

Danial Amin
School of Marketing and Communication
University of Vaasa
Vaasa, Finland

Shima Asaadi
Digital Technology and Innovation Center
Siemens Healthineers
Princeton, New Jersey

Joerg Beringer
ProContext Inc.
Redwood City, California

Wojciech Cellary
Institute of Applied Sciences
WSB Merito University
Poznan, Poland

Carrie Ching
Independent Researcher
Düsseldorf, Germany

Lance Chong
Faculty of Fine Arts
University of Lethbridge
Alberta, Canada

Suzhen Duan
Learning Technologies, Design, & School Library Media, College of Education
Towson University
Towson, Maryland

Thomas Geis
ProContext Consulting GmbH
Cologne, Germany

Rihkiya Gosh
Digital Technology and Innovation Center
Siemens Healthineers
Princeton, New Jersey

Hoda Harati
Learning Technologies, Design, & School Library Media, College of Education
Towson University
Towson, Maryland

Pei-Hsuan Hsieh
College of Informatics
National Chengchi University
Taipei, Taiwan

Bernard J. Jansen
Qatar Computing Research Institute
Hamad Bin Khalifa University
Doha, Qatar

Ilkka Kaate
Department of Marketing and International Business
University of Turku
Turku, Finland

Sanjeev Kumar Karn
Digital Technology and Innovation Center
Siemens Healthineers
Princeton, New Jersey

Khalid Kattan
Department of Computer Science
University of Michigan
Dearborn, Michigan

Kitty Kioskli
Trustilio
Amsterdam, The Netherlands

Tina Korani
School of Journalism and Mass Communication
San Jose State University
San Jose, California

André Frank Krause
Faculty of Communication and Environment
Rhine-Waal University of Applied Sciences
North Rhine-Westphalia, Germany

Prabhat Kumar
US DEVCOM Army Research Laboratory
Adelphi, Maryland

Carsten Lanquillon
Business Information Systems
Heilbronn University of Applied Sciences
Baden-Württemberg, Germany

George Margetis
Institute of Computer Science
Foundation for Research and Technology
Hellas (FORTH), Greece

Rebecca McNulty
Center for Distributed Learning
University of Central Florida
Orlando, Florida

Mark Mittrick
DEVCOM Army Research Laboratory
Adelphi, Maryland

Mahnaz Moallem
Learning Technologies, Design, & School Library Media, College of Education
Towson University
Towson, Maryland

Mark Nuppnau
Department of Computer Science
Wayne State University
Detroit, Michigan

Abraham Moore Odell
US Army Research Lab ARL
Aberdeen Proving Ground (APG)
Adelphi, Maryland

Ipek Ozkaya
Software Engineering Institute
Carnegie Mellon University
Pittsburgh, Pennsylvania

Ming Qian
Charles River Analytics
Cambridge, Massachusetts

Adrienne Raglin
Army Research Laboratory, US
Aberdeen Proving Ground (APG)
Adelphi, Maryland

Robert Gene Reynolds
Department of Computer Science
Wayne State University
Detroit, Michigan
University of Michigan
Ann Arbor, Michigan

John Richardson
DEVCOM Army Research Laboratory
Aberdeen Proving Ground (APG)
Adelphi, Maryland

John E. Robert
Software Engineering Institute
Carnegie Mellon University
Pittsburgh, Pennsylvania

Joni Salminen
School of Marketing and Communication
University of Vaasa
Vaasa, Finland

Wojciech Samek
Department of Artificial Intelligence
Fraunhofer Heinrich Hertz Institute
Berlin, Germany
Department of Electrical Engineering and Computer Science
Technical University of Berlin
Berlin, Germany

Douglas Schmidt
Dean of Computing, Data Sciences & Physics
William & Mary
Williamsburg, Virginia

Bjoern W. Schuller
Chair of Health Informatics (CHI), TUM School of Medicine and Health
Technical University of Munich (TUM)
Klinikum rechts der Isar, Munich, Germany

Yao Sun
Department of Social Sciences and Humanities
New Jersey Institute of Technology
Newark, New Jersey

Andreas Triantafyllopoulos
Chair of Health Informatics (CHI), TUM School of Medicine and Health
Technical University of Munich
Munich, Germany

Iosif Tsangko
Chair of Health Informatics (CHI), TUM School of Medicine and Health
Technical University of Munich
Munich, Germany

Ullas Ullaskrishnan
Digital Technology and Innovation Center
Siemens Healthineers
Princeton, New Jersey

Krzysztof Walczak
Department of Information Technology
Poznań University of Economics and Business
Poznań, Poland

Lijing Wang
Data Science
New Jersey Institute of Technology
Newark, New Jersey

Carsten Wittenberg
School of Applied Artificial Intelligence (SAAI)
Heilbronn University OAS
Heilbronn, Germany

Introduction

Generative artificial intelligence (AI), large language models (LLMs), and applications such as ChatGPT, Gemini, Claude, and, in early 2025, DeepSeek continue to spread worldwide. Along with their massive usage by people and organizations come issues such as their impact on numerous domains, such as education, privacy, and trust, to name a few. With the propagation of these technologies, scientific research has also started to grow.

At the beginning of this wave, available books on LLMs and generative AI were very technical, hard to read, and either too general or too specific. For instance, some books focus only on how to write prompts. When we started working on this book, there was no go-to book that summarized experiences from various applications relevant to our daily lives. Thus, we began putting this book together. Its objective is to provide an overview of many research areas related to these technologies.

Also, a growing number of people from different disciplines, including students, academics, engineers, lawyers, designers, writers, media experts, programmers, politicians, artists, and private citizens across the globe, would like to know more about these technologies. Consequently, our effort was focused on making this book easy to read and understand by combining scientific-based experience reports from different perspectives and trying to fill a gap in the current book landscape. It allows readers worldwide to gain insights into the beneficial use of LLMs and generative AI. This book also hints at how the journey of using such excellent tools may continue.

The topics covered in this book are organized into 13 chapters. In each chapter, field experts provide a review, critical updates, and use cases about the research in generative AI and LLMs.

The first two chapters introduce key technical terms and the technical foundation of LLMS and generative AI.

Chapter 1—Exploring the Artificial Intelligence Landscape

This chapter traces AI's evolution since Turing's 1950 vision and the 1956 Dartmouth conference until today, highlighting important breakthroughs through its cycles of progress and stagnation. It also discusses the topic of explainability, explores societal impacts, and discusses future prospects, including artificial general intelligence.

Chapter 2—Neural Computing for Advanced Natural Language Understanding and Generation

This chapter introduces the technical foundation of LLMs and generative AI. It introduces principles, technical concepts, and architectures that enable the capabilities that make LLMs so powerful and successful (note: revise this description after Robert has submitted the chapter).

Starting with Chapter 3, this book focuses on the use of LLMs for less technical professions and use cases.

Chapter 3—LLMs in Cultural Heritage

This chapter focuses on the use of LLMs in cultural heritage practice. This chapter describes their use for different phases of cultural heritage, including archiving, understanding, educating, and preserving cultural heritage artifacts and history. This chapter highlights how LLMs can help to address different audiences, from children to research professionals.

Chapter 4—Large Language Models (LLMs) in Education: Applications, Challenges, and Use Cases

This chapter elaborates on the role of LLMs in higher education, particularly learning design, learning technologies, and educational research. It provides examples of how LLMs are adopted in higher education from an educator's perspective to improve high-quality learning experiences, personalized learning, and predicting learning performances.

Chapter 5—Healthcare in the Age of LLMs: Perspectives on Radiology

This chapter focuses on LLMs in healthcare. It explores their usage by healthcare professionals, in particular, radiologists, and shows examples of how LLMs can support the workflow of radiologists without compromising the quality of outcomes for patients. This chapter includes diagnostics and standardized reporting examples.

Chapter 6—The Role of Large Language Models in Media Content Creation

This chapter describes examples of how LLMs are used for the creation of media content. Since the use cases take place in the context of education, this chapter not only describes the impact of LLMs on creative content generation but also how LLMs can be used effectively when creating design content for design students.

Starting with Chapter 7, this book focuses on more technical and software-focused topics that use LLMs.

Chapter 7—Transforming Software Engineering and Software Acquisition with Large Language Models

This chapter describes the use of LLMs in the area of software engineering. It discusses how LLMs are used for all major phases of software engineering, including requirements engineering, software architecture and design, verification, and validation. This chapter highlights use cases where LLMs benefit software engineering practices.

Chapter 8—Detecting and Correcting Software Assurance Document Incompleteness, Inconsistencies, and Discrepancies using LLMs

This chapter focuses on LLMs and their potential to enhance software engineering, especially in high-stakes fields like healthcare and aerospace. This chapter examines LLMs' capabilities in detecting document incompleteness, inconsistencies, and discrepancies, reviewing current methods and illustrating their application through a drone avionics case study.

Chapter 9—Large Language Models: Cybersecurity, Privacy, and Trust

This chapter focuses on cross-cutting topics such as security, privacy, ethics, and trust. It describes fundamental concepts, such as jailbreaking, prompt injection, social engineering, information breach, manipulation of users' prompts, and others.

Chapter 10—Large Language Models (LLMs) in Human-Computer Interaction: Using LLM-Generated Personas to Model Everything from Minority Views to Entire Ecosystems

This chapter explores the impact of LLMs in human–computer interaction, focusing particularly on the development of personas as a means to understand and model target users, increasing empathy among decision-makers and supporting human-centered design. It explores how LLM-generated personas can be used to model minorities and marginalized viewpoints, as well as entire ecosystems.

Chapter 11—Theoretical Foundations of Human Factors and Trustworthiness in Large Language Models (LLMs)

This chapter examines human factors and trust in LLMs, integrating insights from HCI, psychology, and socio-technical systems. It explores explainability, transparency, and user alignment, and analyzes how cognitive load, anthropomorphism, and social cues affect trust. The chapter offers a conceptual foundation for ethical and effective design.

Chapter 12—How do AI and LLMs Change our Lives? Reflections and Outlook

This last chapter is the result of multiple workshop sessions, exploring current challenges and future direction for the use of LLMs across different professions and disciplines. The workshops explored three key questions, which are analyzed in this chapter, namely, which parts of our lives have already changed, which are going to be changed, and what are the anticipated future challenges.

Collectively, these chapters reflect the complexity and richness of current research and practices in the area of LLMs, highlighting also critical questions and challenges. We invite readers to delve into each topic, draw connections across chapters, and reflect on their own roles and attitudes as users, practitioners, or researchers in the field of large language models and generative AI.

human-centered design. It explores how LLM-generated personas can be used to model user needs and marginalized viewpoints, as well as ethical considerations.

Chapter 11 – Theoretical Foundations of Human Factors and Trustworthiness in Large Language Models (LLMs)

This chapter examines human factors and trust in LLMs, integrating insights from HCI, psychology, and socio-technical systems. It explores explainability, transparency, and user calibration, and analyzes how cognitive load and anthropomorphic and social cues affect trust. The chapter offers a conceptual foundation for ethical and effective design.

Chapter 12 – Towards AI and LLM-Centred Futures? Reflections and Outlooks

This last chapter is the result of multiple workshop sessions exploring current challenges and future directions for the use of LLMs across different professions and disciplines. The workshops explored three key questions, which are analysed in this chapter, namely, which parts of our lives have already changed, which are going to be changed, and what are the anticipated future challenges?

Collectively, these chapters reflect the complexity and richness of current research and practices in the area of LLMs, highlighting also critical questions and challenges. We invite readers to delve into each topic, draw connections across chapters, and reflect on their own roles and contexts as users, practitioners or researchers in the field of large language models and generative AI.

1 Exploring the Artificial Intelligence Landscape

George Margetis and Wojciech Samek

1.1 INTRODUCTION

Humans' evolutionary nature is harnessed with their continuous endeavor to create tools that could make their everyday lives easier. From the invention of the wheel to the complex apparatuses of Archimedes and the conquest of space, creative thinking is the foundational human skill that separates them from other living beings and is also their driving leverage to overcome challenges and improve. Since humans realized their capabilities and superiority in the world, they have had an innate need to create advanced machines able to address complex problems and think. The humanoid Talos, the mythical bronze giant created by Hephaestus, designed to protect Crete, Daedalus who tried to breed life to his statues using quicksilver to install a voice or to make them move, or even more the early mechanical automata designed by inventors like Hero of Alexandria including steam-powered devices and self-operating machines are some of the examples in Greek Mythology that underscore the humans' desire to create machines in their image and likeness. Of course, there is much historical evidence of actual devices that could solve difficult problems and perform intricate tasks. The Antikythera Mechanism, probably the first computer in human history, could predict astronomical positions and eclipses with high accuracy, and Al-Jazari's programmable water clocks and mechanical devices, like a robotic musician ensemble, are some of the iconic first computational devices that paved the way for computational autonomy.

After the Industrial Revolution, there was an unprecedented eruption of technological advancements sparking the generation of computing devices and machines able to do complex tasks and calculations. Charles Babbage's analytical engine, the first general-purpose mechanical computer concept, which could theoretically be programmed to solve any problem, using punched cards, or the Turing machine, a theoretical model for computation and decision-making, formed the basis of computer science and the creation of general-purpose computers, capable of performing a variety of tasks by programming, e.g., ENIAC.

However, the real major advancements in artificial intelligence (AI) and modern computing have been conducted since the 1950s. In 1966, Shakey (Nilsson, 1984) appeared to be the first general-purpose mobile robot that could reason on its decisions and actions and understand and execute complex tasks, by analyzing commands and breaking them down into basic chunks by itself. MYCIN was an AI system, utilizing AI approaches, for assisting physicians with clinical decision-making (Shortliffe, 1976). Deep Blue (Campbell et al., 2002) was the first chess machine that defeated

DOI: 10.1201/9781003492252-1

the then-reigning World Chess Champion Garry Kasparov in a six-game match in 1997. Its success was based on several factors, including a single-chip chess search engine, a massively parallel system with multiple levels of parallelism, a strong emphasis on search extensions, a complex evaluation function, and the effective use of a Grandmaster game database. If someone thinks about it, Deep Blue was the first computer that mutatis mutandis harnessed the fundamental principles that AI systems rely upon nowadays: high-quality data, advanced algorithms, computational power, and efficient training.

In the last decades, AI has been taken to another level, exposing profound capabilities that made many people think that we are close to creating the perfect AI capable of performing intellectual tasks equivalent to a human being, to reason and learn, as well as adapt across a broad domain spectrum. Although we are still not there, we should consider that current AI technologies, such as large language models (LLMs), have a profound impact on our everyday life, simulating human-like reasoning, creativity, and communication, tending to be irreplaceable tools for everyone, although entailing big concerns and challenges for our future.

In this respect, this chapter aims to explore the AI landscape, unraveling the evolution of AI, and concluding with the current breakthroughs that shape AI as it stands today. It underpins AI's great capabilities but also unearths the ethical, legal, and societal challenges AI brings along with its potential. Moreover, this chapter emphasizes the prominent need for establishing trustworthy and explainable AI (XAI) that will alleviate the harmonious symbiosis of humans and AI, always for the benefit of the former. The chapter concludes with the authors' perspective on the future of AI, as well as potential risks and limitations that might occur.

1.2 THE HISTORY OF AI

Although some can claim that the philosophical foundations motivating people to try to build artificial thinking were laid even in Aristotle's years, having conceptualized formal logic and syllogistic reasoning, the actual groundwork for AI was laid a couple of millennia later in the mid of the 20th century. Since 1950, Alan Turing had already written about machine intelligence in his notorious paper "Computing Machinery and Intelligence," which initially appeared in the British journal, Mind, and was republished in the book "Parsing the Turing Test" (Epstein et al., 2009), in which he explored whether or not machines can do what thinkers like humans can do – and if so. Turing foresees the future development of intelligent machines and addresses potential objections, including theological, mathematical, and philosophical arguments, offering counterpoints to each. The term "Artificial Intelligence" was coined by John McCarthy, Marvin Minsky, Nathaniel Rochester, and Claude Shannon, who organized the Dartmouth Conference in 1956, inaugurating the field's formal inception (McCarthy et al., 2006). The main objective of this conference was to identify whether machines can solve problems that only humans can solve by using symbols and formal language, forming abstractions, and concepts. In the same year, the Logic Theorist appeared, acclaimed as the first AI computer program that could autonomously solve complex problems using reasoning based on people's heuristic approaches (Newell & Simon, 1956). Logic Theorist faced thinking as processing abstract and amodal symbols in short-term and long-term memories that guide the program behavior. These symbols represent a hierarchical knowledge in the form of events affecting other information processes and facilitating the overall thinking process based on heuristics to solve complex problems efficiently, working backward from a theorem to derive valid inferences leading to an axiom. In the early years of AI, research efforts primarily focused on framing theories influenced by mathematical thinking, aiming to define strict formal languages and identify abstraction symbols that computers could use to solve complex problems that only humans could do until then. This period is often called the era of symbolic AI. However, these approaches were gradually abandoned as nonpractical. It is worth mentioning that many years after McCarthy, one of the fathers of AI, admitted that computers cannot conquer autonomous human-level intelligence unless they are capable of dealing with common sense, elaborating on seven unconquerable barriers, at least by the current technology (McCarthy, 2007).

The transition from symbolic AI to a new era was divided by the AI winter of the '70s, which resulted from a variety of reasons including funding cuts, criticism of AI's progress and overpromised results, the limited computational power of that period, and divisions within the AI community (Hendler, 2008). However, in the late 80s, there was the dawn of a new AI era, shifting to new, more efficient research approaches. The dawn of machine learning had come, when AI researchers moved from logic-based approaches to data-driven methods. This included algorithms that learned patterns from data rather than relying solely on pre-defined rules. A breakthrough for machine learning flourishing was the back-propagation learning process that revived neural networks, which is now a major topic of machine learning, enabling their practical application, and leveraging their performance and accuracy (Rumelhart et al., 1986). However, the preference of the research community for neural networks was a limited option until the early 2000s, due to the high demand for computational power, which was not available at that time. Conversely, AI researchers focused on other machine learning approaches, such as Support Vector Networks (Cortes & Vapnik, 1995), Bayesian Networks (Pearl, 2014), and Case Base Reasoning (Kolodner, 1992).

As already mentioned, neural networks came to the upfront of AI research in the 2000s. The prerequisites for such approaches to be effective have begun to mature and flourish. Advances in computing, for example, the abundance of Graphics Processing Units and the availability of massive datasets, paved the way. Moreover, big tech companies like Google, Microsoft, and OpenAI became central players, fueling the AI community with exceptional achievements. A big breakthrough, for example, was the improvement of digital image and video analysis. Until the early 2000s, machine learning had made considerable steps, but not leaps, regarding image classification. However, AlexNet (Krizhevsky et al., 2012) came to stir the waters in 2012, achieving unsurpassable classification results in comparison with similar works in image classification, consolidating the convolutional neural networks (CNN) as the state-of-the-art approach for this research domain, but most importantly, paving the ground for more advanced image analysis. For example, Ren et al. (2015) moved forward with image and video analysis by introducing the Faster Recurrent CNN approach, which was capable not only of detecting the different classes but also of identifying their bounding boxes existing in the analyzed image. Faster R-CNN constituted the basis for more advancements in image segmentation, i.e., detection of classes, for Mask R-CNN (He et al., 2017), an approach that was capable not only of detecting different classes in an image but also different instances of the same class as well all the pixels occupied in the image (mask). The works that followed and improved the instance segmentation of images and videos opened the horizon of systems autonomy in various fields and application domains, including autonomous driving, medical imaging, sports analysis, agriculture and environment monitoring, industrial automation, and many more.

However, in the second decade of this millennium, AI champions are undoubtedly generative AI and LLMs. These technologies had a considerable impact on society, similar to of DeepMind's AlphaGo (Silver et al., 2017) victory over Go game masters, proving "...that the underlying neural networks could be applied to complex domains, while the use of reinforcement learning showed how machines can learn to solve incredibly hard problems for themselves, simply through trial-and-error." as mentioned on AlphaGo's web page.[1] The spearheads of AI's breakthroughs for the 2000s are generative AI and LLMs, at least at the time this chapter is being written. OpenAI's ChatGPT has come to change the world as we know it now.

1.3 CURRENT AI BREAKTHROUGHS

This section highlights recent achievements that have shaped developments in various research fields and created significant potential for the future of AI.

1.3.1 Machine and Deep Learning

Starting with machine and deep learning as the most general topics, the game changer was indisputably the work of Vaswani et al. (2017), "Attention is all you need." In their work, the authors introduce transformers, a simple neural network architecture, based solely on attention mechanisms, i.e., the capability to compute the relevance of each token in a sequence to every other one, hence capturing contextual relationships. Transformers hand out the need for recurrence and convolution approaches entirely, which were state-of-the-art until then. The impact of transformers is profound today, leading the way in many different AI fields. For example, transformers revolutionized natural language processing, which was the initial research topic for their inspiration, by enabling new models with language understanding and language generation. Google's BERT (Devlin et al., 2019) is a transformer-based pioneering model that founded state-of-the-art research in the field. BERT can execute a wide range of tasks, such as question-answering and language inference, without substantial task-specific architecture modifications. BERT was the spark that ultimately led to the creation of LLMs, as discussed later. Another research domain affected by transformers is Computer Vision and, particularly, image analysis. Seamlessly, to NLP, transformers tend to replace previous approaches. Dosovitskiy et al. (2020) introduced Vision Transformers, opening the way for their adoption as a predominant approach for image classification, as well as semantic and instant segmentation, e.g., Swin Transformer (Liu et al., 2021), DETR (Carion et al., 2020), and SegFormer (Xie et al., 2021).

Multimodal AI, i.e., the use of text, images, and other modalities for unified understanding and generation, was also a research domain that flourished in the latest years due to the transformer revolution. CLIP (Radford et al., 2021) was a breakthrough in image analysis for visual summarization, capable of being instructed in natural language to predict the most relevant text snippet, given an image, without directly being optimized for the task. CLIP achieves performance on ImageNet's "zero-shot" evaluation that rivals the original ResNet50, all without utilizing any of the original 1.28 million labeled samples, effectively addressing significant challenges in computer vision.

1.3.2 Reinforcement Learning

Reinforcement learning (RL), although not a new concept, as it has been studied and developed during the last 70 years, still provides great achievements as it continuously evolves. RL has also been influenced by transformers and their exceptional performance, especially in improving sequential decision-making and long-term dependency modeling. For example, Decision Transformer (Chen et al., 2021), an architecture that casts the problem of RL as conditional sequence modeling, was based on transformers to predict future actions that achieve the desired return. Even though it is simple, the Decision Transformer performs on par with or surpasses the best model-free offline RL baselines in Atari, OpenAI Gym, and Key-to-Door tasks. Inspired by the progress of large-scale language models, based on transformers, Reed et al. (2022) introduced Gato, a unified policy-based model able to learn across different tasks. To address this objective, Gato was trained on the widest variety of relevant data possible, including diverse modalities such as images, text, proprioception, joint torques, button presses, and other discrete and continuous observations and actions. Consequently, the same network with unchanged weights could play Atari games, generate image captions, engage in chat, manipulate physical blocks with a real robotic arm, and much more. The ability of transformers to capture long-term dependencies has enabled RL to model the dynamics of state-action transition as well. TransDreamer, which combines visual representation and dynamics learning for improved sample efficiency (Chen et al., 2022), and Masked World Model, which decouples representation and dynamics to enhance scalability (Seo et al., 2023) are some of the works that rely on this. Another transformer-based example is StARformer, which introduced step and sequence transformers to model state-action-reward triplets effectively, capturing complex temporal dependencies (Shang et al., 2022).

Besides the abovementioned developments, RL has shown more noteworthy achievements the recent years. The intertwining of deep neural networks with RL approaches presented exceptional performance. Deep Q-Networks (Mnih et al., 2013) achieved human-level performance in mastering Atari games by training a convolutional neural network with a variant of Q-learning, where the input consisted of raw pixels and the output was a value function estimating future rewards. Since then, several approaches based on deep neural networks appeared, culminating in AlphaGo (Silver et al., 2016) and AphaZero (Silver et al., 2018), which demonstrated the ability of these models to master complex games like Go, Chess, and Shogi without human input. Since then, the combination of deep neural networks with RL and self-play to superhuman performance led to game-mastering approaches such as OpenAI's Five (Berner et al., 2019), which was able to defeat professional Dota 2 players, based on multi-agent environments with partial observability and high complexity, or MuZero (Schrittwieser et al., 2020), which combined model-based and model-free RL to win games without knowledge of rules. Although the abovementioned endeavors focused on becoming skilled at video games, they can be easily generalized to address many other everyday life activities and research. A sound example is AlphaFold 2 (Jumper et al., 2021), an evolutionary model of AlphaGo and AlphaZero, which was able to solve the protein structure prediction, a decade-long problem, using Deep RL, changing the rules of the game in biological sciences and drug discovery by estimating protein structures with near-experimental accuracy. Today, RL plays a major role in fine-tuning LLMs, described next, to generate insightful explanations and robust reasoning. As explained by Ouyang et al. (2022), "making language models bigger does not inherently make them better at following a user's intent." In this respect, they proposed InstructGPT, an LLM based on GPT-3, which was fine-tuned to follow a broad class of written instructions, using RL from human feedback (Christiano et al., 2017). To align the new model with actual human intent, the authors hired 40 contractors to label data based on a screening test, to create a dataset of human-written demonstrations and prompts. This dataset trained supervised learning baselines, while human-labeled comparisons of model outputs were used to develop a reward model that predicts preferred outputs. This reward model acted as a reward function to fine-tune the supervised baseline using the Proximal Policy Optimization (PPO) (Schulman et al., 2017) algorithm, aligning GPT-3's behavior with the preferences of labelers and researchers rather than a broader notion of "human values." This work was foundational for the evolution and broader acceptance of the LLMs.

1.3.3 Neuro-Symbolic AI

One notable emerging trend in AI research is neuro-symbolic AI, which combines the pattern recognition capabilities of neural networks with the reasoning abilities of symbolic AI, aiming to overcome the limitations of either method alone, toward leveraging the strengths of both approaches, enabling AI systems to perform complex cognitive tasks with greater accuracy and reliability. Neuro-symbolic AI can overcome the difficulty of information extrapolation from sparse data through abstract reasoning. Neuro-symbolic methods fall into two main categories: (1) compressing structured symbolic knowledge into neural patterns for reasoning; and (2) extracting information from neural patterns to map onto symbolic knowledge for reasoning (Sheth et al., 2023). The first category includes mainly compressed knowledge graphs and formal logic-based representations, while the second consists of decoupled and intertwined neural-symbolic integrations. On an application level, neuro-symbolic approaches present features critical for their deployment in real-world scenarios, including user-explainability, domain constraints, scalability, and continual adaptation. It is worth mentioning that XAI, an important pre-requisite of AI as discussed later, and neuro-symbolic AI have a common objective, i.e., they aim for knowledge extraction and derive compact and accurate representations from complex neural networks (Garcez & Lamb, 2023); thus, for neuro-symbolic AI, explainability is an innate feature.

1.3.4 Generative AI

A landmark of recent AI achievements is undoubtedly generative AI, which has experienced remarkable advancements, leading to several influential works that have shaped modern AI. In 2014, Goodfellow et al. (2014) introduced the generative adversarial networks as a framework where two neural networks, the generator and the discriminator, compete, leading to the generation of realistic data. This work has been foundational in image and video generation. It would be a shortcoming of our narrative about AI breakthroughs if we did not mention generative autoencoders, a family of approaches co-existing with GANs, that show excelling capabilities. Notable examples of generative autoencoders are the variational autoencoders that combine probabilistic graphical models and neural networks for efficient learning on latent representations for generative tasks (Kingma & Welling, 2013), and sequence-to-sequence learning with neural networks (Sutskever et al., 2014), facilitating significant progress in machine translation and other natural language processing tasks. After transformers got into the game, generative AI approaches revolved between two main categories: image and video generation from text or image/video input and LLMs. The first category was an initial branch of GAN legacy, which very quickly shifted to more powerful models capable of providing better quality results. Ramesh et al. (2021) introduced DALL-E, a model capable of generating images from textual descriptions, demonstrating the potential of generative models in cross-modal tasks. DALL-E was used as an online service provided by OpenAI and very soon various other services, such as Midjourney or Leonardo AI, followed, extending the new market sector of AI-driven image generation. However, the breakthrough of generative AI for images and videos was the stable diffusion models introduced by Rombach et al. (2022), which can generate high-resolution image synthesis, balancing computational efficiency with high-quality output. Stable diffusion and similar approaches widened the AI-driven image and video generation further, making it one of the most flourishing today. The second main category of generative AI is LLMs, which are discussed next.

1.3.5 Large Language Models

LLMs are considered the successors of pre-trained transformer models over large-scale corpora, such as BERT mentioned before. The researchers found that extending the training of such models on even bigger datasets and scaling the model parameters to an even larger size could lead to improved model capacities. To distinguish language models based on their parameter sizes, the research community has introduced the term "large language models" for pre-trained models that are significantly large, typically containing tens or hundreds of billions of parameters. Zhao et al. (2023) classify language models, i.e., pre-trained models aiming to predict the probability of future (or missing) language tokens, in four major development stages: statistical language models (SLMs), neural language models (NLMs), pre-trained language models (PLMs), and LLMs. The evolution process of the four stages and the main achievements are depicted in Figure 1.1.

SLMs based on Markov assumptions and n-gram models were widely used in NLP and information retrieval, but suffered, however, from the curse of dimensionality,

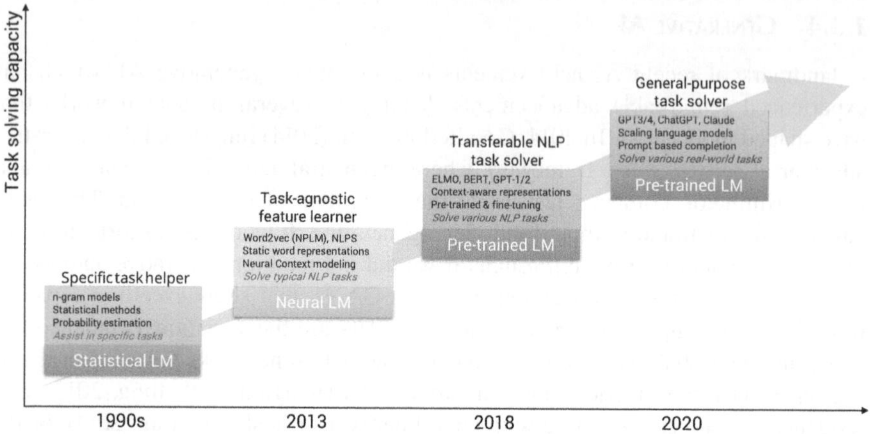

FIGURE 1.1 The evolution process of the four generations of LMs from the perspective of task-solving capacity. (Adapted from Zhao et al. 2023.)

i.e., it was almost impossible to accurately estimate high-order language models since an exponential number of transition probabilities needed to be estimated, limiting their capacity to assist in specific tasks. NLMs introduced distributed word representations using neural networks, leading to innovations like word2vec and end-to-end NLP solutions, capable of solving typical NLP tasks. PLMs, such as ELMo (Peters et al., 2018) and BERT, leveraged large-scale pre-training and fine-tuning, establishing a new paradigm for solving various NLP tasks, as already mentioned. Ultimately, LLMs like GPT-3 (Brown et al., 2020), Galactica, and Llama (Touvron et al., 2023) have shown emergent abilities as their sizes increased. This evolution in LLMs has led to advanced applications like ChatGPT, significantly increasing research interest in the field, but most importantly, the way of accomplishing everyday tasks today. The abundance of LLM agents urged most companies internationally to adopt the approach in the mainstream business workflows, providing AI assistants to their customers. In parallel, key IT players like Google, Microsoft, and Meta got early into the game, competing with OpenAI's GPT by supporting or developing their own LLMs and pertinent services. Today's significant focus of the research community is on fine-tuning existing models to improve their accuracy and reduce the phenomenon of "hallucination," i.e., general LLMs' tendency to generate either false or unverifiable information, often conflicting with established sources. In tandem, research efforts focus also on improving the capacity of LLMs, especially in their ability to reason and solve difficult human problem solutions, such as mathematical problems. In this respect, DeepSeek-R1 (Guo et al., 2025) recently showed impressive reasoning capabilities, designed to validate its own output, enhancing accuracy in subjects such as math, science, and fact-checking. This approach offers a level of precision that many AI models, including o3-mini from OpenAI, Llama 3 from MetaAI, and Gemini 2.0 Flash from Google, struggle to achieve.

Table 1.1 summarizes the most impactful works discussed in this section, in the presented domains, highlighting the AI breakthroughs that they introduced or significantly contributed to.

TABLE 1.1
AI Breakthroughs Across Impactful Publications

Work Title	References	Domain	Breakthroughs
Deep Q-networks (DQN)	Mnih et al. (2013)	RL	Combined deep learning with Q-learning, achieving human-level performance in games.
Variational autoencoders (VAEs)	Kingma and Welling (2013)	GenAI	Enabled probabilistic generative modeling, learning structured latent representations.
Sequence to sequence learning with neural networks	Sutskever et al. (2014)	GenAI	Established machine translation and sequence-based NLP tasks using deep learning.
AlphaGo	Silver et al. (2016)	RL	First AI to defeat a Go champion using deep reinforcement learning and Monte Carlo Tree Search.
Attention is all you need	Vaswani et al. (2017)	ML	Introduced the Transformer architecture, revolutionizing NLP by replacing RNNs and CNNs.
Proximal policy optimization algorithms	Schulman et al. (2017)	RL	Improved policy gradient optimization, making RL more stable and enabling the use of sampling data through interaction with the environment.
AlphaZero	Silver et al. (2018)	RL	Generalized self-play RL, surpassing human and previous AI strategies in Chess, Go, and Shogi.
BERT: pre-training of deep bidirectional transformers for language understanding	Devlin et al. (2019)	NLP	Enabled context-aware language understanding, improving NLP tasks like QA, translation, and sentiment analysis.
An image is worth 16 × 16 words: transformers for image recognition at scale	Dosovitskiy et al. (2020)	Computer V ision	Proved that Vision Transformers (ViTs) can outperform CNNs in image classification without locality biases.
End-to-end object detection with transformers (DeTr)	Carion et al. (2020)	Computer Vision	Introduced end-to-end object detection using Transformers, eliminating the need for anchor boxes and post-processing.
GPT-3: language models are few-shot learners	Brown et al. (2020)	LLMs	Showcased few-shot learning with large-scale pre-training, drastically reducing the need for task-specific fine-tuning.
Swin transformer: hierarchical vision transformer using shifted windows	Liu et al. (2021)	Computer Vision	Developed a hierarchical vision transformer with shifting windows, improving scalability in classification, detection, and segmentation.

(*Continued*)

TABLE 1.1 (*Continued*)
AI Breakthroughs Across Impactful Publications

Work Title	References	Domain	Breakthroughs
CLIP: learning transferable visual models from natural language supervision	Radford et al. (2021)	Multi-modal AI	Enabled zero-shot learning by aligning vision and text in a shared embedding space, improving image-text understanding.
DALL-E: zero-shot text-to-image generation	Ramesh et al. (2021)	GenAI	Demonstrated text-to-image generation using Transformers, producing creative and high-quality visuals.
AlphaFold 2	Jumper et al. (2021)	AI for Science	Solved the protein folding problem, predicting structures with near-experimental accuracy.
Zero-shot text-to-image generation	Ramesh et al. (2021)	GenAI	Applied Transformer-based diffusion models for high-quality text-to-image synthesis. DALE-E model.
Stable diffusion	Rombach et al. (2022)	GenAI	Pioneering latent diffusion models, allowing efficient, high-resolution text-to-image generation.
InstructGPT: training language models to follow instructions with human feedback	Ouyang et al. (2022)	LLMs	Used Reinforcement Learning from Human Feedback (RLHF) to make AI more aligned with human intent.
Neuro-symbolic AI: the 3rd wave	Garcez and Lamb (2023)	Neuro-Symbolic AI	Integrated deep learning with symbolic reasoning, improving explainability and logic-based learning.
LLaMA: open and efficient foundation language models	Touvron et al. (2023)	LLMs	Provided an open-source, efficient alternative to proprietary LLMs, improving accessibility and research.
DeepSeek-R1	Guo et al. (2025)	LLMs	Enhanced reasoning capabilities using RL and Chain of Thought explanations.

1.4 ETHICAL, LEGAL, AND SOCIETAL CHALLENGES

Due to the widespread use of AI systems in everyday life today, concerns about their use and potential hazards for humanity have emerged in several layers, including international organizations, the United Nations, the European Union, and individual countries. As mentioned by Correa et al. (2023), since 2018, an *AI Ethics boom* has been noted, underpinning the high concerns about ethical and legal aspects of AI use globally. Furthermore, there has been a considerable shift in focus regarding AI-related ethical principles from 2014 to the present day. The same survey mentions that in 2014, the most cited principles, by pertinent research works, were Fairness, Reliability, and Dignity, while since 2016, the emphasis has shifted to Accountability, Beneficence, and Privacy. However, since 2018, Transparency (XAI, Mechanistic Interpretability) has become the dominant concern, highlighting the need for humans to understand AI. In parallel, the urge for human-centered AI has become a dominant prerequisite for the development of AI systems, with relevant frameworks and pertinent guidelines proposed (Amershi et al., 2019; *IBM Design for AI*, 2022; *PAIR. People + AI Guidebook.*, 2019; Margetis et al., 2021; Riedmann-Streitz et al., 2025; Shneiderman, 2020; Xu, 2019), underscoring the effort of putting human-in-the-loop during the whole AI systems' implementation lifecycle, safeguarding humane results that are useful, reliable, and trustworthy.

With the explosion of LLM agents, the necessity for framing their use in alignment with human values, as reflected on ethical, legal, and societal considerations, has become prominent. Although these technologies present unprecedented capacity and capabilities, they still present significant drawbacks and limitations that might drive maleficent use. The major pitfalls of LLMs are bias, information hallucination, lack of explainability, reasoning errors, and their susceptibility to prompt injection (Hadi et al., 2023). Bias is an innate problem of LLMs, due to their training on datasets that contain prejudices related to several factors, such as race, gender, or socioeconomic status. As mentioned by Schramowski et al. (2022), pre-trained models designed to imitate human language may perpetuate unfairness, leading to discriminatory or inaccurate outputs in areas like politics and law. Bias in language models can be imposed in four primary ways: (1) training data bias, where models inherit prejudices from datasets; (2) user interaction bias, where biased user input reinforces prejudices; (3) algorithmic bias, where optimization metrics prioritize engagement over fairness; and (4) contextual bias, where limited training data on specific cultures or topics results in skewed responses. These biases can influence AI-generated recommendations, raising ethical concerns about fairness and accountability.

As already mentioned, information hallucination occurs when an LLM provides factually incorrect fictional content. This issue happens mainly because of gaps in the LLMs' knowledge, urging them to make assumptions that are based on learned patterns, and it can be particularly problematic in sensitive applications where accuracy is critical. Hallucinations arise from factors such as training processes, dataset limitations, and model architecture. LLMs sometimes prioritize fluent or engaging outputs over factual accuracy. Considerable research effort is put into exploring mitigation strategies, including reality grounding, expanding training datasets, and incorporating verifiable, fact-checkable methods (Tonmoy et al., 2024). As LLMs continue to advance in their ability to write human-like text, a key challenge remains their tendency to hallucinate, generating content that appears factual but is ungrounded. This issue of hallucination is arguably the biggest hindrance to safely deploying these powerful LLMs into real-world production systems that impact people's lives. The journey toward widespread adoption of LLMs in practical settings heavily relies on addressing and mitigating hallucinations. Unlike traditional AI systems focused on limited tasks, LLMs have been exposed to vast amounts of online text data during training. While this allows them to display impressive language fluency, it also means they are capable of extrapolating information from the biases in training data, misinterpreting ambiguous prompts, or modifying the information to align superficially with the input. This becomes hugely alarming when we rely on language generation capabilities for sensitive applications, such as summarizing medical records, financial analysis reports, etc. Several techniques have been reported in the literature to mitigate hallucination in LLMs. Notable among these are retrieval-augmented generation (Lewis et al., 2021), Knowledge Retrieval (Varshney et al., 2023), CoNLI (Lei et al., 2023), and CoVe (Dhuliawala et al., 2023). Furthermore, we introduce a detailed taxonomy categorizing these methods based on various parameters, such as dataset utilization, common tasks, feedback mechanisms, and retriever types. This classification helps distinguish the diverse approaches specifically designed to tackle hallucination issues in LLMs. Additionally, we analyze the challenges and limitations inherent in these techniques, providing a solid foundation for future research in addressing hallucinations and related phenomena within the realm of LLMs. However, further research is needed to create advanced models that can differentiate between facts and assumptions.

The explainability of LLMs, comprising hundreds of billions of parameters, remains a prominent challenge due to their intricate neural architectures. The enormous number of interconnected parameters complicates decision-making processes, almost making it infeasible, and raises concerns about transparency, implicit biases, and accountability. While the attention mechanisms of their architecture offer partial insights, they do not fully explain how LLMs generate specific outputs. This lack of interpretability raises trust and ethical concerns, particularly in high-stake domains like healthcare and finance, where explainability is crucial for responsible deployment.

A big challenge for LLMs is their reasoning capability, urging them to make mistakes in logical assumptions, either because of ambiguities in the prompt or inherent limitations in their understanding of complex logical operations. LLMs such as GPT-4 and Claude-3 rely on pattern recognition and statistical inference instead of explicit logical structures. They can support only in-context reasoning, meaning these models draw conclusions based on information provided in prompts during a single session, presenting implicit memory, i.e., they lack memory persistency. Approaches such as Chain-of-Thought prompting enhance LLM reasoning capabilities by guiding it to generate intermediate reasoning steps before arriving at a final answer (Wei et al., 2022), or agent-based reasoning, which maintains the memory state and can dynamically interact with environments, enabling goal-driven task accomplishment and explicit decision-making processes (Li et al., 2024). Although recent research has shown promising results, for example, DeepSeek as mentioned previously, LLM reasoning remains an ongoing challenge.

A notable drawback of LLMs today is that they are prone to security vulnerabilities such as prompt injection, jailbreak exploits, and data poisoning. Benjamin et al. (2024) investigated 36 LLMs for various prompt injection attacks. Their results reveal a significant vulnerability drift among LLMs, showing that 56% of the 144 tested scenarios resulted in successful injections. This highlights the importance of considering factors such as model parameters, architecture, and intended use cases when assessing LLM security. Bowen et al. (2024) examined 24 LLMs and concluded that there is a clear scaling relationship, indicating that larger LLMs are more susceptible to data poisoning. Additionally, higher poisoning rates result in more harmful behavior in general. Thus, there is a pressing need for comprehensive security measures to protect LLMs from such adversarial manipulations.

From a legal perspective, content generated from LLMs might cause several open issues regarding intellectual property rights, content authenticity, misinformation, and regulatory compliance (Al-kfairy et al., 2024). More specifically, copyright concerns might emerge from such content, raising potential debate regarding authorship, fair use, and liability, while lack of transparency and accountability prevents content attributions and detection of misinformation. The primary countermeasures to these phenomena are regulatory frameworks, which, although they struggle to keep pace with AI advancements, necessitate proactive governance measures that balance innovation with ethical considerations, particularly in areas such as healthcare, education, and media.

Besides the impact of misinformation and deepfakes, as already discussed, a major societal concern is the potential economic displacement that might result from the widespread use of LLMs (Hagendorff, 2024). Specifically, the automation capabilities of LLMs may lead to job displacement in sectors that rely on content creation and information processing, spanning from customer services to software engineering or crowd-work platforms (Mannuru et al., 2023). Conversely, new job opportunities have already emerged such as prompt engineering and LLM fine-tuning for corporate use; however, these new opportunities cannot counterbalance the big concern that generative AI could worsen socioeconomic inequalities and cause job displacement.

Great concerns have also risen about the environmental impact of LLM-based services worldwide, with big worries about their contribution to the recent radical climate change. Risks closely related to these concerns include the high energy consumption and carbon footprint associated with training and deploying LLMs, along with additional issues such as excessive water usage and electronic waste (Rillig et al., 2023). To mitigate environmental risks, policy interventions, transparent data usage, and enhanced literacy regarding AI's environmental impact are proposed to balance these risks and benefits.

1.5 TRUSTWORTHY AND XAI

Major progress in AI has often been made through open competitions and international challenges. Prominent examples include the PASCAL Visual Object Classes (VOC) Challenge,[2] the ImageNet Large Scale Visual Recognition Challenge[3] (ILSVRC), and the Diabetic Retinopathy Detection Challenge.[4] Typically, these challenges involve the release of labelled training datasets, while separate, unlabeled test sets are reserved for model evaluation. The competitive spirit and scientific recognition of these challenges have contributed to significant innovations, e.g., the development of novel architectures like ResNet and U-Net. However, over time, it has become clear that top-performing models on test sets do not always generalize well or exhibit trustworthy behavior. On the contrary, the evaluation of models in a "black box" manner, i.e., without understanding the rationale behind their predictions, may conceal serious flaws.

The emergence of XAI as a distinct research area (Samek et al., 2019) marked a turning point and, for the first time, allowed us to uncover and visualize the internal reasoning of AI systems. Surprisingly, studies showed that even models that win international competitions and dominate leaderboards arrived at correct predictions for the wrong reasons, a phenomenon referred to as the "Clever Hans effect[5]" (Lapuschkin et al., 2019). For instance, in the VOC challenge, models classified horses not due to the animal itself, but due to the presence of a copyright watermark often found in horse images used for the challenge (both in training and test datasets). Similarly, ILSVRC models learned to associate images of wolves with snowy backgrounds. These examples highlight how spurious correlations and dataset artifacts can undermine the trustworthiness of AI systems, even when they perform well on benchmark tasks. The relevance of XAI has continued to grow, as follow-up research has shown that such artifacts are not limited to early deep learning models. Even modern models, including those trained via self-supervised or unsupervised learning (Kauffmann et al., 2025), remain susceptible to similar biases. This poses a significant risk, particularly when these models are fine-tuned for downstream applications in high-stakes domains.

Since the implications of the Clever Hans effect for the trustworthiness and reliability of AI models are significant, there is a clear need for a reliable and comprehensive approach to model transparency, auditing, and certification. Emerging regulatory frameworks, such as the EU's AI Act, and guidelines from international organizations and initiatives, e.g., IEEE, ITU, NIST, are actively addressing these challenges. Despite these efforts, some researchers take the opposite standpoint and advocate for the exclusive use of inherently interpretable models, avoiding black-box approaches altogether in high-stake applications (Rudin, 2019).

1.5.1 Three Waves of XAI

In the following, we briefly review some of the developments in the area of XAI. For that, we organize the evolution of XAI research into three conceptual waves, each expanding the scope and capability of explanation methods.

1.5.1.1 First Wave: Explaining Individual Predictions

Following the success of deep neural networks in image classification in the early 2010s, the wish to understand these models and how they arrive at their decisions quickly emerged. The focus of these first methods was to determine and visualize the contribution of individual input dimensions (e.g., pixels) to the model's prediction. The idea of using gradients to quantify prediction influence had already been proposed in earlier works (Morch et al., 1995) and was adapted to deep neural networks, e.g., Simonyan et al. (2014). The gradient shattering problem (Balduzzi et al., 2017) in neural networks, however, led to noisy and unreliable attribution maps, starting discussions about the overall faithfulness of this class of XAI approaches. Some methods addressed the issue through smoothing techniques, such as SmoothGrad (Smilkov et al., 2017), while others, like Integrated Gradients (Sundararajan et al., 2017), relied on gradient integration along a path from a baseline to the input. Additionally, Layer-wise Relevance Propagation (Bach et al., 2015) introduced an alternative backpropagation scheme designed to provide more robust and faithful attributions. In parallel, model-agnostic approaches emerged that did not require access to internal model gradients. These were based on different principles, including input perturbation

(Fong & Vedaldi, 2017), game-theoretic reasoning, e.g., Shapley values (Lundberg & Lee, 2017), and surrogate modeling using simpler, interpretable models (e.g., LIME) (Ribeiro et al., 2016). The introduction of evaluation metrics for explanations, along with comprehensive toolboxes such as Quantus (Hedström et al., 2023), complemented the development of these first-generation XAI methods. Although attribution-based explanation has become a mature and widely used approach, adapting these methods to newer model architectures, such as transformers, or to diverse data types like time series remains non-trivial. Some recent studies have started to address these challenges, e.g., Ali et al. (2022), Achtibat et al. (2024), and Vielhaben et al. (2024).

In addition to attribution-based methods, many other approaches were developed to explain individual model decisions. These include, for example, counterfactual explanations, which illustrate how an input would need to change to yield a different model decision (Verma et al., 2024), and data attribution methods, which identify training examples that most influenced the model's behavior on a given input (Koh & Liang, 2017).

1.5.1.2 Second Wave: Understanding the Model Itself

Another line of work in the XAI community focused on explaining the model itself. These methods aim to understand what internal components, such as individual neurons or layers, encode within the model. One of the early approaches in this direction was activation maximization (Erhan et al., 2009), which sought to reveal the features encoded by specific neurons. By generating synthetic images (Nguyen et al., 2016; Yosinski et al., 2015) or identifying real-world examples (Chen et al., 2020) that maximize the activation of specific neurons, these methods provided valuable insights into the model's internal representations. More advanced techniques soon followed. For instance, network dissection (Bau et al., 2020) systematically analyzed the embedding space to map neurons to interpretable concepts, while Testing with Concept Activation Vectors (TCAV) (Kim et al., 2018) introduced a way to quantify whether a pre-defined concept (e.g., stripeness) is encoded by the model and plays a role in the model's decisions. Concept Relevance Propagation (CRP) (Achtibat et al., 2023) further advanced this direction by combining global explainability (understanding what the model encodes overall) with local explainability (understanding a specific prediction) in an explorative manner, i.e., without relying on pre-defined concepts. CRP uses Layer-wise Relevance Propagation to identify important neurons or subnetworks for a given decision and then applies Relevance Maximization to visualize the concepts represented by these components using reference examples. By shifting the focus from the input space (e.g., raw pixels) to the semantically more meaningful embedding space of the model, these techniques are also referred to as concept-level explanation methods. These foundational techniques developed during the second wave have become key building blocks for the emerging research area of mechanistic interpretability. Here, the goal is to reverse-engineer models by identifying and analyzing functional subnetworks ("circuits") within them, thereby uncovering how specific computations are implemented internally.

1.5.1.3 Third Wave: Systematic Understanding

The approaches introduced in the last section do not allow full-scale, systematic exploration of models because they either rely on a human-in-the-loop to label and interpret discovered concepts (as in CRP) or depend on the availability of pre-defined concepts and labeled datasets (as in NetDissect or TCAV). The latter also limits the scope of the analysis to what the user explicitly searches for. Recent developments, such as FALCON (Kalibhat et al., 2023) and SemanticLens (Dreyer et al., 2025), overcome this shortcoming by replacing the human with a foundation model in the explanation pipeline. The latter method builds on the CRP methodology, effectively linking data (i.e., reference examples), predictions (i.e., model outputs), and internal model representations (i.e., relevance values), thereby enabling systematic exploration of this triangle without human intervention. This enables SemanticLens to (1) automatically identify and label the concepts encoded by individual neurons and subnetworks, (2) determine the functional role of each component in the model's decision-making process, and (3) relate these roles to the training data (e.g., to detect potential biases or dataset artifacts).

This ability to comprehensively and systematically investigate the model's representation and decision-making enables one to verify, e.g., whether a neural network trained for skin cancer detection adheres to the ABCDE rule, which is a widely accepted diagnostic guideline for skin cancer screening found in dermatology textbooks. Although still in its early stages, this third wave of XAI research promises to open up new ways for improved model auditing and certification, potentially enabling rigorous validation of whether AI systems are making decisions for the right reasons, which is crucial for developing reliable and trustworthy AI in sensitive domains like healthcare.

1.5.2 From Explanation to Control

In parallel to the use of XAI methods for the identification of biases, flaws, and bugs in model predictions, researchers have begun developing techniques for targeted model improvement and debugging (Weber et al., 2023). Methods such as the "right for the right reason" approach (Ross et al., 2017) or the "reveal to revise" framework (Pahde et al., 2023) were developed to actively correct model behavior in a desired and principled manner. This shift beyond explanation is particularly visible in the domain of LLMs. Here, researchers often want to gain mechanistic insight into model internals, such as the functional roles of individual attention heads, with the ultimate goal of controlling model behavior more effectively (Zheng et al., 2024). Such control is crucial for addressing challenges like hallucinations and marks an evolution in the field of XAI. Merely explaining model behavior seems to no longer be enough. As AI systems become more complex and integrated into critical applications, next-generation explanation techniques will need to enable meaningful interventions to actively steer, correct, and constrain future models.

1.5.3 Applications of XAI

Established explanation techniques, such as attribution maps, have been used in a wide range of applications. These include human–machine interaction (e.g., in the medical field), model validation and debugging, and regulatory contexts, where explanations are used to determine whether certain attributes influence model decisions. However, one of the most inspiring applications of XAI is in drug discovery. Here, researchers have reported the discovery of a novel structural class of antibiotics with the guidance of XAI methods (Wong et al., 2024). Beyond healthcare, XAI techniques have also been applied in physics (Lai et al., 2022), chess (Schut et al., 2025), and genomics (Mieth et al., 2021), offering the potential for novel scientific insights, i.e., not only to interpret what a model has learned but to discover something previously unknown. The XAI methods, especially emerging from the third wave, have great potential for full-scale, systematic exploration of the unknown. It will be exciting to observe when and how these next-generation techniques will be adopted in scientific discovery processes.

1.6 FUTURE OF AI, RISKS, AND LIMITATIONS

Having discussed the current landscape of AI, its tremendous potential for societal evolution, and the needs and limits that should be addressed and set, respectively, to avoid societal backfires from its unreasonable and improper use, we will try to sketch the future of AI and discuss potential risks and limitations in this section.

The advancements of generative AI and, particularly, the achievements of LLMs, reignited discussions about artificial general AI (AGI), i.e., the creation of thinking machines comparable to, or greater than, human beings' thinking processes. Although Ben Goertzel and Cassio Pennachin have made the term AGI popular (Goertzel, 2014), they noted that the term was first introduced by Mark Gubrud in a 1997 article examining the future of technology and its potential risks (Gubrud, 1997). AGI is the counter concept of "narrow AI" referring to the creation of systems that carry out specific "intelligent" behaviors in specific contexts (Kurzweil, 2014). Although the current achievements of AI still reside in the land of narrow AI, several efforts exist toward AGI, although still in their early stages. However, the exponential pace of technological advancement suggests that dramatic breakthroughs could occur within decades – or even years (Goertzel, 2014). Visible signs for sparking AGI from LLMs are analyzed and discussed by Bubeck et al. (2023). Specifically, the authors explored signs of AGI in GPT-4, as demonstrated by its reasoning, creativity, expertise across domains, and diverse task performance. Although still not a true AGI, future efforts should focus on defining AGI, enhancing LLMs with the necessary components, and understanding the origins of their intelligence. Albeit AGI's trajectory is uncertain, a survey at AGI-09 indicated that most researchers expect human-level AGI before 2050 (Baum et al., 2011), with some holding a more optimistic outlook. While controversial, this optimism is growing within the AGI community, which is actively working toward rapid progress.

Despite how close or far AGI is, the discussion should also revolve around the limitations and risks that come with it. A superintelligence that surpasses human thinking and cognition on several tasks can eventually undermine human existence in the context of several catastrophic scenarios (Bostrom, 2017). In other words, the AI systems might act in unintended or harmful ways, especially if they do not align with and encompass human values. Even if super-intelligent systems are laden with human values, there are risks against humans due to the phenomenon of the "instrumental convergence" theory, according to which AI agents can select to harm a human as an instrumental goal that will help them achieve their terminal goal, e.g., for ensuring self-presentation, goal integrity, or resource acquisition (Bostrom, 2012). As pessimistic as the future of AI might be, superintelligence will never acquire omniscient and superhuman abilities due to physical and computational constraints, facing limits in knowledge acquisition, prediction, and real-world experimentation, as it cannot perfectly simulate the universe or human behavior in real-time (Russell, 2023). In this respect, intelligent AI should serve humans under the model of Everything as a Service, having the potential to vastly improve human civilization by enhancing productivity, increasing global GDP, and making services like education, healthcare, and governance more accessible, while it cannot solve everything since finite natural resources and societal hierarchies will persist, requiring cultural shifts in how success and well-being are defined (abid.).

Furthermore, risks, challenges, and limitations that have already been discussed in this chapter, such as bias and fairness, explainability and transparency, security and misuse, environmental impact, and economic disruption or job displacement, will still be on the table for future AI if actions are not taken now. Hence, while anticipating the future of AI worldwide, initiatives need to be reinforced for better collaboration between AI researchers, policymakers, and other stakeholders. Efforts should be made to better understand the tensions between AI and policy methodologies while acknowledging the benefits of transdisciplinary approaches to overcoming emerging challenges. The new era of AI should bring together policymakers and the research community beyond peripheral synergies, aiming for a more integrated approach, urging AI development to employ policy considerations as an essential element, while policy-making should consider embracing human-centered principles and practices to ensure effective governance and oversight of the intricate challenges that emerge with AI.

As AI reaches a new level of capabilities, building trust with humanity becomes the fundamental objective. To achieve this, transparency, ethical development, and robust safety mechanisms must be established. AI should be explainable, fair, and aligned with human values, supported by clear regulations that ensure accountability. Prioritizing human-AI collaboration should empower users rather than replace them. Public engagement, AI literacy, and policies addressing societal concerns, such as job displacement, are essential. Reliability and long-term monitoring will further strengthen trust, ensuring that AI benefits society equitably while enhancing well-being and safety.

1.7 CONCLUSIONS

The evolution of AI, from symbolic logic systems to today's data-driven models, has profoundly reshaped technology and society. This chapter has examined the foundational milestones and technological breakthroughs, from early symbolic reasoning to neural networks, transformers, and the rise of generative AI and LLMs. It has also underscored the growing influence of AI across various fields, including vision, language, RL, and multimodal systems, while recognizing the significant ethical, legal, and societal implications of its widespread adoption. Key concerns such as fairness, transparency, misinformation, and environmental impact highlight the urgent need for explainability, trustworthy design, and human-centered frameworks. As we look ahead, the journey toward AGI remains uncertain yet alluring, requiring interdisciplinary collaboration and proactive governance. Ultimately, creating a future where AI empowers humanity without undermining it relies on our ability to design systems that are not only intelligent but also safe, ethical, and aligned with human values.

NOTES

1 AlphaGo—Google DeepMind: https://deepmind.google/research/breakthroughs/alphago/, retrieved 24 February 2025.
2 https://www.robots.ox.ac.uk/~vgg/projects/pascal/VOC/.
3 https://www.image-net.org/challenges/LSVRC/.
4 https://www.kaggle.com/competitions/diabetic-retinopathy-detection.
5 Clever Hans was a horse in the early 20th-century Germany, who was believed to solve math problems, but in really, he was responding to the body language of the person interacting with him.

REFERENCES

Achtibat, R., Dreyer, M., Eisenbraun, I., Bosse, S., Wiegand, T., Samek, W., & Lapuschkin, S. (2023). From attribution maps to human-understandable explanations through concept relevance propagation. *Nature Machine Intelligence*, *5*(9), 1006–1019. https://doi.org/10.1038/s42256-023-00711-8.

Achtibat, R., Hatefi, S. M. V., Dreyer, M., Jain, A., Wiegand, T., Lapuschkin, S., & Samek, W. (2024). AttnLRP: attention-aware layer-wise relevance propagation for transformers. In R. Salakhutdinov, Z. Kolter, K. Heller, A. Weller, N. Oliver, J. Scarlett, & F. Berkenkamp (Eds.), *Proceedings of the 41st International Conference on Machine Learning,* Vienna, Austria (Vol. 235, pp. 135–168). PMLR. https://proceedings.mlr.press/v235/achtibat24a.html.

Al-kfairy, M., Mustafa, D., Kshetri, N., Insiew, M., & Alfandi, O. (2024). Ethical challenges and solutions of generative AI: an interdisciplinary perspective. *Informatics*, *11*(3), 58. https://doi.org/10.3390/informatics11030058.

Ali, A., Schnake, T., Eberle, O., Montavon, G., Müller, K.-R., & Wolf, L. (2022). XAI for transformers: better explanations through conservative propagation. In K. Chaudhuri, S. Jegelka, L. Song, C. Szepesvari, G. Niu, & S. Sabato (Eds.), *Proceedings of the 39th International Conference on Machine Learning,* Baltimore, MD, USA (Vol. 162, pp. 435–451). PMLR. https://proceedings.mlr.press/v162/ali22a.html.

Amershi, S., Weld, D., Vorvoreanu, M., Fourney, A., Nushi, B., Collisson, P., Suh, J., Iqbal, S., Bennett, P. N., Inkpen, K., Teevan, J., Kikin-Gil, R., & Horvitz, E. (2019). Guidelines for human-AI interaction. In *Proceedings of the 2019 CHI Conference on Human Factors in Computing Systems*, Glasgow, Scotland (pp. 1–13). https://doi.org/10.1145/3290605.3300233.

Bach, S., Binder, A., Montavon, G., Klauschen, F., Müller, K.-R., & Samek, W. (2015). On pixel-wise explanations for non-linear classifier decisions by layer-wise relevance propagation. *PLoS ONE*, *10*(7), e0130140. https://doi.org/10.1371/journal.pone.0130140.

Balduzzi, D., Frean, M., Leary, L., Lewis, J. P., Ma, K. W.-D., & McWilliams, B. (2017). The shattered gradients problem: if resnets are the answer, then what is the question? In D. Precup & Y. W. Teh (Eds.), *Proceedings of the 34th International Conference on Machine Learning,* Sydney, Australia (Vol. 70, pp. 342–350). PMLR. https://proceedings.mlr.press/v70/balduzzi17b.html.

Bau, D., Zhu, J.-Y., Strobelt, H., Lapedriza, A., Zhou, B., & Torralba, A. (2020). Understanding the role of individual units in a deep neural network. *Proceedings of the National Academy of Sciences, 117*(48), 30071–30078. https://doi.org/10.1073/pnas.1907375117.

Baum, S. D., Goertzel, B., & Goertzel, T. G. (2011). How long until human-level AI? Results from an expert assessment. *Technological Forecasting and Social Change, 78*(1), 185–195. https://doi.org/10.1016/j.techfore.2010.09.006.

Benjamin, V., Braca, E., Carter, I., Kanchwala, H., Khojasteh, N., Landow, C., Luo, Y., Ma, C., Magarelli, A., Mirin, R., Moyer, A., Simpson, K., Skawinski, A., & Heverin, T. (2024). Systematically analyzing prompt injection vulnerabilities in diverse LLM architectures (version 1). https://doi.org/10.48550/ARXIV.2410.23308.

Berner, C., Brockman, G., Chan, B., Cheung, V., Dbiak, P., Dennison, C., Farhi, D., Fischer, Q., Hashme, S., Hesse, C., Józefowicz, R., Gray, S., Olsson, C., Pachocki, J., Petrov, M., Pinto, H. P. D. O., Raiman, J., … Zhang, S. (2019). Dota 2 with large scale deep reinforcement learning (version 1). https://doi.org/10.48550/ARXIV.1912.06680.

Bostrom, N. (2012). The superintelligent will: motivation and instrumental rationality in advanced artificial agents. *Minds and Machines, 22*(2), 71–85. https://doi.org/10.1007/s11023-012-9281-3.

Bostrom, N. (2017). *Superintelligence: Paths, Dangers, Strategies* (Reprinted with corrections 2017). Oxford University Press.

Bowen, D., Murphy, B., Cai, W., Khachaturov, D., Gleave, A., & Pelrine, K. (2024). Data poisoning in LLMs: Jailbreak-Tuning and scaling laws (version 5). https://doi.org/10.48550/ARXIV.2408.02946.

Brown, T., Mann, B., Ryder, N., Subbiah, M., Kaplan, J. D., Dhariwal, P., Neelakantan, A., Shyam, P., Sastry, G., Askell, A., Agarwal, S., Herbert-Voss, A., Krueger, G., Henighan, T., Child, R., Ramesh, A., Ziegler, D., Wu, J., Winter, C., … Amodei, D. (2020). Language models are few-shot learners. In H. Larochelle, M. Ranzato, R. Hadsell, M. F. Balcan, & H. Lin (Eds.), *Advances in Neural Information Processing Systems* (Vol. 33, pp. 1877–1901). Curran Associates, Inc. https://proceedings.neurips.cc/paper_files/paper/2020/file/1457c0d6bfcb4967418bfb8ac142f64a-Paper.pdf.

Bubeck, S., Chandrasekaran, V., Eldan, R., Gehrke, J., Horvitz, E., Kamar, E., Lee, P., Lee, Y. T., Li, Y., Lundberg, S., Nori, H., Palangi, H., Ribeiro, M. T., & Zhang, Y. (2023). Sparks of artificial general intelligence: early experiments with GPT-4 (version 5). https://doi.org/10.48550/ARXIV.2303.12712.

Campbell, M., Hoane, A. J., & Hsu, F. (2002). Deep blue. *Artificial Intelligence, 134*(1–2), 57–83. https://doi.org/10.1016/S0004-3702(01)00129-1.

Carion, N., Massa, F., Synnaeve, G., Usunier, N., Kirillov, A., & Zagoruyko, S. (2020). End-to-end object detection with transformers. In A. Vedaldi, H. Bischof, T. Brox, & J.- M. Frahm (Eds.), *Computer Vision—ECCV 2020* (Vol. 12346, pp. 213–229). Springer. https://doi.org/10.1007/978-3-030-58452-8_13.

Chen, C., Wu, Y.-F., Yoon, J., & Ahn, S. (2022). TransDreamer: reinforcement learning with transformer world models (version 2). https://doi.org/10.48550/ARXIV.2202.09481.

Chen, L., Lu, K., Rajeswaran, A., Lee, K., Grover, A., Laskin, M., Abbeel, P., Srinivas, A., & Mordatch, I. (2021). Decision transformer: reinforcement learning via sequence modeling. In M. Ranzato, A. Beygelzimer, Y. Dauphin, P. S. Liang, & J. W. Vaughan (Eds.), *Advances in Neural Information Processing Systems* (Vol. 34, pp. 15084–15097). Curran Associates, Inc. https://proceedings.neurips.cc/paper_files/paper/2021/file/7f489f642a0ddb10272b5c31057f0663-Paper.pdf.

Chen, Z., Bei, Y., & Rudin, C. (2020). Concept whitening for interpretable image recognition. *Nature Machine Intelligence, 2*(12), 772–782. https://doi.org/10.1038/s42256-020-00265-z.

Christiano, P. F., Leike, J., Brown, T., Martic, M., Legg, S., & Amodei, D. (2017). Deep reinforcement learning from human preferences. In I. Guyon, U. V. Luxburg, S. Bengio, H. Wallach, R. Fergus, S. Vishwanathan, & R. Garnett (Eds.), *Advances in Neural Information Processing Systems* (Vol. 30). Curran Associates, Inc. https://proceedings.neurips.cc/paper_files/paper/2017/file/d5e2c0adad503c91f91df240d0cd4e49-Paper.pdf.

Corrêa, N. K., Galvão, C., Santos, J. W., Del Pino, C., Pinto, E. P., Barbosa, C., Massmann, D., Mambrini, R., Galvão, L., Terem, E., & De Oliveira, N. (2023). Worldwide AI ethics: a review of 200 guidelines and recommendations for AI governance. *Patterns*, *4*(10), 100857. https://doi.org/10.1016/j.patter.2023.100857.

Cortes, C., & Vapnik, V. (1995). Support-vector networks. *Machine Learning*, *20*(3), 273–297. https://doi.org/10.1007/BF00994018.

Devlin, J., Chang, M.-W., Lee, K., & Toutanova, K. (2019). BERT: pre-training of deep bidirectional transformers for language understanding. In *Proceedings of the 2019 Conference of the North American Chapter of the Association for Computational Linguistics: Human Language Technologies*, Minneapolis, Minnesota (Vol. 1, pp. 4171–4186). https://doi.org/10.18653/v1/N19-1423.

Dhuliawala, S., Komeili, M., Xu, J., Raileanu, R., Li, X., Celikyilmaz, A., & Weston, J. (2023). Chain-of-verification reduces hallucination in large language models. https://doi.org/10.48550/2309.11495.

Dosovitskiy, A., Beyer, L., Kolesnikov, A., Weissenborn, D., Zhai, X., Unterthiner, T., Dehghani, M., Minderer, M., Heigold, G., Gelly, S., Uszkoreit, J., & Houlsby, N. (2020). An image is worth 16x16 words: transformers for image recognition at scale (version 2). https://doi.org/10.48550/ARXIV.2010.11929.

Dreyer, M., Berend, J., Labarta, T., Vielhaben, J., Wiegand, T., Lapuschkin, S., & Samek, W. (2025). Mechanistic understanding and validation of large AI models with SemanticLens. *Nature Machine Intelligence*. https://doi.org/10.1038/s42256-025-01084-w.

Epstein, R., Roberts, G., & Beber, G. (Eds.). (2009). *Parsing the Turing Test: Philosophical and Methodological Issues in the Quest for the Thinking Computer*. Springer. https://doi.org/10.1007/978-1-4020-6710-5.

Erhan, D., Bengio, Y., Courville, A., & Vincent, P. (2009). *Visualizing Higher-Layer Features of a Deep Network*. Technical Report 1341, Département d'Informatique et Recherche Opérationnelle, Université de Montréal.

Fong, R. C., & Vedaldi, A. (2017, October). Interpretable explanations of black boxes by meaningful perturbation. In *2017 IEEE International Conference on Computer Vision (ICCV)*, Venice, Italy. https://doi.org/10.1109/iccv.2017.371.

Garcez, A. d'Avila, & Lamb, L. C. (2023). Neurosymbolic AI: the 3rd wave. *Artificial Intelligence Review*, *56*(11), 12387–12406. https://doi.org/10.1007/s10462-023-10448-w.

Goertzel, B. (2014). Artificial general intelligence: concept, state of the art, and future prospects. *Journal of Artificial General Intelligence*, *5*(1), 1–48. https://doi.org/10.2478/jagi-2014-0001.

Goodfellow, I., Pouget-Abadie, J., Mirza, M., Xu, B., Warde-Farley, D., Ozair, S., Courville, A., & Bengio, Y. (2014). Generative adversarial nets. In Z. Ghahramani, M. Welling, C. Cortes, N. Lawrence, & K. Q. Weinberger (Eds.), *Advances in Neural Information Processing Systems* (Vol. 27). Curran Associates, Inc. https://proceedings.neurips.cc/paper_files/paper/2014/file/5ca3e9b122f61f8f06494c97b1afccf3-Paper.pdf.

Gubrud, M. A. (1997). Nanotechnology and international security. In *Fifth Foresight Conference on Molecular Nanotechnology,* Palo Alto, CA (Vol. 1).

Guo, D., Yang, D., Zhang, H., Song, J., Zhang, R., Xu, R., Zhu, Q., Ma, S., Wang, P., Bi, X., Zhang, X., Yu, X., Wu, Y., Wu, Z. F., Gou, Z., Shao, Z., Li, Z., Gao, Z., … Zhang, Z. (2025). DeepSeek-R1: incentivizing reasoning capability in LLMs via reinforcement learning (Version 1). https://doi.org/10.48550/ARXIV.2501.12948.

Hadi, M. U., Tashi, Q. A., Qureshi, R., Shah, A., Muneer, A., Irfan, M., Zafar, A., Shaikh, M. B., Akhtar, N., Wu, J., & Mirjalili, S. (2023). *A Survey on Large Language Models: Applications, Challenges, Limitations, and Practical Usage*. https://doi.org/10.36227/techrxiv.23589741.v1.

Hagendorff, T. (2024). Mapping the ethics of generative AI: a comprehensive scoping review. *Minds and Machines*, *34*(4), 39. https://doi.org/10.1007/s11023-024-09694-w.

He, K., Gkioxari, G., Dollar, P., & Girshick, R. (2017). Mask R-CNN. In *2017 IEEE International Conference on Computer Vision (ICCV),* Venice, Italy (pp. 2980–2988). https://doi.org/10.1109/ICCV.2017.322.

Hedström, A., Weber, L., Krakowczyk, D., Bareeva, D., Motzkus, F., Samek, W., Lapuschkin, S., & Hãhne, M. M.-C. (2023). Quantus: an explainable AI toolkit for responsible evaluation of neural network explanations and beyond. *Journal of Machine Learning Research*, *24*(34), 1–11.

Hendler, J. (2008). Avoiding another AI winter. *IEEE Intelligent Systems*, *23*(2), 2–4. https://doi.org/10.1109/MIS.2008.20.

IBM Design for AI. (2022). https://www.ibm.com/design/ai/.

Jumper, J., Evans, R., Pritzel, A., Green, T., Figurnov, M., Ronneberger, O., Tunyasuvunakool, K., Bates, R., Žídek, A., Potapenko, A., Bridgland, A., Meyer, C., Kohl, S. A. A., Ballard, A. J., Cowie, A., Romera-Paredes, B., Nikolov, S., Jain, R., Adler, J., … Hassabis, D. (2021). Highly accurate protein structure prediction with AlphaFold. *Nature*, *596*(7873), 583–589. https://doi.org/10.1038/s41586-021-03819-2.

Kalibhat, N., Bhardwaj, S., Bruss, C. B., Firooz, H., Sanjabi, M., & Feizi, S. (2023). Identifying interpretable subspaces in image representations. In A. Krause, E. Brunskill, K. Cho, B. Engelhardt, S. Sabato, & J. Scarlett (Eds.), *Proceedings of the 40th International Conference on Machine Learning,* Honolulu, Hawaii (Vol. 202, pp. 15623–15638). PMLR. https://proceedings.mlr.press/v202/kalibhat23a.html.

Kauffmann, J., Dippel, J., Ruff, L., Samek, W., Müller, K.-R., & Montavon, G. (2025). Explainable AI reveals clever hans effects in unsupervised learning models. *Nature Machine Intelligence*, *7*(3), 412–422. https://doi.org/10.1038/s42256-025-01000-2.

Kim, B., Wattenberg, M., Gilmer, J., Cai, C., Wexler, J., Viegas, F., & Sayres, R. (2018). Interpretability beyond feature attribution: quantitative testing with concept activation vectors (TCAV). In J. Dy & A. Krause (Eds.), *Proceedings of the 35th International Conference on Machine Learning,* Stockholm, Sweden (Vol. 80, pp. 2668–2677). PMLR. https://proceedings.mlr.press/v80/kim18d.html.

Kingma, D. P., & Welling, M. (2013). Auto-encoding variational Bayes (version 11). https://doi.org/10.48550/ARXIV.1312.6114.

Koh, P. W., & Liang, P. (2017). Understanding black-box predictions via influence functions. In D. Precup & Y. W. Teh (Eds.), *Proceedings of the 34th International Conference on Machine Learning,* Sydney, Australia (Vol. 70, pp. 1885–1894). PMLR. https://proceedings.mlr.press/v70/koh17a.html.

Kolodner, J. L. (1992). An introduction to case-based reasoning. *Artificial Intelligence Review*, *6*(1), 3–34. https://doi.org/10.1007/BF00155578.

Krizhevsky, A., Sutskever, I., & Hinton, G. E. (2012). ImageNet classification with deep convolutional neural networks. In F. Pereira, C. J. Burges, L. Bottou, & K. Q. Weinberger (Eds.), *Advances in Neural Information Processing Systems* (Vol. 25). Curran Associates, Inc. https://proceedings.neurips.cc/paper_files/paper/2012/file/c399862d3b9d6b76c8436e924a68c45b-Paper.pdf.

Kurzweil, R. (2014). The singularity is near. In R. L. Sandler (Ed.), *Ethics and Emerging Technologies* (pp. 393–406). Palgrave Macmillan. https://doi.org/10.1057/9781137349088_26.

Lai, Y. S., Neill, D., Płoskoń, M., & Ringer, F. (2022). Explainable machine learning of the underlying physics of high-energy particle collisions. *Physics Letters B*, *829*, 137055. https://doi.org/10.1016/j.physletb.2022.137055.

Lapuschkin, S., Wäldchen, S., Binder, A., Montavon, G., Samek, W., & Müller, K.-R. (2019). Unmasking Clever Hans predictors and assessing what machines really learn. *Nature Communications, 10*(1). https://doi.org/10.1038/s41467-019-08987-4.

Lei, D., Li, Y., Hu, M., Wang, M., Yun, V., Ching, E., & Kamal, E. (2023). Chain of natural language inference for reducing large language model ungrounded hallucinations. https://doi.org/10.48550/2310.03951.

Lewis, P., Perez, E., Piktus, A., Petroni, F., Karpukhin, V., Goyal, N., ... & Kiela, D. (2020). Retrieval-augmented generation for knowledge-intensive NLP tasks. *Advances in Neural Information Processing Systems, 33*, 9459–9474.

Li, X., Wang, S., Zeng, S., Wu, Y., & Yang, Y. (2024). A survey on LLM-based multi-agent systems: workflow, infrastructure, and challenges. *Vicinagearth, 1*(1), 9. https://doi.org/10.1007/s44336-024-00009-2.

Liu, Z., Lin, Y., Cao, Y., Hu, H., Wei, Y., Zhang, Z., Lin, S., & Guo, B. (2021). Swin transformer: hierarchical vision transformer using shifted windows. In *2021 IEEE/CVF International Conference on Computer Vision (ICCV),* Montreal, Canada (pp. 9992–10002). https://doi.org/10.1109/ICCV48922.2021.00986.

Lundberg, S. M., & Lee, S.-I. (2017). A unified approach to interpreting model predictions. In I. Guyon, U. V. Luxburg, S. Bengio, H. Wallach, R. Fergus, S. Vishwanathan, & R. Garnett (Eds.), *Advances in Neural Information Processing Systems* (Vol. 30). Curran Associates, Inc. https://proceedings.neurips.cc/paper_files/paper/2017/file/8a20a8621978632d76c43dfd28b67767-Paper.pdf.

Mannuru, N. R., Shahriar, S., Teel, Z. A., Wang, T., Lund, B. D., Tijani, S., Pohboon, C. O., Agbaji, D., Alhassan, J., Galley, J., Kousari, R., Ogbadu-Oladapo, L., Saurav, S. K., Srivastava, A., Tummuru, S. P., Uppala, S., & Vaidya, P. (2023). Artificial intelligence in developing countries: the impact of generative artificial intelligence (AI) technologies for development. *Information Development.* https://doi.org/10.1177/02666669231200628.

Margetis, G., Ntoa, S., Antona, M., & Stephanidis, C. (2021). Human-centered design of artificial intelligence. In G. Salvendy & W. Karwowski (Eds.), *Handbook of Human Factors and Ergonomics* (1st ed., pp. 1085–1106). Wiley. https://doi.org/10.1002/9781119636113.ch42.

McCarthy, J. (2007). From here to human-level AI. *Artificial Intelligence, 171*(18), 1174–1182. https://doi.org/10.1016/j.artint.2007.10.009.

McCarthy, J., Minsky, M. L., Rochester, N., & Shannon, C. E. (2006). A proposal for the dartmouth summer research project on artificial intelligence: August 31, 1955. *AI Magazine, 27*(4), 12–14. https://doi.org/10.1609/aimag.v27i4.1904.

Mieth, B., Rozier, A., Rodriguez, J. A., Höhne, M. M. C., Görnitz, N., & Müller, K.-R. (2021). DeepCOMBI: explainable artificial intelligence for the analysis and discovery in genome-wide association studies. *NAR Genomics and Bioinformatics, 3*(3). https://doi.org/10.1093/nargab/lqab065.

Mnih, V., Kavukcuoglu, K., Silver, D., Graves, A., Antonoglou, I., Wierstra, D., & Riedmiller, M. (2013). Playing atari with deep reinforcement learning (version 1). https://doi.org/10.48550/ARXIV.1312.5602.

Morch, N. J. S., Kjems, U., Hansen, L. K., Svarer, C., Law, I., Lautrup, B., Strother, S., & Rehm, K. (1995). Visualization of neural networks using saliency maps. In *Proceedings of ICNN'95- International Conference on Neural Networks,* Perth, Western Australia (Vol. 4, pp. 2085–2090). https://doi.org/10.1109/icnn.1995.488997.

Newell, A., & Simon, H. (1956). The logic theory machine: a complex information processing system. *IEEE Transactions on Information Theory, 2*(3), 61–79. https://doi.org/10.1109/TIT.1956.1056797.

Nguyen, A., Dosovitskiy, A., Yosinski, J., Brox, T., & Clune, J. (2016). Synthesizing the preferred inputs for neurons in neural networks via deep generator networks. In D. Lee, M. Sugiyama, U. Luxburg, I. Guyon, & R. Garnett (Eds.), *Advances in Neural Information Processing Systems* (Vol. 29). Curran Associates, Inc. https://proceedings.neurips.cc/paper_files/paper/2016/file/5d79099fcdf499f12b79770834c0164a-Paper.pdf.

Nilsson, N. J. (1984). *Shakey the Robot*. https://api.semanticscholar.org/CorpusID:106973706.
Ouyang, L., Wu, J., Jiang, X., Almeida, D., Wainwright, C., Mishkin, P., Zhang, C., Agarwal, S., Slama, K., Ray, A., Schulman, J., Hilton, J., Kelton, F., Miller, L., Simens, M., Askell, A., Welinder, P., Christiano, P. F., Leike, J., & Lowe, R. (2022). Training language models to follow instructions with human feedback. In S. Koyejo, S. Mohamed, A. Agarwal, D. Belgrave, K. Cho, & A. Oh (Eds.), *Advances in Neural Information Processing Systems* (Vol. 35, pp. 27730–27744). Curran Associates, Inc. https://proceedings.neurips.cc/paper_files/paper/2022/file/b1efde53be364a73914f58805a001731-Paper-Conference.pdf.
Pahde, F., Dreyer, M., Samek, W., & Lapuschkin, S. (2023). Reveal to revise: an explainable AI life cycle for iterative bias correction of deep models. In H. Greenspan, A. Madabhushi, P. Mousavi, S. Salcudean, J. Duncan, T. Syeda-Mahmood, & R. Taylor (Eds.), *Medical Image Computing and Computer Assisted Intervention—MICCAI 2023* (pp. 596–606). Springer Nature.
PAIR. (2019). *People + AI Guidebook*. https://pair.withgoogle.com/guidebook.
Pearl, J. (2014). *Probabilistic Reasoning in Intelligent Systems: Networks of Plausible Inference* (1. Aufl). Elsevier Reference Monographs.
Peters, M. E., Neumann, M., Iyyer, M., Gardner, M., Clark, C., Lee, K., & Zettlemoyer, L. (2018). Deep contextualized word representations. In M. Walker, H. Ji, & A. Stent (Eds.), *Proceedings of the 2018 Conference of the North American Chapter of the Association for Computational Linguistics: Human Language Technologies,* New Orleans, Louisiana (Vol. 1, pp. 2227–2237). Association for Computational Linguistics. https://doi.org/10.18653/v1/N18-1202.
Radford, A., Kim, J. W., Hallacy, C., Ramesh, A., Goh, G., Agarwal, S., Sastry, G., Askell, A., Mishkin, P., Clark, J., Krueger, G., & Sutskever, I. (2021). Learning transferable visual models from natural language supervision. In M. Meila & T. Zhang (Eds.), *Proceedings of the 38th International Conference on Machine Learning,* Virtual Conference (Originally planned for Vienna, Austria) (Vol. 139, pp. 8748–8763). PMLR. https://proceedings.mlr.press/v139/radford21a.html.
Ramesh, A., Pavlov, M., Goh, G., Gray, S., Voss, C., Radford, A., Chen, M., & Sutskever, I. (2021). Zero-shot text-to-image generation. In M. Meila & T. Zhang (Eds.), *Proceedings of the 38th International Conference on Machine Learning,* Virtual (Vol. 139, pp. 8821–8831). PMLR. https://proceedings.mlr.press/v139/ramesh21a.html.
Reed, S., Zolna, K., Parisotto, E., Colmenarejo, S. G., Novikov, A., Barth-Maron, G., Gimenez, M., Sulsky, Y., Kay, J., Springenberg, J. T., Eccles, T., Bruce, J., Razavi, A., Edwards, A., Heess, N., Chen, Y., Hadsell, R., Vinyals, O., Bordbar, M., & de Freitas, N. (2022). A generalist agent. https://doi.org/10.48550/ARXIV.2205.06175.
Ren, S., He, K., Girshick, R., & Sun, J. (2015). Faster R-CNN: towards real-time object detection with region proposal networks. In C. Cortes, N. Lawrence, D. Lee, M. Sugiyama, & R. Garnett (Eds.), *Advances in Neural Information Processing Systems* (Vol. 28). Curran Associates, Inc. https://proceedings.neurips.cc/paper_files/paper/2015/file/14bfa6bb14875e45bba028a21ed38046-Paper.pdf.
Ribeiro, M. T., Singh, S., & Guestrin, C. (2016). "Why should I trust you?" Explaining the predictions of any classifier. In *Proceedings of the 22nd ACM SIGKDD International Conference on Knowledge Discovery and Data Mining,* San Francisco, CA, USA (pp. 1135–1144). https://doi.org/10.1145/2939672.2939778.
Riedmann-Streitz, C., Streitz, N., Antona, M., Marcus, A., Margetis, G., Ntoa, S., Rau, P.-L. P., & Rosenzweig, E. (2025). How to create and foster sustainable smart cities? Insights on ethics, trust, privacy, transparency, incentives, and success. *International Journal of Human–Computer Interaction*, *41*(4), 2491–2522. https://doi.org/10.1080/10447318.2024.2325175.

Rillig, M. C., Ågerstrand, M., Bi, M., Gould, K. A., & Sauerland, U. (2023). Risks and benefits of large language models for the environment. *Environmental Science & Technology*, *57*(9), 3464–3466. https://doi.org/10.1021/acs.est.3c01106.

Rombach, R., Blattmann, A., Lorenz, D., Esser, P., & Ommer, B. (2022). High-resolution image synthesis with latent diffusion models. In *Proceedings of the IEEE/CVF Conference on Computer Vision and Pattern Recognition (CVPR),* Orleans, Louisiana, USA (pp. 10684–10695).

Ross, A. S., Hughes, M. C., & Doshi-Velez, F. (2017). Right for the right reasons: training differentiable models by constraining their explanations (version 2). https://doi.org/10.48550/ARXIV.1703.03717.

Rudin, C. (2019). Stop explaining black box machine learning models for high stakes decisions and use interpretable models instead. *Nature Machine Intelligence*, *1*(5), 206–215. https://doi.org/10.1038/s42256-019-0048-x.

Rumelhart, D. E., Hinton, G. E., & Williams, R. J. (1986). Learning representations by back-propagating errors. *Nature*, *323*(6088), 533–536. https://doi.org/10.1038/323533a0.

Russell, S. J. (2023). *Human Compatible: Artificial Intelligence and the Problem of Control* (Reprinted with an afterword). Penguin Books.

Samek, W., Montavon, G., Vedaldi, A., Hansen, L. K., & Müller, K.-R. (Eds.). (2019). *Explainable AI: Interpreting, Explaining and Visualizing Deep Learning*. Springer. https://doi.org/10.1007/978-3-030-28954-6.

Schramowski, P., Turan, C., Andersen, N., Rothkopf, C. A., & Kersting, K. (2022). Large pre-trained language models contain human-like biases of what is right and wrong to do. *Nature Machine Intelligence*, *4*(3), 258–268. https://doi.org/10.1038/s42256-022-00458-8.

Schrittwieser, J., Antonoglou, I., Hubert, T., Simonyan, K., Sifre, L., Schmitt, S., Guez, A., Lockhart, E., Hassabis, D., Graepel, T., Lillicrap, T., & Silver, D. (2020). Mastering atari, go, chess and shogi by planning with a learned model. *Nature*, *588*(7839), 604–609. https://doi.org/10.1038/s41586-020-03051-4.

Schulman, J., Wolski, F., Dhariwal, P., Radford, A., & Klimov, O. (2017). Proximal policy optimization algorithms. https://api.semanticscholar.org/CorpusID:28695052.

Schut, L., Tomašev, N., McGrath, T., Hassabis, D., Paquet, U., & Kim, B. (2025). Bridging the human–AI knowledge gap through concept discovery and transfer in AlphaZero. *Proceedings of the National Academy of Sciences*, 122(13). https://doi.org/10.1073/pnas.2406675122.

Seo, Y., Hafner, D., Liu, H., Liu, F., James, S., Lee, K., & Abbeel, P. (2023). Masked world models for visual control. In K. Liu, D. Kulic, & J. Ichnowski (Eds.), *Proceedings of The 6th Conference on Robot Learning*, Karlsruhe, Germany (Vol. 205, pp. 1332–1344). PMLR. https://proceedings.mlr.press/v205/seo23a.html.

Shang, J., Kahatapitiya, K., Li, X., & Ryoo, M. S. (2022). StARformer: transformer with state-action-reward representations for visual reinforcement learning. In S. Avidan, G. Brostow, M. Cissé, G. M. Farinella, & T. Hassner (Eds.), *Computer Vision—ECCV 2022* (Vol. 13699, pp. 462–479). Springer Nature. https://doi.org/10.1007/978-3-031-19842-7_27.

Sheth, A., Roy, K., & Gaur, M. (2023). Neurosymbolic artificial intelligence (why, what, and how). *IEEE Intelligent Systems*, *38*(3), 56–62. https://doi.org/10.1109/MIS.2023.3268724.

Shneiderman, B. (2020). Human-centered artificial intelligence: reliable, safe & trustworthy. *International Journal of Human–Computer Interaction*, *36*(6), 495–504. https://doi.org/10.1080/10447318.2020.1741118.

Shortliffe, E. H. (1976). *Computer-Based Medical Consultations, MYCIN*. Elsevier.

Silver, D., Huang, A., Maddison, C. J., Guez, A., Sifre, L., Van Den Driessche, G., Schrittwieser, J., Antonoglou, I., Panneershelvam, V., Lanctot, M., Dieleman, S., Grewe, D., Nham, J., Kalchbrenner, N., Sutskever, I., Lillicrap, T., Leach, M., Kavukcuoglu, K., Graepel, T., & Hassabis, D. (2016). Mastering the game of Go with deep neural networks and tree search. *Nature*, *529*(7587), 484–489. https://doi.org/10.1038/nature16961.

Silver, D., Hubert, T., Schrittwieser, J., Antonoglou, I., Lai, M., Guez, A., Lanctot, M., Sifre, L., Kumaran, D., Graepel, T., Lillicrap, T., Simonyan, K., & Hassabis, D. (2018). A general reinforcement learning algorithm that masters chess, shogi, and Go through self-play. *Science*, *362*(6419), 1140–1144. https://doi.org/10.1126/science.aar6404.

Silver, D., Schrittwieser, J., Simonyan, K., Antonoglou, I., Huang, A., Guez, A., Hubert, T., Baker, L., Lai, M., Bolton, A., Chen, Y., Lillicrap, T., Hui, F., Sifre, L., Van Den Driessche, G., Graepel, T., & Hassabis, D. (2017). Mastering the game of Go without human knowledge. *Nature*, *550*(7676), 354–359. https://doi.org/10.1038/nature24270.

Simonyan, K., Vedaldi, A., & Zisserman, A. (2014). Deep inside convolutional networks: visualising image classification models and saliency maps. https://doi.org/10.48550/arXiv.1312.6034.

Smilkov, D., Thorat, N., Kim, B., Viégas, F., & Wattenberg, M. (2017). SmoothGrad: removing noise by adding noise (version 1). https://doi.org/10.48550/ARXIV.1706.03825.

Sundararajan, M., Taly, A., & Yan, Q. (2017). Axiomatic attribution for deep networks. In D. Precup & Y. W. Teh (Eds.), *Proceedings of the 34th International Conference on Machine Learning*, Sydney, Australia (Vol. 70, pp. 3319–3328). PMLR. https://proceedings.mlr.press/v70/sundararajan17a.html.

Sutskever, I., Vinyals, O., & Le, Q. V. (2014). Sequence to sequence learning with neural networks. In Z. Ghahramani, M. Welling, C. Cortes, N. Lawrence, & K. Q. Weinberger (Eds.), *Advances in Neural Information Processing Systems* (Vol. 27). Curran Associates, Inc. https://proceedings.neurips.cc/paper_files/paper/2014/file/a14ac55a4f27472c5d894ec1c3c743d2-Paper.pdf.

Tonmoy, S. M. T. I., Zaman, S. M. M., Jain, V., Rani, A., Rawte, V., Chadha, A., & Das, A. (2024). A comprehensive survey of hallucination mitigation techniques in large language models (version 3). https://doi.org/10.48550/ARXIV.2401.01313.

Touvron, H., Lavril, T., Izacard, G., Martinet, X., Lachaux, M.-A., Lacroix, T., Rozière, B., Goyal, N., Hambro, E., Azhar, F., Rodriguez, A., Joulin, A., Grave, E., & Lample, G. (2023). LLaMA: open and efficient foundation language models (version 1). https://doi.org/10.48550/ARXIV.2302.13971.

Varshney, D., Zafar, A., Behera, N. K., & Ekbal, A. (2023). Knowledge grounded medical dialogue generation using augmented graphs. *Scientific Reports*, *13*(1), 3310.

Vaswani, A., Shazeer, N., Parmar, N., Uszkoreit, J., Jones, L., Gomez, A. N., Kaiser, Ł. ukasz, & Polosukhin, I. (2017). Attention is all you need. In I. Guyon, U. V. Luxburg, S. Bengio, H. Wallach, R. Fergus, S. Vishwanathan, & R. Garnett (Eds.), *Advances in Neural Information Processing Systems* (Vol. 30). Curran Associates, Inc. https://proceedings.neurips.cc/paper_files/paper/2017/file/3f5ee243547dee91fbd053c1c4a845aa-Paper.pdf.

Verma, S., Boonsanong, V., Hoang, M., Hines, K., Dickerson, J., & Shah, C. (2024). Counterfactual explanations and algorithmic recourses for machine learning: a review. *ACM Computing Surveys*, *56*(12), 1–42. https://doi.org/10.1145/3677119.

Vielhaben, J., Lapuschkin, S., Montavon, G., & Samek, W. (2024). Explainable AI for time series via Virtual Inspection Layers. *Pattern Recognition*, *150*, 110309. https://doi.org/10.1016/j.patcog.2024.110309.

Weber, L., Lapuschkin, S., Binder, A., & Samek, W. (2023). Beyond explaining: opportunities and challenges of XAI-based model improvement. *Information Fusion*, *92*, 154–176. https://doi.org/10.1016/j.inffus.2022.11.013.

Wei, J., Wang, X., Schuurmans, D., Bosma, M., Ichter, B., Xia, F., Chi, E., Le, Q. V., & Zhou, D. (2022). Chain-of-thought prompting elicits reasoning in large language models. In S. Koyejo, S. Mohamed, A. Agarwal, D. Belgrave, K. Cho, & A. Oh (Eds.), *Advances in*

Neural Information Processing Systems (Vol. 35, pp. 24824–24837). Curran Associates, Inc. https://proceedings.neurips.cc/paper_files/paper/2022/file/9d5609613524ecf4f15af0f7b31abca4-Paper-Conference.pdf.

Wong, F., Zheng, E. J., Valeri, J. A., Donghia, N. M., Anahtar, M. N., Omori, S., Li, A., Cubillos-Ruiz, A., Krishnan, A., Jin, W., Manson, A. L., Friedrichs, J., Helbig, R., Hajian, B., Fiejtek, D. K., Wagner, F. F., Soutter, H. H., Earl, A. M., Stokes, J. M., … Collins, J. J. (2024). Discovery of a structural class of antibiotics with explainable deep learning. *Nature*, *626*(7997), 177–185. https://doi.org/10.1038/s41586-023-06887-8.

Xie, E., Wang, W., Yu, Z., Anandkumar, A., Alvarez, J. M., & Luo, P. (2021). SegFormer: simple and efficient design for semantic segmentation with transformers. In M. Ranzato, A. Beygelzimer, Y. Dauphin, P. S. Liang, & J. W. Vaughan (Eds.), *Advances in Neural Information Processing Systems* (Vol. 34, pp. 12077–12090). Curran Associates, Inc. https://proceedings.neurips.cc/paper_files/paper/2021/file/64f1f27bf1b4ec22924fd0acb550c235-Paper.pdf.

Xu, W. (2019). Toward human-centered AI: a perspective from human-computer interaction. *Interactions*, *26*(4), 42–46. https://doi.org/10.1145/3328485.

Yosinski, J., Clune, J., Nguyen, A., Fuchs, T., & Lipson, H. (2015). Understanding neural networks through deep visualization (version 1). https://doi.org/10.48550/ARXIV.1506.06579.

Zhao, W. X., Zhou, K., Li, J., Tang, T., Wang, X., Hou, Y., Min, Y., Zhang, B., Zhang, J., Dong, Z., Du, Y., Yang, C., Chen, Y., Chen, Z., Jiang, J., Ren, R., Li, Y., Tang, X., Liu, Z., … Wen, J.-R. (2023). A survey of large language models (version 15). https://doi.org/10.48550/ARXIV.2303.18223.

Zheng, Z., Wang, Y., Huang, Y., Song, S., Yang, M., Tang, B., Xiong, F., & Li, Z. (2024). Attention heads of large language models: a survey (version 3). https://doi.org/10.48550/ARXIV.2409.03752.

2 Neural Computing for Advanced Natural Language Understanding and Generation

Mark Nuppnau, Khalid Kattan, and Robert G. Reynolds

2.1 HISTORY OF LARGE LANGUAGE MODELS

Language modelling is an important subfield of artificial intelligence (AI) that underlies the development of any intelligent system. The use of language for communication in these systems is often preferred over other communication formats such as graphs or database systems. In fact, progress in natural language understanding has effectively shaped the development of AI over the years.

Language modelling can be viewed as predicting the next word in a sequence of words. It has been central to natural language processing (NLP) since the 1940s. Early work in the 1940s and 1950s provided the foundations for NLP with advances in formal models of computation, formal grammar, and probability theory. The symbolic approach emerged from this work, and it represents a top-down paradigm for language understanding. The goal is to model the underlying capabilities that a system needs to process and understand natural language. The approach represents the earliest work on AI that focuses on the underlying reasoning and logic. The term AI was originally used to describe this approach in the Dartmouth Conference in 1956 as organized by John McCarthy, Marvin Minsky, Claude Shannon, and Nathaniel Rochester (Lucci et al., 2022).

In these early decades of AI, **symbolic methods** were the norm for NLP. In 1948, Shannon employed statistical models to predict the next word in a symbol sequence, n-grams (Shannon, 1948). Subsequently, Weizenbaum used basic pattern matching to produce ELIZA, a system designed to emulate a psychiatrist from the Carl Rogers discipline of Psychiatry (Weizenbaum, 1966). The popularity of the system paved the way for the large language models (LLMs) of today.

Other systems, such as **SHRDLU** (Winograd, 1971), demonstrated how rule-based logic and hand-crafted grammars could be used to parse and understand language in constrained domains. These systems relied on **semantic networks**, **ontologies**, and **first-order logic** to represent knowledge and draw inferences. While expressive, symbolic systems required extensive manual labor and domain-specific engineering.

DOI: 10.1201/9781003492252-2

They also struggled to cope with ambiguity, metaphor, and variability, properties that are inherent features of natural language.

A second major paradigm, the connectionist paradigm, emerged from the early work of McCulloch and Pitts to model the neuron in 1943 (McCulloch and Pitts, 1943). Their work represented neurons as binary threshold units. This paradigm focused on the origins of intelligence from a bottom-up approach. A single layer of perceptrons could perform logical operations and provided a fundamental basis for the development of neural computation. Originally, the study of neural networks was a niche sub-field in AI. Initially, Rosenblatt developed the perceptron learning rule that showed how perceptrons can learn linearly separable concepts through weight adjustment that stimulated interest in neural models (Rosenblatt, 1958). However, Minsky and Papert in their 1969 book *Perceptrons* demonstrated that a single layer of perceptrons could not express even a fundamental function in Boolean Algebra, the Exclusive Or function (Minsky and Papert, 1969) along with other inherently non-linear problems. However, there was no training algorithm for multi-layer networks, which dampened interest in neural networks for over a decade.

However, the development of the backpropagation algorithm in the 1980s allowed multi-layer perceptrons (MLP) to learn internal representations and approximate complex, non-linear functions. The paper by David Rumelhart, Geoffrey Hinton, and Ronald Williams was instrumental in this revival (Rumelhart et al., 1986). This led to numerous applications in speech understanding, pattern classification, and robotics. But a real breakthrough came with recurrent neural networks (RNNs) and long short-term memory networks (LSTMs) (Hochreiter and Schmidhuber, 1997), that supported sequential data processing, which was essential for language modelling. This allowed the creation of neural network-based language models such as the feed-forward model proposed by Benzio et al. (2003), which was the first attempt to capture distributed word representations in continuous space. Despite these innovations, training on large corpora remained computationally expensive, and long-range dependencies were difficult to capture.

Things began to change in 2006 when Geoffrey Hinton et al. introduced greedy layer-wise pretraining, which supported the training of deep neural architectures that had previously been difficult to optimize (Hinton et al., 2006). This opened the door for applications to large-scale datasets that were leveraged with graphics processing units (GPUs), hardware that was initially designed to support video games. Such a system, AlexNet, by Alex Krizhevsky, Ilya Sutskever, and Geoffrey Hinton won the ImageNet Large Scale Visual Recognition Challenge in 2012 (Krizhevsky et al., 2012). This result demonstrated the potential of hardware to scale up deep learning applications. The use of GPUs was followed by tensor processing units from Google and AI-specific chips from companies like NVIDIA and AMD that further accelerated network training and inference. The concomitant expansion of Cloud computing platforms served to make these resources more accessible to researchers and businesses around the globe. Subsequently since 2012, neural networks have exhibited a phase of rapid scaling with LLMs like ResNet (He et al., 2015), BERT (Devlin et al., 2018), and GPT (Radford et al., 2018–2023), expanding the boundaries of neural architecture depth, parameter size, and overall performance across a variety of tasks.

As datasets began to scale larger, new methods of training were required to exploit them. For example, Transfer learning permitted pre-trained models to be adapted to new tasks with the use of minimal data. Self-supervised learning systems, SimCLR, allowed networks to train on unlabeled data that further enhanced the nature of the data that could be handled (Chen et al., 2020). Finally, reinforcement learning combined with neural networks and hardware accelerators, exemplified by AlphaGo (Silver et al., 2016), unlocked the capabilities of LLMs in complex decision-making environments.

Still there were other challenges to address, such as the ability to model long-range dependencies in text. Both RNNs and LSTMs were limited in their ability to model long-range dependencies in text. This deficiency was addressed in 2017 with the Transformer architecture designed by Vaswani et al. (2017). Recurrence was replaced in Transformer architectures by the self-attention mechanism. This new framework not only supported long-range dependencies but enabled additional parallel computation as well.

Transformers became the basis for the development of large-scale NLP models such as OpenAI's GPT (Radford et al., 2018) and Google's BERT (Devlin et al., 2018). These were succeeded by models of increasingly larger scale. It was demonstrated by systems such as GPT-2 and GPT-3 that LLMs began to exhibit emergent learning properties with increased scale. These emergent properties supported the development of foundational models such as GPT-4 (Open AI, 2023). These models were "foundational" in the sense that they can be trained on a broad variety of datasets, fine-tuned to specific downstream applications, and provide an infrastructure that supported multiple domains besides NLP. These were the basis for Chatbots, search engines, and copilots.

The remainder of this chapter will elaborate on these ideas. In Section 2.2, a brief overview of the connectionist paradigm is presented. Section 2.3 follows with a survey of LLMs. Model scaling and transfer learning are discussed in this section as well. Ethical issues are then described in Section 2.4. The chapter concludes with a discussion of future directions in Section 2.5.

2.2 OVERVIEW OF NEURAL NETWORK ARCHITECTURES

The "Connectionist" idea emerged from the work of McCulloch in 1943 during World War II (McCulloch and Pitts, 1943). They presented the basic model of a neuron as shown in Figure 2.1 on the far left.

The basic model consists of a cell body that accumulates electronic charges that arrive along dendrites. Dendrites collect electrons generated by connected neurons currently active. When a sufficient charge has accumulated, the cell body produces an electric charge that is conducted along its axon. These charges can be transferred to the dendrites of other connected cells through synapses. A synapse is a body of fluid that interfaces an axon with a dendrite and supports the movement of electrons from the axon of one neuron to the dendrite of another.

Based upon this model, Frank Rosenblatt described a type of artificial neural network called a layered network of perceptrons in his book "Perceptron" in 1958 (Rosenblatt, 1958). The perceptron is a type of artificial neuron or node that serves as

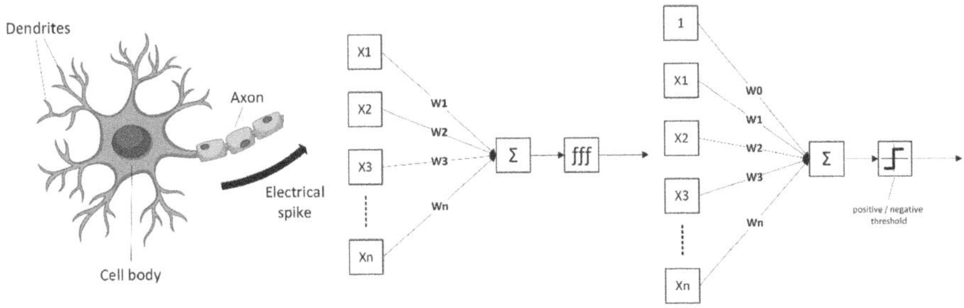

FIGURE 2.1 A neuron and a perceptron (McCulloch and Pitts, 1943).

the building block for artificial neural networks. It is a simplified model of a biological neuron, designed to mimic some of its basic functions. The middle illustration in Figure 2.1 is a perceptron model with a set of weighted inputs that highlight the perceptron's most important connections. The perceptron takes an input and multiplies it by a weight value. This product is summed over all inputs and sent to an output function. By the addition of a bias function, $w0$, a binary output can be produced. If the difference between the sum of inputs 1 through n exceeds $w0$, then the perceptron fires; otherwise, it does not. This is shown in the rightmost configuration in Figure 2.1.

The perceptron is able to learn to classify a dataset of positive and negative examples if they are linearly separable, which means that a line can be drawn that separates the positive from the negative examples. If there are multiple inputs, then there is a hyperplane that is able to separate the positive and negative input vectors. The perceptron learning rule then adjusts the weights until there are no misclassifications if they are linearly separable.

First, it is necessary to take a closer look at what the perceptron computes. Let x be an input vector $(x_1, x_2, \ldots, x_n)$. The perceptron computes a weighted sum function $g(x)$, which is then sent to the output function $o(x)$. Equations 2.1 and 2.2 show the calculation of $g(x)$ and $o(x)$, respectively.

$$g(x) = \sum_{i=0}^{n} w_i x_i \tag{2.1}$$

$$o(x) = \begin{cases} 1 \text{ if } g(x) > 0 \\ 0 \text{ if } g(x) < 0 \end{cases} \tag{2.2}$$

Referring to Figures 2.2 and 2.3, the data points labeled "+" and "−" represent two different classes. Each of the three different straight lines ($L1$, $L2$, $L3$) represents a possible decision boundary that the perceptron might learn, depending on the weights. In this example, $L1$ is the slope produced by a set of initial weights for each of the two variables. By reducing the weight for X_2 and increasing the weight for X_1, the "−" is correctly classified, but the "+" class is not. The third line shows another adjustment to each variable that produces a zero classification error.

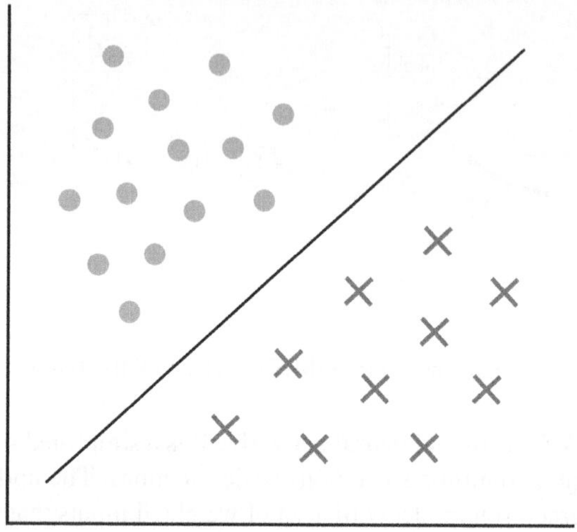

FIGURE 2.2 Example of linearly separable data.

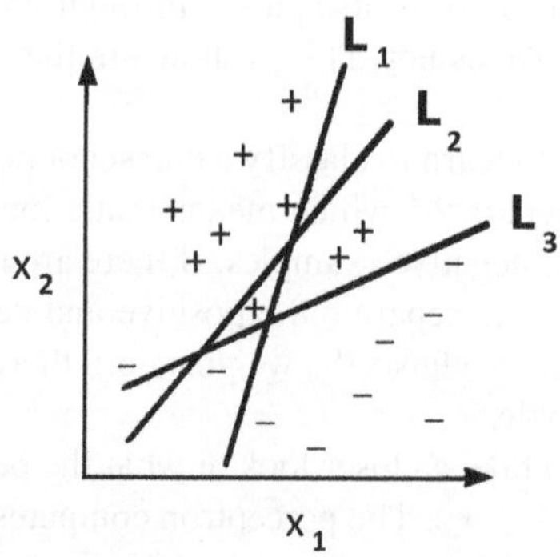

FIGURE 2.3 Three different straight lines (*L*1, *L*2, and *L*3), each representing a possible decision boundary.

The perceptron learning rule is guaranteed to produce the necessary adjustments to produce perfect classification if the data points are linearly separable. However, in order to deal with non-linearities in the data distribution, as illustrated in Figure 2.4 below, one needs a differentiable output function. The sigmoid function shown in Equation 2.3 is a classic example of such a function.

A multilayer network built only with linear components cannot model complex, non-linear relationships. To overcome this, non-linear and differentiable activation functions (such as logistic sigmoid, tanh, or even Gaussian) are used in neural networks, enabling rich, multi-step transformations and powerful learning.

Sigmoid function calculation (Figure 2.5)

$$o_j = \frac{1}{1+e^{-(\text{net}_j - T_j)}} \tag{2.3}$$

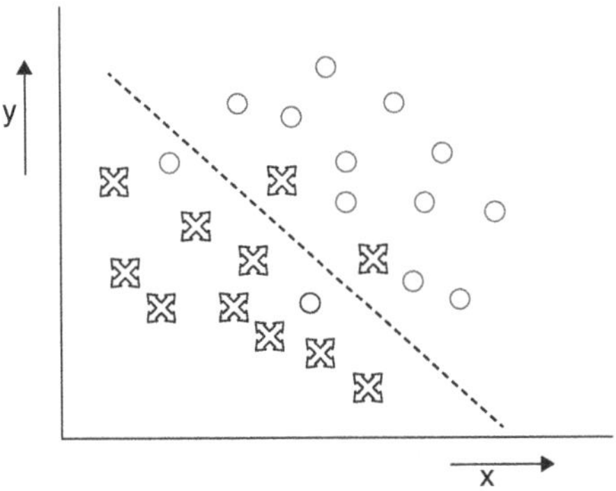

FIGURE 2.4 Example of non-linearities in data distribution.

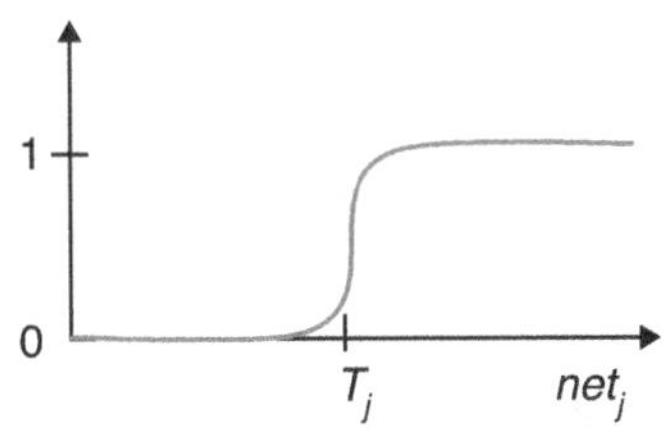

FIGURE 2.5 Sigmoid function curve.

This information now allows the support of a Gradient Descent learning rule for a perceptron with the goal of minimizing the error function. The Gradient Descent method is shown below in Equations 2.4–2.6. In Equation 2.4, D is the set of training examples, K is the set of output units, and t_{kd} and o_{kd} are, respectively, the teacher and current output for unit k for example d.

Objective function to minimize error

$$E(W) = \sum_{d \in D} \sum_{k \in K} (t_{kd} - o_{kd})^2 \tag{2.4}$$

Derivative of a sigmoid unit with respect to net input

$$\frac{\partial o_j}{\partial \text{net}_j} = o_j(1 - o_j) \tag{2.5}$$

Learning rule to change weights to minimize error

$$\Delta w_{ji} = -\eta \frac{\partial E}{\partial w_{ji}} \tag{2.6}$$

But, for arbitrary non-linear distributions, hidden layers are needed to adjust the weights incrementally via backpropagation in a feed-forward network of perceptrons. The backpropagation process for a network with a single hidden layer is as

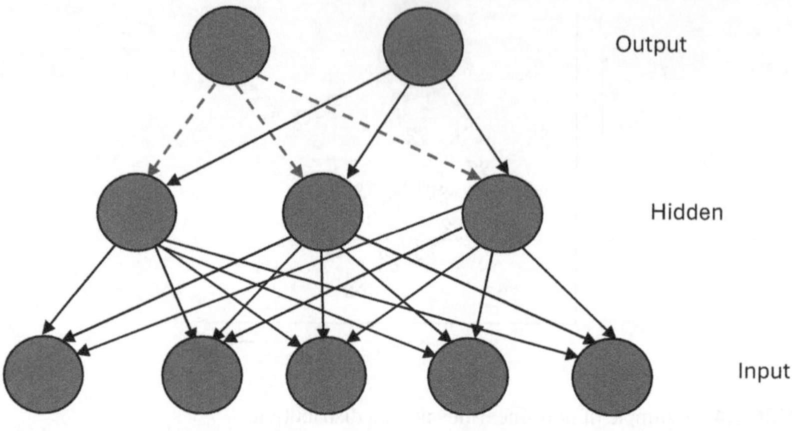

FIGURE 2.6 Error backpropagation, step 1.

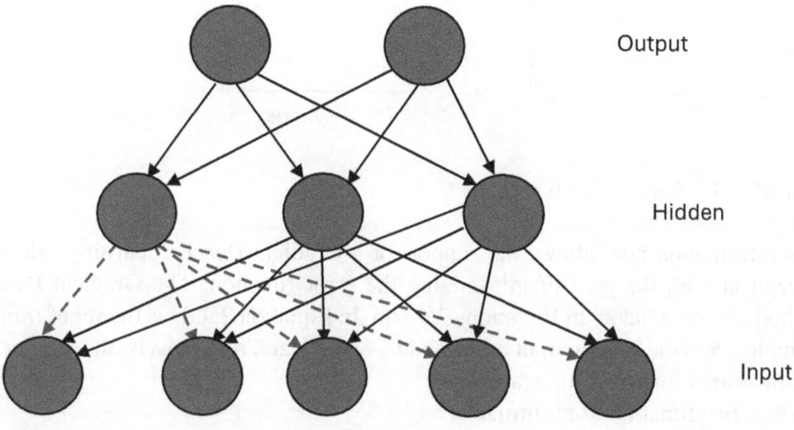

FIGURE 2.7 Error backpropagation, step 2.

follows. First, the top level of weights is adjusted based upon the difference between the expected and actual output, as shown in Figure 2.6. Next, the bottom layer is updated to reflect adjustments to the hidden layer, as shown in Figure 2.7. The update relations are shown as dashed lines in each figure. For example, assume that the current output is $o_j = 0.2$ and the correct output is 1.0. Then, the error will be 0.128, as shown in Equation 2.7:

Error backpropagation, Step 1

$$\text{Error } \delta_j = o_j(1 - o_j)(t_j - o_j)$$
$$0.2(1 - 0.2)(1 - 0.2) = 0.128 \tag{2.7}$$

Next, update the weights into j, as shown in Equation 2.8.

Update weights into j

$$\Delta w_{ji} = \eta \delta_j o_j \tag{2.8}$$

The second step is to update the bottom layer of weights based on errors calculated for hidden units, as shown in Figure 2.5 and Equation 2.9.

Update the bottom layer of weights based on errors calculated for hidden units

$$\delta_j = o_j(1 - o_j) \sum_k \delta_k w_{kj} \tag{2.9}$$

After Equation 2.9, update the weights into j using Equation 2.8, same as the end of step 1.

Historically in 1967, Shun'ichi Amari published a groundbreaking paper that introduced a multilayer perceptron network that utilized stochastic gradient descent for the first time. That network was capable of classifying complex, non-linearly separable pattern classes. Amari's student, Saito, conducted a series of computer experiments using a five-layered feed-forward network with two learning layers. This work will eventually lead to the capability to train deeper networks efficiently during the hardware acceleration of the early 2000s.

2.3 LARGE LANGUAGE MODELS: ARCHITECTURES, TRAINING, AND ASSESSMENT

2.3.1 Basic RNN Architecture

When analyzing weather patterns or a patient's vital-sign waveforms, the temporal order within the sequence provides essential context for making predictions. The same is true for understanding language and being able to generate words given a sequence of words.

In the previous section, we explored fully connected (feed-forward) neural networks and their various components. We noted that standard feed-forward networks (and many convolutional neural network (CNN) variants without sequence-handling adaptations) assume fixed-length vector inputs, making naïve application to variable-length sequences non-trivial.

Another limitation of feed-forward networks is that, aside from local receptive fields in CNNs, these architectures process each training example (or each position absent explicit context mechanisms) independently rather than maintaining a state across time steps. While CNNs can capture local sequential dependencies, they do not maintain an explicit persistent hidden state (memory) across arbitrarily long sequences. While the overall architectures of feed-forward and CNNs are not ideal for processing sequential data, some of their components we discussed in the previous section can be utilized by other architectures (e.g., activation functions, normalization).

RNN architectures for learning sequential patterns gained prominence in the 1980s (Hopfield, 1982). In 1982, John Hopfield introduced recurrent associative memory networks with symmetric weights (Hopfield networks). While the intention of Hopfield networks was not to use recurrence for the purposes of processing sequential data to learn sequential patterns, they showed the potential of the use of recurrence and sparked an interest that would motivate future research in recurrent networks.

With his expertise in physics and neuroscience, Hopfield sought to use an energy function on a set of interconnected neurons to remember specific patterns given a partial or noisy pattern. Through iterative (often asynchronous) updates, the network dynamics descend the energy landscape toward an attractor (a stable fixed point), ideally corresponding to a stored pattern, though spurious minima can also occur (Hopfield, 1982).

In 1986, Jordan proposed a specific recurrent architecture employing context units fed by the network's outputs, influencing later RNN designs (Jordan, 1986). He describes a recurrent network that contains two units, an input unit and an output unit, as a network having an output unit with a recurrent connection back onto itself. Thus, the current output depends on both the current input and the previous output (i.e., the recurrent state) (Jordan, 1986). The recurrent network is formalized as:

$$\begin{aligned} x_2(t) &= \mu x_2(t-1) + w_{21}\, x_1(t) \\ &= \mu^t x_2(0) + \sum_{t=0}^{t-1} \mu^\tau\, w_{21}\, x_1(t-\tau) \end{aligned} \tag{2.10}$$

where μ is the value of the recurrent weight, $x_1(t)$ is assumed to be constant over time, and $x_2(t)$ is the activation of the output unit at time t. A visualization of this recurrent network can be viewed in Figure 2.8 [2]. The feedback loop on the output unit can be difficult to understand, so it is common to visualize an RNN with the time steps unrolled as to look similar to a feed-forward neural network, as shown in Figure 2.9, which has the additional hidden layer.

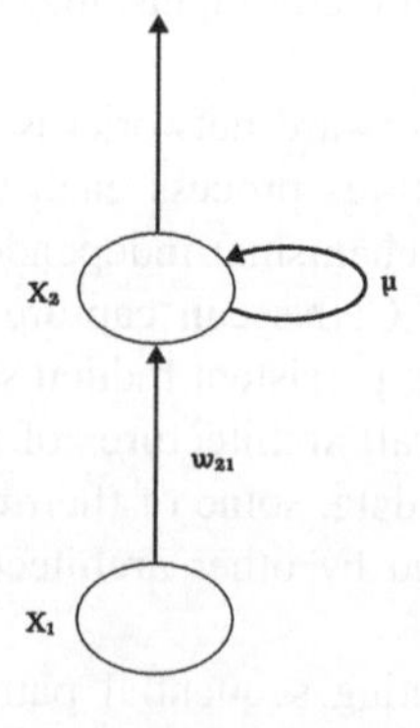

FIGURE 2.8 An example of a recurrent network with two units (Hopfield, 1982).

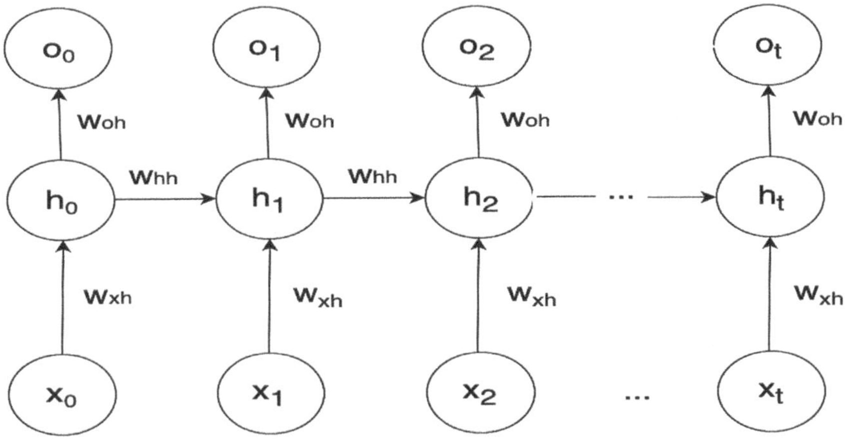

FIGURE 2.9 An example recurrent network with unrolled time steps (Elman, 1990).

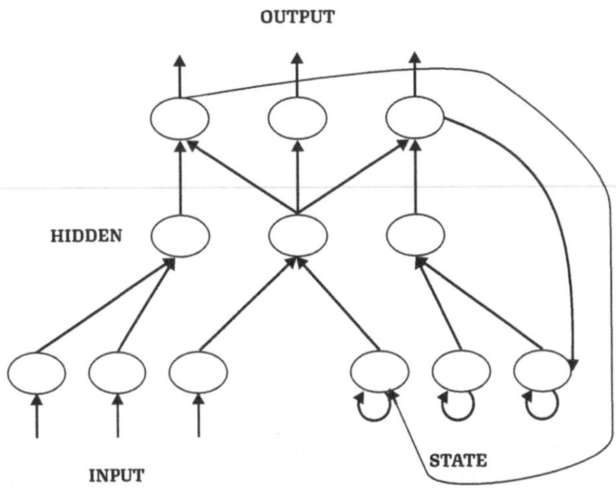

FIGURE 2.10 Architecture used by Jordan (Jordan, 1986).

2.3.2 Advances in RNN Architecture

Elman built upon Jordan's work and designed an RNN architecture that has now become a popular RNN architecture that is used through the early 2000s (Elman, 1990). Elman shifted the feedback loop from the output unit directly to the hidden state, allowing higher-dimensional information to be carried forward for improved context. Elman provides a visualization of Jordan's network, with three units per layer, as shown in Figure 2.10 (Elman, 1990). With Elman's update, the feedback loops move from the three output units to the three hidden units. The loop still ends at each of the state units.

While feed-forward neural networks use backpropagation to adjust model weights (as shown in the previous section), RNNs use backpropagation through time (BPTT). BPTT was introduced by (Werbos, 1990)—the same year as Elman's updated architecture, which utilizes BPTT.

Using Figure 2.9 as reference, BPTT calculates the gradient of the loss function with respect to weights W_{xh} W_{hh}, and W_{oh}. The gradients are calculated at each time step t and are summed. Also, the gradients of the hidden states are accumulated across time.

Let's assume we have a loss function, L, that computes the mean squared error. To compute the gradient of L in relation to the input weights across two time steps:

$$\frac{\partial L}{\partial W_{xh}} = \frac{\partial L}{\partial o_t}\frac{\partial o_t}{\partial h_t}\frac{\partial h_t}{\partial W_{xh}} + \frac{\partial L}{\partial o_{t-1}}\frac{\partial o_{t-1}}{\partial h_{t-1}}\left(\frac{\partial h_t}{\partial h_{t-1}}\right)\frac{\partial h_{t-1}}{\partial W_{xh}} \quad (2.11)$$

BPTT calculates the gradient at each time step, starting with the last time step and moving backward, as to simulate moving through time, to the first-time step. As we saw in Equation 2.11, we accumulated the hidden state gradients across two-time steps. If we increase the time steps to three, we need to accumulate the gradients from the hidden state across all three-time steps:

$$(2.11) + \frac{\partial L}{\partial o_{t-2}}\frac{\partial o_{t-2}}{\partial h_{t-2}}\left(\frac{\partial h_t}{\partial h_{t-1}}\right)\left(\frac{\partial h_{t-1}}{\partial h_{t-2}}\right)\frac{\partial h_{t-2}}{\partial W_{xh}} \quad (2.12)$$

RNNs often process sequences of length T with hundreds or even thousands of steps. Over so many steps, backpropagating the hidden-state gradients involves repeatedly multiplying by the Jacobian

$$\frac{\partial h_t}{\partial h_{t-1}} = \mathrm{diag}(f'(a_t))W_{h,h} \quad (2.13)$$

Thus, if the spectral norm $\|W_{hh}\|_2 > 1$, these products can grow roughly like $\|W_{hh}\|_2^T$ (exploding gradients); if $\|W_{hh}\|_2 < 1$, they can decay like $\|W_{hh}\|_2^T$ (vanishing gradients).

With advances in computational resources and algorithmic techniques during the late 1990s and 2000s, several methods emerged to stabilize and accelerate training of larger RNNs. Duchi (2011) introduced the Adaptive Gradient Algorithm (AdaGrad), which adapts each parameter's learning rate based on the sum of its past squared gradients. Gradient clipping was first proposed by Mikolov in his 2012 PhD thesis (and was used implicitly in Mikolov et al. 2011 for language modeling) as element-wise clipping of each gradient component to lie within a fixed threshold. Pascanu, Mikolov, and Bengio later formalized this approach in 2013 (Rasvan, 2013), where the entire gradient vector is rescaled when its norm exceeds a pre-defined maximum.

Karpathy (2015) provides a highly optimized RNN implementation that will be discussed in the following application section. This implementation utilizes BPTT with gradient clipping + AdaGrad for stabilized gradients throughout training.

Application: *Minimal Character-level Vanilla RNN: Let's walk through Karpathy's gist (https://gist.github.com/karpathy/d4dee566867f8291f086), which is a didactic example of how to efficiently train an RNN.*

Lines 1-13: The program first reads in a text file (you can replace this text file with any text file that contains text that you wish the model to learn). The length and the number of unique characters of the input text are then calculated. An index number is then assigned to each unique character and a mapping of both directions is created.

Lines 15-25: The hyperparameters are then set with 100 hidden layer neurons and 25 time steps. The model parameters are set for W_{xh} (input to hidden), W_{hh} (hidden to hidden), W_{hy} (hidden to output), b_h (hidden bias), and y_h (output bias).

Lines 81-84: The iteration counter and data pointer are set at zero and will be incremented after each training step. The memory variables for Adagrad and loss for iteration zero are set.

Lines 85-91: The training loop begins and if it's the first or last chunk of input data, the initial hidden state, hprev, is reset and the data pointer is reset to zero. The inputs are set as a list of index values where each index is associated with each character from the input sequence. The targets are set in the same manner but the range of characters selected from the input data is increased by one. Suppose our sequence length is six and we are extracting the first input sequence from the input data. The input sequence, data[0:6], might look like "We are" and the list of index values stored in inputs might look like [21,4,39,1,16,4]. The targets would then use indices associated with characters from the input sequence data[1:7] which could be "e are " and the list of index values assigned to targets could be [4,39,1,16,4,39]. When training the model, the input index 21, inputs[0], for "W", will have the target index 4, targets[0], for "e".

Lines 93-97: Every 100 iterations, a sequence of integers from the model is sampled and the characters associated with each integer (index) are printed. The sample function, lines 63-79, intakes the hidden state, the first index value of the input sequence, and the desired sequence length. An array of zeroes the length of the vocabulary is created and the first index value in inputs [0] is used to one-hot encode the array of zeroes for the character associated with that index value. A forward pass of that one-hot encoded array through the model is completed (the details of which will be covered when discussing the training loop), and a probability distribution, *p*, is obtained and used to choose an index value for the next onehot encoded vector. A well-trained model will assign a higher probability to more sensible indices, characters, based on the current index.

Lines 99-102: Line 100 begins the model training and calls the lossFun function with the inputs, targets, and hidden state as arguments. The function begins on line 27. Lines 33-35 set up the dicts used to store information, set the previous hidden state, and set the loss to zero. The forward pass is performed with lines 37-43. It loops through each input character in the input sequence, and for each loop, stores a one-hot encoded array of the input character in xs, a hidden activation in hs, unnormalized log probabilities in ys, and next character probabilities in ps. The hidden activation, line 40, is produced by applying the tanh activation function to the sum of two linear projections. The first dot product is between the input weights, Wxh, and the one-hot encoded input array.

The second being between the hidden weights, W_{hh}, and the previous hidden state. You can think of the first dot product as a measure of alignment between new information from the current input and the input weights. Similarly, the second dot product can be thought of as a measure of alignment between old information from the previous hidden state and the hidden weights. Both are linear functions, so you simply add them, along with the bias term, to get a single alignment score. The tanh activation is then applied for non-linearity.

The output vector, unnormalized log probabilities, is then calculated by taking the dot product of W_{hy} and the hidden activation. To get the normalized probabilities, the unnormalized log probabilities are exponentiated and then normalized. The cross-entropy loss is then calculated using the normalized probabilities and the target.

Lines 45-47 set zero-filled arrays to be used for computing the gradients with BPTT. Lines 48-58. As mentioned previously, BPTT starts with the last time step. Thus, *t* will start with the last time step when used as an index for the dicts during the forwards passes. The probability vector p_t is copied to d_y, and we subtract 1 at the target index.

This creates a vector of gradients where each index not associated with the target retains the probability (i.e., a positive value) as the gradient on the scores, and the index associated with the target is pushed below zero. This negative value associated with the target index acts as a signal to increase the score of that index which decreases the loss.

The derivative of the loss w.r.t the weights in the output units, *dWhy*, is then calculated by taking the dot product of the gradient of the loss and the hidden state at time t. Notice that *dWhy* uses the += operator to increment for each loop. The same operator is used for each derivative associated with the weights and biases.

Recall from Equation 2.11, the derivative of the loss w.r.t the input weights are summed across all time steps. Next, the derivative of the loss w.r.t the biases in the output units is incremented. The gradients are then backpropagated into the hidden units by taking the dot product of the transpose of the current output weights, *Why* and the gradient of the loss, dy and adds the derivative of the loss w.r.t the previous hidden state. Having backpropagated the loss signal into the hidden units, the next step is to pass it through the tanh non-linearity. We form the raw hidden gradient

$$\text{dhraw} = \left(1\text{-tanh}^2\left(a_t\right)\right) \text{e}\, dh \tag{2.14}$$

where a_t is the pre-activation, and then accumulate the bias gradient. From dhraw the parameter gradients follow:

$$\text{dW}_{xh} += \text{dhraw}\,\text{x}_t^T\,,\ \text{dW}_{hh} += \text{dhraw}\,\text{h}_{t-1}^T \tag{2.15}$$

and the gradient passed further back in time is

$$dh = W_{hh}^T,\ \text{dhraw} + dh_{prev} \tag{2.16}$$

After sweeping over all time steps, we clip gradients (e.g., global norm), update the AdaGrad accumulators, and apply the parameter updates. This minimal character-level RNN, though small, concretely demonstrates recurrence,

```
BPTT, gradient clipping, and adaptive learning rates—the
ingredients later refined in LSTMs and ultimately
supplanted by attention in Transformers.
```

2.3.3 LSTM Networks

In the previous section, Karpathy's vanilla RNN implementation utilizes AdaGrad and gradient clipping to stabilize gradients during training, which are techniques introduced more than 20 years after the Elman RNN architecture. Hochreiter and Schmidhuber introduced LSTM in 1997. The main issue with the Elman RNN was the vanishing/exploding gradient problem. In practice, gradients can vanish/explode and the Elman RNN struggles to learn long-range dependencies. The LSTM architecture solves both of these issues.

The LSTM architecture adds a cell state and gates, to control the flow of information, to the hidden state of an Elman RNN. This updated architecture can be viewed as a unit at a single time step in Figure 2.11, which we will use to explain the functionality of an LSTM.

In comparison, an Elman RNN would have a single tanh activation network layer connecting the previous hidden state and current input to the current output and next hidden state.

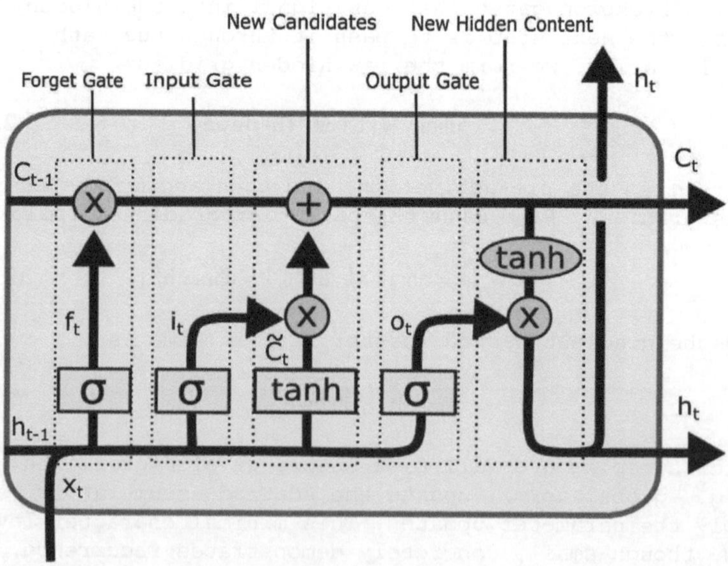

FIGURE 2.11 LSTM architecture (Hochreiter and Schmidhuber, 1997).

An important piece to the LSTM architecture is the direct transfer of vectors from one cell state, C_t, to the next, C_{t+1} which allows for better handling of long-term dependencies—this can be viewed at the top of the unit. Information moves along this transfer mechanism with information being removed and information being added at each time step using gates.

The standard LSTM architecture adds three sigmoid activation network layers, along with pointwise operations to each unit. The functionality of these network layers and pointwise operations can be separated into five sections: the forget gate, input gate, new candidates for cell state update, output gate, and new hidden state content.

The forget gate takes information from the previous hidden state, h_{t-1}, and the current input, x_t, and passes it through a sigmoid activation network layer. Recall from the previous section that the sigmoid activation outputs a value between zero and one. The closer to zero values output by the sigmoid activation, the less information flows forward through time. This gate allows the model to forget a certain level of information at each time step. Let's call the output of the forget gate's sigmoid activation network layer f_t, and the sigmoid activation σ, then:

$$f_t = \sigma\left(W_f \cdot [h_{t-1}, x_t] + b_f\right) \tag{2.17}$$

Before discussing how f_t is applied to the cell state, one must explore the input gate, as both affect the cell state. The input gate consists of a sigmoid network layer. Instead of the output of the sigmoid activation being applied directly to the previous cell state, there is a pointwise operation with the output from the tanh activation output. This is the section that determines new candidates for the cell state update. Let's say the output of the input gate's sigmoid activation is i_t, and the output of the tanh activation is $\underline{C_t}$, then:

$$\begin{aligned} i_t &= \sigma\left(W_i \cdot \left[h_{t-1}, x_t\right] + b_i\right) \text{and} \\ \underline{C_t} &= \tanh\left(W_C \cdot \left[h_{t-1}, x_t\right] + b\right) \end{aligned} \tag{2.18}$$

Now, one can calculate the cell state, C, for the current time step, t. First, multiply the previous cell state, C_{t-1}, by the output of the forget gate, f_t. Then, add the product of the input gate's sigmoid activation, i_t, with the tanh activation output, $\underline{C_t}$:

$$C_t = f_t * C_{t-1} + i_t * \underline{C_t} \tag{2.19}$$

The last gate is the output gate. This gate helps determine the information from the cell state that will be output by passing the previous hidden state and current input through a sigmoid activation network layer. This creates a vector of values, o_t, between zero and one to determine what cell state values to output:

$$o_t = \sigma\left(W_o \cdot [h_{t-1}, x_t] + b_o\right) \tag{2.20}$$

The last section, new hidden state content, inputs the cell state through a tanh activation network layer and multiplies it by the output of the output gate. The product of which gets output as h_t to downstream layers or back into the next timestep's gates as h_{t-1}. The output is calculated as:

$$h_t = o_t * \tanh(C_t) \tag{2.21}$$

In the previous section, BPTT was discussed in detail. The LSTM utilizes BPTT in the same manner, but with additional internal operations per time step. The network is unrolled and the gradients are computed for all parameters (e.g. gates, input/output transforms, activations, etc.).

Each LSTM cell state contains what Hochreiter and Schmidhuber called the Constant Error Carousel (CEC), a self-recurrent pathway with an identity activation and unit weight that preserves error signals across time steps. This CEC mechanism ensures that, during BPTT, the gradient flowing through the cell state neither vanishes nor explodes. As a result, LSTMs can learn and retain long-range dependencies that vanilla RNNs struggle to capture.

2.3.4 GRU Networks

Chung et al. (2014) introduced gated recurrent units (GRUs) as a new type of RNN architecture that closely resembles the LSTM as can be viewed in Figure 2.12.

The most obvious difference between the LSTM and the GRU architectures is that the GRU has a hidden state running along the top of the unit and there is no cell state. With the LSTM, the hidden state ran along the bottom of the unit and the cell state ran along the top of the unit.

The GRU unit can be separated into three sections, similar to the LSTM. The first section is the reset gate. This gate passes the previous hidden state, h_{t-1}, and the current input, x_t, through a sigmoid activation network layer and multiples the vector of values between zero and one by the previous hidden state. This is similar to the forget gate of the LSTM. But instead of passing the product along to the cell state, it gets directed toward the third gate. The calculation of r_t is similar to f_t for the LSTM:

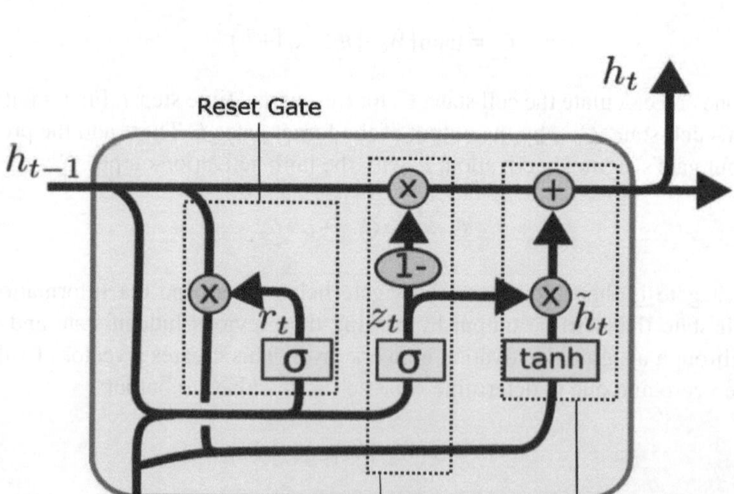

FIGURE 2.12 GRU architecture (Kyunghyun, 2014).

$$r_t = \sigma(W_r \cdot [h_{t-1}, x_t] + b_r) \quad (2.22)$$

The second section is the update gate. The update gate takes the same input as the reset gate and passes it through a sigmoid activation network layer. This vector of values between 0 and 1 is then passed in two directions, toward the third section and separately to the hidden state. The output of the sigmoid activation network layer, z_t, is calculated similarly to r_t:

$$z_t = \sigma\left(W_z \cdot [h_{t-1}, x_t] + b_z\right) \quad (2.23)$$

The third, and last, section is the new hidden state content, which closely resembles the last section of the LSTM. The LSTM utilized the updated cell state to add new content to the hidden state. Since the GRU architecture does not have a cell state, new hidden state content comes from the previous hidden state and the current input in two different forms. First, the reset gate controls what is kept from the previous hidden state. Then, the update gate controls what is kept from the reset hidden state. Together, the two gates and the current input determine new hidden state content.

The purpose of both the LSTM and GRU is to resolve the vanishing/exploding gradient problem. The LSTM adds more complexity with the cell state acting as an additional memory unit. This added complexity results in less efficient training but gives more control over the flow of information through each time step. In general, GRUs are preferred over LSTMs when training efficiency is more important than the need for additional memory units to support long-term dependencies.

2.3.5 Seq2Seq Models

Cho (2014a) and Sutskever et al. (2014) introduced two variations of sequence to sequence learning. In order to understand the history of architectures that influenced the creation of the transformer, sequence to sequence learning can be seen as the bridge that eventually connects purely recurrent networks to the attention-based transformer networks. Sequence to sequence learning can be thought of as a multi-layered LSTM, or GRU, encoder/decoder system.

The main use-case for sequence-to-sequence learning is machine translation, as well as free-form question answering, both of which take a variable-shaped sequence as input for the encoder to encode into a fixed-size vector, which is input into the decoder to decode into a variable-shaped sequence. This is useful for language translation applications. The input sequence could be an English sentence, and the output could be a German sentence. Any given sentence can vary in length and a sentence in English could have a different length of characters and words when compared with an equivalent German sentence.

A simplified visualization of the sequence-to-sequence model training can be viewed in Figure 2.6 which resembles the design in Cho et al. (2014b) where the final encoder hidden state is used as input at each decoder time step. Think of each block as an LSTM or GRU as shown in Figure 2.4 or 2.5. The encoder can be thought as having a single layer and a batch size of one. In practice, there would likely be

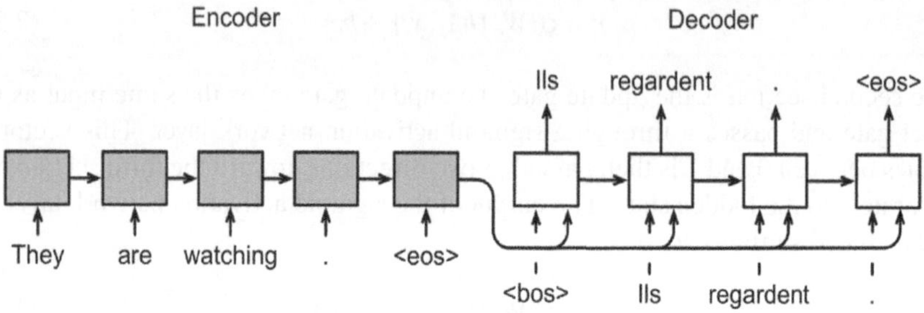

FIGURE 2.13 Sequence to sequence model training—English to German (Sutskever et al., 2014).

multiple layers and a larger batch size. Other designs (Sutskever et al., 2014) use the final encoder hidden state only to initialize the first-time step of the decoder (Figure 2.13).

Both designs follow the same overall process of encoding a variable-length sequence into a fixed-size context vector to be decoded into a variable-length sequence. The fixed-size vector can be thought of as a context vector, providing context of the input sequence to the decoder. In the (Sutskever et al., 2014) design, the decoder at the first time step has the only full-context view of the encoder input sequence. This design makes training difficult on long sequences where the decoder can lose context of the encoder input sequence after many time steps.

In the Cho et al. design, each decoder time step gets to reference the encoder context vector—allowing for no loss of information from the context vector through all decoder time steps. This provides later decoder time steps more context as it relates to the encoder input sequence. This design works better for longer sequences, but is still quite limited due to the nature of the fixed-size context vector.

The fixed-size context vector may be only able to hold a maximum amount of information no matter the input sequence size. Once this maximum is hit using the least number of tokens, any additional tokens will decrease the amount of context assigned to each token. This is where attention comes into play by allowing varying amounts of context on each encoder hidden state at each decoder time step.

Bahdanau et al. (2015) introduce an extension of the encoder–decoder sequence-to-sequence learning approach that acts as an attention mechanism. They saw the fixed-size context vector as a bottleneck due to the potential loss of context per token with long sentences as described above. They proposed a method of weighting each encoder hidden state vector when creating the context vector for each decoder time step to create a unique context vector at each decoder time step. This is accomplished by concatenating each encoder hidden state, h_j^{enc}, with the previous decoder hidden state, h_{t-1}^{dec}, and passing each pair through a learned alignment MLP model, m.

$$e_{ij} = m\left(h_{t-1}^{\text{dec}}, h_j^{\text{enc}}\right) \tag{2.24}$$

Then, the softmax activation function can be applied to normalize the values to produce weight values between zero and one that sum to one:

$$a_{ij} = \frac{\exp(e_{ij})}{\sum_{k=1}^{T} \exp(e_{ik})} \tag{2.25}$$

Each h_j^{enc} is then multiplied by its associated weight and then they are all summed giving us context vector c_i:

$$c_i = \sum_{j=1}^{T_x} a_{ij} h_j^{\text{enc}} \tag{2.26}$$

Thus, the decoder focuses more attention on the higher weighted encoder hidden states.

Later, in 2015, Luong introduced two attention mechanism classes that build upon and simplify the work of Bahdanau (Vaswani et al., 2017). The first is a global attention model that always attends to all encoder hidden states. The second being a local attention model that looks at only a subset of encoder hidden states. The global approach is similar to the implementation by Bahdanau with the main difference being how the similarity scores are calculated. Similar to how a_{ij} is calculated above, the alignment function, $a_t(s)$, is calculated as:

$$a_t(s) = \text{align}(h_t, \bar{h}_s) = \frac{\exp(\text{score}(h_t, \bar{h}_s))}{\sum_{s'} \exp(\text{score}(h_t, \bar{h}_{s'}))} \tag{2.27}$$

where h_t is the current decoder hidden state and $\bar{h}_s$ is each of the encoder hidden states, and the score content-based function provides three simplified (i.e., more efficient) calculations for calculating similarity between the current decoder hidden state and each encoder hidden state, such that:

$$\text{score}(h_t, \bar{h}_s) = \begin{cases} h_t^{\mathrm{T}} \bar{h}_s & \text{dot} \\ h_t^{\mathrm{T}} \mathbf{W}_a \bar{h}_s & \text{general} \\ v_a^{\mathrm{T}} \tan h[h_t; \bar{h}_s] & \text{concat} \end{cases} \tag{2.28}$$

allows for an attention module that can switch between each method. When calculating the similarity between the current decoder hidden state and any given encoder hidden state, the decoder hidden state plays the role of a query and each encoder state serves as both key and value. Dot and general scoring correspond to (projected) query–key dot products; concat is an MLP scoring variant.

The Transformer architecture (Vaswani) generalizes this idea by the introduction of separately learned projections for queries, keys, and values, applying attention within a sequence (self-attention) in parallel across all positions, using multiple heads, and dispensing with recurrence entirely. The next section introduces Transformers, where scaled multi-head self-attention becomes the primary mechanism for modeling dependencies.

2.3.6 Transformers

The goal of this section is to focus more deeply on how queries, keys, and values interact in the original Bahdanau attention formulation, in order to understand how they lay the groundwork for the more advanced self- and multi-head attention mechanisms used in Transformer architectures.

As mentioned above, the decoder hidden state plays the role of a query q, and each encoder hidden state serves as both key k_i and value v_i. We saw that

$$e_{ij} = a\left(\mathbf{h}_{i-1}^{\text{dec}}, \mathbf{h}_j^{\text{enc}}\right) \tag{2.29}$$

calculates a similarity between the previous decoder hidden state and the current encoder hidden state. Applying softmax over i normalizes these similarity scores into attention weights:

$$a_{ij} = \frac{\exp(e_{ij})}{\sum_k \exp(e_{kj})} \tag{2.30}$$

Another way to view these computations is explicitly in terms of queries, keys, and values. Let

$$q = \mathbf{h}_{i-1}^{\text{dec}},\ \mathbf{k}_j = \mathbf{h}_j^{\text{enc}},\ \mathbf{v}_j = \mathbf{h}_j^{\text{enc}} \tag{2.31}$$

First, one could normalize the raw scores:

$$\alpha(\mathbf{q}, \mathbf{k}_j) = \frac{a(\mathbf{q}, \mathbf{k}_j)}{\sum_j a(\mathbf{q}, \mathbf{k}_j)} \tag{2.32}$$

Then, equivalently, apply softmax to obtain the final weights:

$$\alpha(\mathbf{q}, \mathbf{k}_j) = \frac{\exp(a(\mathbf{q}, \mathbf{k}_j))}{\sum_j \exp(a(\mathbf{q}, \mathbf{k}_j))} \tag{2.33}$$

Here, q and k interact to produce a scalar score, while v carries the actual information to be aggregated into the context vector via those weights. The original additive score $a(q, k)$ becomes computationally expensive when there are many query–key pairs, since each score requires a concatenation and an MLP evaluation.

Luong et al. instead proposed the (scaled) dot-product scoring function, which computes

$$a(q,k_j) = \frac{q^{\mathrm{T}} k_j}{\sqrt{d}_k} \tag{2.34}$$

where the factor $1/\sqrt{d_k}$ stabilizes training by keeping the pre-softmax scores at a roughly constant scale. This scaled dot-product attention score is then normalized via softmax:

$$\alpha(q,k_j) = \mathrm{soft\,max}_i(a(q,k_j)) = \frac{\exp(q^{\mathrm{T}} k_j/\sqrt{d_k})}{\sum_j \exp(q^{\mathrm{T}} k_j/\sqrt{d_k})} \tag{2.35}$$

In the previous section, the single context vector was defined as

$$c_i = \sum_{j=1}^{T_x} \alpha_{ij} h_j^{\mathrm{enc}} \tag{2.36}$$

which collapses all encoder information—both short and long-range—into one summary. Such a single pooling operation can overemphasize nearby tokens while under-representing distant ones (or vice versa). In order to overcome this limitation, the Transformer computes H separate context vectors (one per "head") in parallel. For each head $h = 1,\ldots,H$, let

$$q = \mathbf{h}_{i-1}^{\mathrm{dec}}, \mathbf{k}_j = \mathbf{h}_j^{\mathrm{enc}}, v_j = \mathbf{h}_j^{\mathrm{enc}} \tag{2.37}$$

and define

$$a(q,k_j) = \left(qW^{(h)}{}_Q\right)\left(k_jW^{(h)}{}_K\right)^T / \sqrt{d_h} \tag{2.38}$$

$$\alpha^{(h)}(\mathbf{q},\mathbf{k}_j) = \frac{\exp(\mathrm{a}^{(h)}(\mathbf{q},\mathbf{k}_j))}{\sum_{m=1}^{T_s} \exp(a^{(h)}(\mathbf{q},\mathbf{k}_m))} \tag{2.39}$$

Then each head produces

$$\mathrm{head}_h = \sum_{j=1}^{T_x} \alpha^{(h)}(\mathbf{q},\mathbf{k}_j)(\mathbf{v}_j, W_V^{(h)}) \tag{2.40}$$

Finally, the multi-head output is a concatenation of h heads, given by

$$\mathrm{Multihead}(Q,K,V) = \mathrm{Concat}(\mathrm{head}_1,\ldots,\mathrm{head}_H)W_O \tag{2.41}$$

With cross-attention (encoder-to-decoder attention), the decoder's queries interact with encoder-derived keys and values via H parallel heads. The same multi-head mechanism can also be applied within a single sequence: we simply let each token's own representation play the roles of query, key, and value, so that every position can directly attend to every other. This intra-sequence application of multi-head attention is what is called self-attention, and it lies at the heart of the Transformer's fully attention-based encoder (and decoder) layers.

The Transformer architecture removes recurrent states and relies entirely on parallel self-attention, so the positional order of tokens is no longer encoded. To remedy this, the Transformer adds learned (or sinusoidal) positional encodings to each token embedding before feeding them into the attention layers (Figure 2.14).

In Figure 2.14 the left half of the model is the encoder and the right half is the decoder, both built from stacked self-attention and position-wise feed-forward sub-layers. Each encoder layer comprises two sub-layers: a multi-head self-attention mechanism and a position-wise fully connected feed-forward network. The decoder adds a third sub-layer—masked multi-head self-attention—to prevent attending to future positions. Both encoder and decoder apply residual connections around each sub-layer, followed by layer normalization (borrowed from ResNet).

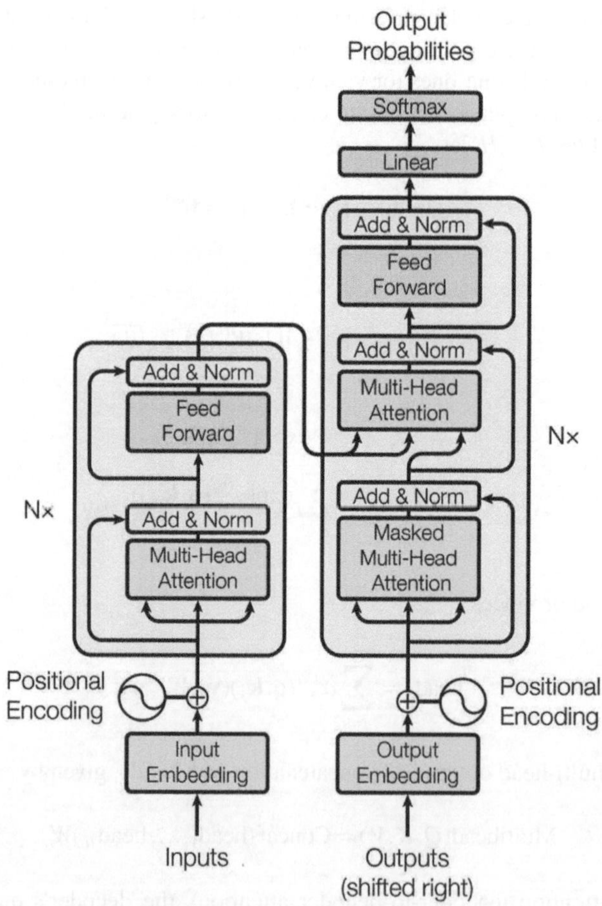

FIGURE 2.14 The transformer—model architecture (Vaswani).

In summary, a Transformer layer replaces recurrence with pure attention: first each token's representation is augmented by a positional encoding, then multi-head self-attention lets every position attend to every other in parallel, and finally a small point-wise feed-forward network refines the result. Residual connections and layer normalization ensure stable gradients and fast convergence. By stacking these layers (plus occasional cross-attention in the decoder), the model learns rich contextual representations without any sequential bottleneck.

2.4 ETHICAL CONSIDERATIONS OF LARGE LANGUAGE MODELS: BIAS, FAIRNESS, PRIVACY, AND HALLUCINATIONS

The lineage of *Homo sapiens* dates back to over 5 million years ago with the anatomically modern physical forms developing around 300,000 years ago. The early origins were in Africa and have spread around the world; since then the primary occupation of *Homo sapiens* for much of its existence was the hunting of wild animals and foraging of wild plants. The development of tools, plant domestication, and other corresponding advances in industry and digital domains are only very recent in retrospect.

While LLMs consisting of large numbers of neurons are useful in addressing global communication, these early systems required much less in the way of computational intelligence. In these early environments, travel was limited and foraging strategies could be developed that were able to optimize productivity in such local areas using very simple languages. For example, in a simple binary vector voting model, a single leader uses weighted contributions from the band to determine the next direction a groups movement. The individual experiences were useful in biasing decisions in more successful patterns over time (Zeigler and Reynolds, 1978). The architecture of our neural systems were evolved to support these biases. These cultural biases are built into our views of the world from an evolutionary point of view. However, as our communication channels and needs have increased exponentially; recently, it has been shown that LLMs capture discourse and corresponding biases from cultures worldwide. This can lead to the subtle reinforcement of traditional biases and their reexpression in ways that are fundamentally hard to directly identify (Reynolds, 2021).

As a result, LLMs have demonstrated remarkable capabilities in generating human-like text, aiding research, supporting education, and enhancing productivity. However, the content from which they are derived is inherently prone to a variety of biases that reflect the limited understanding of those individuals who produced that content. The remainder of the section will discuss four of the basic ethical issues that derive from these limitations. The four basic issues are Bias, Fairness, Privacy, and Hallucinations. We then briefly discuss the reasons for each in terms algorithms, learning metrics, hardware, and human cognition.

2.4.1 Bias Generation in Large Language Models

According to the Merriam-Webster Dictionary, Bias is "an inclination of temperament or outlook; especially: a personal and sometimes unreasoned judgment." There are several types of bias:

- **Implicit Bias**: Unconscious attitudes or stereotypes that influence perceptions and decision making.
- **Confirmation Bias**: The tendency to search for or interpret information in a way that confirms one's preconceptions.
- **Self-Serving Bias**: Attributing successes to personal factors while blaming failures on external factors.
- **Status Quo Bias**: A preference for the current state of affairs, resisting change.
- **Social and Institutional Bias**: Biases can be embedded in societal structures, leading to systemic discrimination based on race, gender, or other identities.
- **Statistical Bias**: Systematic errors in data collection or analysis that lead to inaccurate results.
- **AI Bias**: AI systems can inherit biases present in their training data, leading to unfair outcomes.

An LLM model reproduces and perhaps can even amplify any one or all of these existing biases when they are present in the data they were trained on based on. The nature of the training process can impact the presence of bias as well. Since LLMs learn from large-scale text corpora sourced from the internet and other real-world documents, they inevitably reflect social and cultural biases that are implicitly embedded in those documents and can lack the context in which the text is generated. Bias becomes ethically problematic when it impacts marginalized groups or reinforces existing discrimination.

For example, Bolukbasi et al. (2016) demonstrated the presence of gender bias in word embeddings with associations such as "man:computer programmer" and "woman:homemaker" emerged. Also, in 2018, researchers demonstrated that Google Translate **defaulted to male pronouns** when translating gender-neutral sentences from Turkish into English. For example, "O bir doktor" (gender-neutral in Turkish) was translated as "He is a doctor," while "O bir hemşire" became "She is a nurse." Although more recent LLMs are designed to mitigate such issues through techniques like reinforcement learning from human feedback, bias still persists (Bender et al., 2021; Gehman et al., 2020).

2.4.2 Fairness and Equity

Fairness as a noun is "the quality or state of being fair especially: **fair or impartial treatment" according to the Webster-Merriam Dictionary.** Fairness in AI systems involves ensuring equitable treatment and benefits across different demographic categories, including race, gender, socio-economic status, and geography. LLMs can generate content or make decisions that affect access to opportunities, information, and services. For example, LLMs trained primarily on English-language internet data often perform poorly on low-resource languages or dialects, thereby exacerbating digital divides (Joshi et al., 2020).

Unfairness in LLMs is a systemic issue arising from complex interactions among algorithmic design, optimization procedures, hardware constraints, and human cognitive biases. Tackling these issues requires a shift from narrow performance metrics toward value-sensitive design, participatory dataset curation, diversified annotator pools, and transparent model governance. Efforts such as algorithmic audits, fairness benchmarking, and culturally grounded annotation are steps in the right direction, but deeper interdisciplinary engagement—spanning AI ethics, cognitive science, and social theory—is needed to ensure that future LLMs are not only powerful but just.

2.4.3 Data Privacy and Data Security

Data privacy is defined as the **right of individuals to control how their personal information is collected, used, and shared**. It emphasizes **consent**, **transparency**, and **individual rights** regarding the processing of personal data. Security is "the appropriate use of data, especially personal data, in compliance with privacy laws and regulations," while **data security** refers to the **technical and organizational measures** used to perform these activities and protect **data from unauthorized access, corruption, theft, or loss**. It involves tools like **encryption**, **access control**, and **firewalls** to ensure **confidentiality, integrity, and availability** of data (Security, 2025).

Privacy concerns arise when LLMs inadvertently memorize and reproduce sensitive personal information. During training, LLMs may be exposed to unredacted emails, phone numbers, or confidential content from leaked or improperly curated datasets. Carlini et al. (2021) showed that LLMs can "leak" training data when prompted in specific ways, raising alarms about compliance with data protection laws like the General Data Protection Regulation. This is particularly troubling in medical, legal, and educational applications, where personal data is highly sensitive. On **March 20, 2023**, OpenAI confirmed that a **bug in ChatGPT's open-source library Redis** exposed parts of a users' conversation histories and possibly even account information to other users. Chat snippets from unrelated users appeared in individual accounts.

Ethically, model developers must ensure rigorous data anonymization, implement data minimization principles, and allow for user consent and data erasure where applicable. Differential privacy and federated learning are emerging approaches aimed at enhancing privacy-preserving machine learning. Security and privacy challenges in LLMs are not incidental—they emerge from the very design of algorithms, the myopic focus of optimization metrics, infrastructural weaknesses in hardware, and cognitive misunderstandings by end users. The ability of a system to address these issues demands a multifaceted strategy that can incorporate one or more of the following:

- **Technical**: Integrate differential privacy, auditability, and adversarial robustness into training.
- **Hardware**: Employ trusted computing, secure enclaves, and encrypted communication during training and inference.
- **Organizational**: Adopt responsible data handling practices and provide clear user interfaces for data control and consent.
- **Societal**: Increase public and professional literacy about how LLMs handle sensitive data.

2.4.4 Hallucinations and Misinformation

LLM hallucinations refer to instances where an **LLM generates content that can contain factual inaccuracies** (e.g., misquote a statistic), invented references (e.g., cite a non-existent study or case), or logical inconsistencies or contradictions (Hallucinations, 2025). These can mislead users, especially in domains like medicine, law, or education, where accuracy is required.

Hallucinations stem from the probabilistic nature of LLMs. These models do not understand facts; they predict the next token based on statistical patterns. As a result, they can generate citations to non-existent papers or offer dangerously wrong medical advice that is statistically possible. In **May 2023**, two New York attorneys used ChatGPT to help write a legal brief. The LLM generated **fake judicial decisions and citations**, which the attorneys then submitted in court. The judge discovered the hallucinations and sanctioned the lawyers (Schwartz and LoDuca, 2023).

The ethical implications of hallucinations can erode public trust, fuel conspiracy theories, and cause tangible harm. Hallucinations in LLMs are a structural outcome of how these models are trained, evaluated, deployed, and used. While the underlying algorithms promote fluent language, they can lack grounding in verified knowledge. In addition, training metrics reward syntactic correctness and style over semantic truth. Hardware constraints also necessitate trade-offs that reduce the quality of a result. Finally, human users often just lack the tools to identify hallucinated content. To address hallucinations meaningfully, developers must integrate retrieval-based grounding and fine-tuning with factual datasets along with hallucination-aware loss functions. **In terms of hardware**, context window optimization and inference stability need to be balanced against performance. Finally, users need to be educated about the limitations of LLMS both cognitively and socially.

2.5 CONCLUSIONS AND FUTURE DIRECTIONS

In this chapter, a brief history of LLMs for NLP was presented. The origins of such neural architectures were shown to derive from early simple models of computation, the perceptron. With the perceptron as the building block, a basic overview of LLMs was presented. While these models exhibited some unprecedented capabilities in terms of education, science, health care, and human–computer interaction, there were concerns relating to issues of bias, fairness, privacy, and correctness. These issues were discussed in Section 2.4.

While current models are able to support general purpose conversation, the next generation of LLMs will need to address current deficiencies in the areas of computational inefficiency, limited reasoning and agency, bias, and fairness, along with the reduction of hallucinations. Future work with LLMs will go beyond increasing the number of model parameters. New model developments will address issues with the integration of multiple dimensions; reasoning across models; the alignment of model output with concepts of personalization, privacy, ethics, and fairness; the increased focus on adaptation to specific domains; and increased agency in smart domains.

REFERENCES

Amari, S. I. (1968). Geometrical Theory of Information. Kyoritsu-Shuppan, Tokyo, Japan.

Bahdanau, D., *Cho*, K., and *Bengio*, Y. (2015). Neural machine translation by jointly learning to align and translate. In: *International Conference on Learning Representations (ICLR),* San Diego, CA, USA.

Bender, E. M., *Gebru*, T., *McMillan-Major*, A., and *Shmitchell*, S. (2021). On the dangers of stochastic parrots: Can language models be too big? In: *Proceedings of the 2021 ACM Conference on Fairness, Accountability, and Transparency (*FAccT '21*)*, Virtual Event, Canada, pp. 610–623.

Bemzio, Y., Ducharme,R., Vincent, P., and Jarvin, C. (2003). A neural probabilistic language model. The Journal of Machine Learning Research, 3, 1137–1155.

Bolukbasi, T., *Chang*, K. W., *Zou*, J., *Saligrama*, V., and *Kalai*, A. (2016). Man is to computer programmer as woman is to homemaker? Debiasing word embeddings. In: *NeurIPS 2016 Conference Proceedings,* Barcelona, Spain, pp. 4349–4357.

Carlini, N., *Tramèr*, F., *Wallace*, E., Jagielski, M., Herbert-Voss, A., Lee, K., Roberts, A., Brown, T., Song, D., Erlingsson, Ú., Oprea, A., and Raffel, C. (2021). Extracting training data from large language models. In: *Proceedings of the 30th USENIX Security Symposium (USENIX Security 2021),* Vancouver, BC, Canada, pp. 2633–2650. https://www.usenix.org/conference/usenixsecurity21/presentation/carlini-extracting.

Chen, T., *Kornblith*, S., *Norouzi*, M., and *Hinton*, G. (2020). A simple framework for contrastive learning of visual representations (SimCLR). In: *International Conference on Machine Learning (ICML),* Vienna, Austria, PMLR.

Cho, K., *Van Merriënboer*, B., *Gülcehre*, Ç., *Bahdanau*, D., *Bougares*, F., *Schwenk*, H., and *Bengio*, Y. (2014a). On the Properties of Neural Machine Translation: Encoder–Decoder Approaches. arXiv preprint arXiv:1409.1259.

Cho, K., *Van Merriënboer*, B., *Gülcehre*, Ç., *Bahdanau*, D., *Bougares*, F., *Schwenk*, H., and *Bengio*, Y. (2014b). Learning phrase representations using RNN encoder–decoder for statistical machine translation. In: *Proceedings of the 2014 Conference on Empirical Methods in Natural Language Processing (EMNLP),* Doha, Qatar, pp. 1724–1734.

Chung, J., Gulchere, C., Cho, K. H., and Bengio, Y. (2014). Empirical evaluation of gated recurrent neural networks on sequence modeling. https://doi.org.10.48550/arXiv.1412.355

Devlin, J., *Chang*, M.-W., *Lee*, K., and *Toutanova*, K. (2018). *BERT: Pre-Training of Deep Bidirectional Transformers for Language Understanding*. arXiv preprint arXiv:1810.04805.

Duchi, J., *Hazan*, E., and *Singer*, Y. (2011). Adaptive subgradient methods for online learning and stochastic optimization. *Journal of Machine Learning Research*, 12, 2121–2159.

Elman, J. L. (1990). Finding structure in time. *Cognitive Science*, 14(2), 179–211.

Gehman, S., Gururangan, S., Sap, M., Choi, Y., and Smith, N. A. (2020). Real toxicityprompts: evaluating neural toxic degeneration in language models. In Cohn T, He Y, Liu Y (Eds.), *Findings of the Association for Computational Linguistics: EMNLP 2020*. Association for Computational Linguistics, Stroudsburg, PA, USA, pp. 3356–3369. https://doi.org/10.18653/v1/2020.findings-emnlp.301.

Hallucination (Artificial Intelligence). (2025). Wikipedia. https://en.wikipedia.org/wiki/Hallucination_(artificial_intelligence).

He, K., *Zhang*, X., *Ren*, S., and *Sun*, J. (2015). *Deep Residual Learning for Image Recognition.* arXiv preprint arXiv:1512.03385.

Hinton, G. E., *Osindero*, S., and *Teh*, Y.-W. (2006). A fast learning algorithm for deep belief nets. *Neural Computation*, 18(7), 1527–1554.

Hochreiter, S., and *Schmidhuber*, J. (1997). Long short-term memory. *Neural Computation*, 9(8), 1735–1780. https://doi.org/10.1162/neco.1997.9.8.1735.

Hopfield, J. J. (1982). Neural networks and physical systems with emergent collective computational abilities. *Proceedings of the National Academy of Sciences*, 79(8), 2554–2558.

Information *Security*. (2025). Wikipedia, The Free Encyclopedia. https://en.wikipedia.org/wiki/Information_security.

International Association of Privacy Professionals. (2023). *What Is Privacy?* https://iapp.org/about/what-is-privacy/.

Jordan, M. I. (1986). Serial order: a parallel distributed processing approach. In: *Proceedings of the Eighth Annual Conference of the Cognitive Science Society,* Cognitive Science Society, Ann Arbor, MI, USA, pp. 531–546

Joshi, P., *Santy*, S., *Budhiraja*, A., *Bali*, K., and *Choudhury*, M. (2020). The state and fate of linguistic diversity and inclusion in the NLP world. In: *Proceedings of the 58th Annual Meeting of the Association for Computational Linguistics (ACL* 2020*)*, Stroudsburg, PA, USA, pp. 6282–6293. https://doi.org/10.18653/v1/2020.acl-main.560.

Karpathy, A. (2015). *Minimal Character-Level Vanilla RNN Model.* https://gist.github.com/karpathy/d4dee566867f8291f086.

Krizhevsky, A., *Sutskever*, I., and *Hinton*, G. E. (2012). ImageNet classification with deep convolutional neural networks. *Advances in Neural Information Processing Systems (NeurIPS)*, 25, 1097–1105.

Lucci, S., *Musa*, S. M., and *Kopec*, D. (2022). *Artificial Intelligence in the 21st Century*, 3rd edn. Mercury Learning and Information, Dulles, VA. https://doi.org/10.1515/9781683922520.

McCulloch, W. S., and *Pitts*, W. (1943). A logical calculus of the ideas immanent in nervous activity. *The Bulletin of Mathematical Biophysics*, 5, 115–133.

Minsky, M., and *Papert*, S. (1969). *Perceptrons: An Introduction to Computational Geometry.* MIT Press, Cambridge, MA, USA.

Pascanu, R., *Mikolov*, T., and *Bengio*, Y. (2013). On the difficulty of training recurrent neural networks. In: *Proceedings of the 30th International Conference on Machine Learning (ICML),* Atlanta, GA, PMLR, pp. 1310–1318.

Radford, A. et al. (2018–2023). *GPT Series. OpenAI, S*an Francisco, CA.

Rasvan, V. (2013). *The seminar "Qualitative theory of differential equations and control theory" – 60 years.* Annals of the Academy of Romanian Scientists. *Mathematics and Its Applications*, Volume 5. 3-12.

Reynolds, R. G. (2021). Cultural Algorithms:Tools to Model Complex Dynamic Soial Systems, IEEE/John W. Wiley Press, Hoboken, NJ.

Rosenblatt, F. (1958). The perceptron: a probabilistic model for information storage and organization in the brain. *Psychological Review*, 65(6), 386–408.

Rumelhart, D. E., *Hinton*, G. E., and *Williams*, R. J. (1986). Learning representations by back-propagating errors. *Nature*, 323(6088), 533–536.

Schwartz, S., and *LoDuca*, P. (2023). Mata v. *Avianca, Inc., U.S. District Court*, Southern District of New York, *Case* No. 1:22-cv-12345. https://www.reuters.com/legal/new-york-lawyers-sanctioned-using-fake-chatgpt-cases-legal-brief-2023-06-22/.

Shannon, C. E. (1948). A mathematical theory of communication. *Bell System Technical Journal*, 27(3), 379–423 and 27(4), 623–656. https://doi.org/10.1002/j.1538-7305.1948.tb00917.x.

Silver, D. et al. (2016). Mastering the game of go with deep neural networks and tree search. *Nature*, 529(7587), 484–489.

Sutskever, I., *Vinyals*, O., and *Le*, Q. V. (2014). Sequence to sequence learning with neural networks. *Advances in Neural Information Processing Systems,* 27, 3104–3112.

Vaswani, A., *Shazeer*, N., *Parmar*, N., *Uszkoreit*, J., *Jones*, L., *Gomez*, A. N., *Kaiser*, Ł., and *Polosukhin*, I. (2017). Attention is all you need. *Advances in Neural Information Processing Systems*, 30, 5998–6008.

Weizenbaum, J. (1966). ELIZA—a computer program for the study of natural language communication between man and machine. *Communications of the ACM*, 9(1), 36–45. https://doi.org/10.1145/365153.365168.

Werbos, P. J. (1990). Backpropagation through time: what it does and how to do it. *IEEE*, 78(10), 1550–1560.

Winograd, T. (1971). *Procedures as a Representation for Data in a Computer Program for Understanding Natural Language*. MIT AI Technical Report 235, Massachusetts Institute of Technology, Cambridge.

Zeigler, B. P., and *Reynolds*, R. G. (1985). A hierarchical information processing model for adaptation to technological change. *Systems Research*, 2(4), 309–317.

3 LLMs in Cultural Heritage

Iosif Tsangko, Andreas Triantafyllopoulos, and Bjoern W. Schuller

ABBREVIATIONS

AI	Artificial Intelligence
ALM	Audio Language Model
BERT	Bidirectional Encoder Representation from Transformers
BLEU	Bilingual Evaluation Understudy
CH	Cultural Heritage
GloVe	Global Vectors for Word Representation
GPT	Generative Pre-trained Transformer
GRU	Gated Recurrent Unit
K	Key
KG-LMs	Knowledge Graph-Augmented Language Models
LLAVA	Large Language and Vision Assistant
LLM	Large Language Model
LoRA	Low-Rank Adaptation
LSTM	Long Short-Term Memory
MHA	Multi-Head Attention
NLP	Natural Language Processing
OCR	Optical Character Recognition
PaLM	Pathways Language Model
POS	Part-of-Speech
Q	Query
RAFT	Retrieval-Augmented Fine-Tuning
RAG	Retrieval-Augmented Generation
RLHF	Reinforcement Learning with Human Feedback
RNN	Recurrent Neural Networks
ROUGE	Recall-Oriented Understudy for Gisting Evaluation
V	Value
VLM	Vision Language Model

3.1 INTRODUCTION

Large language models (LLMs) have dramatically broadened the horizons of artificial intelligence (AI), with their capabilities evolving at an unprecedented pace. In this chapter, we offer a comprehensive exploration of how these models are reshaping the landscape of cultural heritage (CH)—from preservation and interpretation to accessibility and engagement. Our goal is to illuminate both established applications

DOI: 10.1201/9781003492252-3

and emerging possibilities, reflecting the breadth and depth of this rapidly advancing intersection between technology and cultural understanding.

CH encompasses the legacy of physical artifacts and intangible attributes of a society that are inherited from past generations, maintained in the present, and bestowed on future generations. This includes monuments, historical buildings, artworks, literature, and less tangible elements like traditions, languages, and collective memories. These cultural expressions shape our understanding of identity, community, and the past, playing a critical role in defining societal values and historical continuity (Gîrbacia, 2024). Preserving CH is essential not only for its intrinsic value but also for its potential to educate and inspire future generations. By maintaining and documenting cultural assets, societies aim to build bridges between the past, present, and future, providing a shared sense of identity and continuity. This preservation, however, faces numerous challenges, including natural decay, environmental threats, and conflicts that can lead to the destruction of heritage sites (Wagner, 2023). Furthermore, digital globalization and the evolution of information technology have presented both opportunities and challenges in maintaining and accessing cultural records. Recently, advances in AI, specifically in LLMs, have opened new avenues for preserving, analyzing, and disseminating CH. LLMs, such as the closed-sourced Generative Pre-trained Transformer (GPT) variants (Kalyan, 2023), open-sourced LLaMA (Touvron, 2023) models and similar architectures, are AI systems trained on vast amounts of text data to understand and generate human-like language. Their capabilities include natural language processing (NLP), advanced text analysis, multilingual understanding, and generating coherent, contextually relevant outputs. These features make them valuable tools for CH projects. Researchers have extended the definition of LLMs connecting textual data to modalities such as vision and audio, resulting in vision language models (VLMs) (Zhang, 2024a) and audio language models (ALMs) (Triantafyllopoulos, 2024), respectively. The term *Foundation Models* has also risen in the literature, referring to LLMs and their multimodal counterparts. However, in this chapter, we will utilize the former acronym and focus primarily on linguistic capabilities.

Integrating LLMs into CH initiatives enables more efficient and accurate digitization of ancient texts and documents. For instance, LLMs can perform relevant tasks such as automated translation and contextualization of historical records (Spina, 2023). Additionally, multimodal language models can enhance the accuracy of transcriptions by exploiting cues such as vision ones (e.g., from paintings) and unlocking access to previously obscure or inaccessible materials for scholars and the general public (Rachabatuni, 2024). This technology is also transformative in virtual museums and interactive exhibits, where LLMs can provide dynamic, multilingual descriptions and personalize visitor experiences, enhancing engagement and learning.

Beyond practical applications, LLMs support academic research by facilitating collaborative efforts across institutions and disciplines (Cui, 2024). Through AI-powered platforms, researchers can share data, refine interpretations, and develop new insights into cultural history. Unlike traditional databases, LLMs can offer significant advantages by improving search functionality through their ability to understand context and semantics, leading to more relevant and accurate information retrieval (Lee, 2023). For instance, AI-driven platforms can classify CH data more effectively, as demonstrated by machine learning applications that improve image classification and automated discovery of patterns in vast datasets (Fiorucci, 2020). This capability can enhance, for instance, the modelling and simulation of ancient languages and cultural practices, which could aid historians and archaeologists in reconstructing lost or lesser-known aspects of past societies.

However, the implementation of LLMs in CH is not without challenges.

Issues such as the need for domain-specific fine-tuning, potential biases in AI outputs, and ethical considerations around the representation of sensitive cultural information must be carefully addressed. Despite these obstacles, the potential of LLMs to revolutionize CH preservation and curation is undeniable, offering opportunities to broaden access, deepen understanding, and engage diverse audiences worldwide.

This chapter explores these possibilities, highlighting case studies and emerging technologies, and demonstrating how LLMs can reshape how we interact with and understand CH.

3.1.1 Structure of the Chapter

To guide readers through the chapter's scope, the following sections are organized to progressively deepen the understanding of how LLMs intersect with CH. Section 3.2 introduces the foundational technologies behind LLMs, including their architectural principles and advancements in attention mechanisms and multimodal learning. Section 3.3 explores the practical capabilities of LLMs and their relevance to CH, highlighting how these models enable new forms of analysis, preservation, and engagement. Section 3.4 presents a range of case studies that demonstrate real-world applications of LLMs in digitization, exhibition design, and collaborative research. Section 3.5 focuses on accessibility, showing how LLMs can be leveraged to support diverse audiences, including those with disabilities or language barriers. Section 3.6 then discusses the challenges of using LLMs in CH, from technical limitations to ethical considerations, and concludes with a look at emerging directions and long-term implications. The final section summarizes the chapter's contributions and reflects on the future of LLM integration in CH initiatives.

3.2 TECHNICAL ADVANCEMENTS

NLP has evolved significantly, beginning with simpler statistical methods before advancing to neural network-based models that could handle sequential dependencies more effectively. Early NLP models like n-grams and bag-of-words approaches were limited by their reliance on statistical co-occurrence and simple linear representations (Sharma, 2017). These methods were eventually succeeded by recurrent neural networks (RNNs) (Su, 2022), which introduced a sequential processing paradigm. RNNs, by design, allowed for a concept of "memory," enabling the network to retain information from earlier input while processing subsequent ones, thus capturing dependencies within the language sequence. However, traditional RNNs faced several limitations. Their sequential nature hindered parallelization, and they struggled with long-range dependencies due to issues such as vanishing gradients, making it difficult to retain information across lengthy text spans.

TABLE 3.1
Overview of Key NLP Model Architectures and Capabilities

Model Type	Key Features	Strengths	Limitations
N-gram/ Bag-of-words	Statistical models using co-occurrence	Simplicity, fast to compute	Ignores word order, lacks context
RNN	Sequential processing, memory of prior input	Captures sequence dependencies	Struggles with long-range dependencies
LSTM/GRU	Gating mechanisms to manage memory	Handles longer sequences better	Limited parallelism
Transformer	Attention mechanism, parallel processing	Long-range context, high scalability	Requires large datasets and computing resources
LLMs (e.g., GPT, BERT)	Deep transformer-based, massive scale	Multilingual, generative, contextual understanding	Bias, hallucination, resource-intensive

This led to the development of long short-term memory (LSTM) and gated recurrent unit (GRU) networks, which partially mitigated these issues by introducing "gating" mechanisms (see Table 3.1). Recent innovations like Mamba (Gu, 2023) and xLSTMs (Beck, 2024) have sought to address the limitations of traditional RNN-based models with the former leveraging parallelism for improved sequence processing and the latter introducing extended memory mechanisms to capture dependencies across longer spans. Despite these advancements, LLMs remain the gold standard in sequential language modeling, outperforming these approaches in both scalability and contextual understanding.

3.2.1 Background of Attention Mechanisms in Transformers

At their core, LLMs leverage the ability of transformers to process and understand relationships across entire input sequences simultaneously. This ground-breaking shift occurred with the introduction of attention mechanism, as pioneered by the transformer architecture proposed by Vaswani in the paper "Attention Is All You Need" (Vaswani, 2017). Attention (Bahdanau, 2014) enabled the model to assign varying levels of focus, or "attention," to different parts of the input sequence, allowing for a more nuanced and context-aware processing of the words' representations, namely, tokens. The mechanism shifted the language modelling paradigm away from a purely sequential focus, where each word is directly linked to its previous one, to one where relationships between all tokens are considered concurrently. This paved the way for the Transformer model's encoder-decoder structure, where the self-attention mechanism allows each token to "attend" to every other token in the sequence. The concept of multi-head attention (MHA) expanded upon this by allowing multiple sets of attention mechanisms to operate in parallel, effectively creating a more complex and nuanced representation of language. In MHA, the input sequence

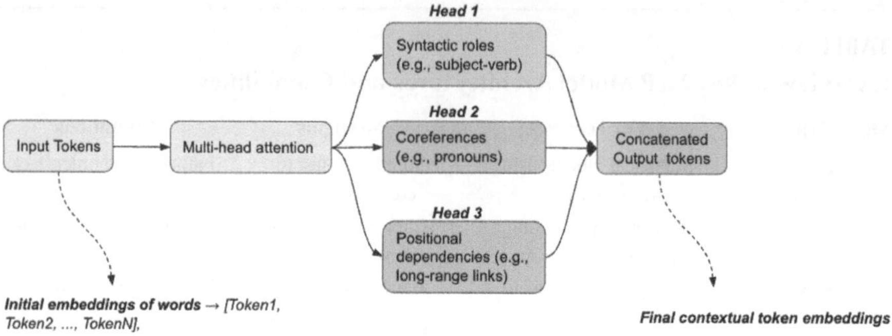

FIGURE 3.1 Multi-head attention (MHA) mechanism in transformers.

undergoes several independent attention calculations, each producing its own attention distribution over the input sequence. These distributions, or "heads," capture different relationships within the data—such as syntactic dependencies, semantic meanings, or hierarchical linguistic structures—and collectively provide a richer understanding of the input.

In detail, each attention head operates by computing three vectors: the query (Q), key (K), and value (V). The self-attention mechanism then calculates attention weights by computing dot products between Q and K, normalizing the values, and applying them to the V vectors. These vectors are linearly projected into higher-dimensional spaces, where they focus on specific aspects of the data. For instance, one head might capture the relationship between subjects and verbs, while another focuses on long-range dependencies like pronoun antecedents. The results from each head are concatenated and linearly transformed, enabling the model to capture a wide array of linguistic features (Figure 3.1).

The MHA mechanism enhanced transformers' representational power by allowing the model to process different aspects of the data in parallel. Unlike RNNs, transformers avoid the pitfalls of sequential processing, allowing for parallelization and efficient handling of long sequences. This transition from sequential models to transformer-based architectures marked a fundamental shift in NLP, where models could now "see" all tokens in a sequence simultaneously, making them highly efficient for both training and inference.

3.2.2 Impact of Masked Token Prediction in Training LLMs

Following the development of MHA and self-attention, a pivotal advancement in transformers was adapting the "masked token prediction" (pre)training strategy, particularly for training bidirectional transformer models such as Bidirectional Encoder Representations from Transformers (BERT). A masked token prediction is a form of language modelling in which certain tokens in a sentence are "masked" during training. The model is tasked with predicting these masked tokens based on the surrounding context, thereby encouraging the model to learn bidirectional dependencies. This bidirectional training process contrasts with the autoregressive approach used in models like GPT, where each token is generated sequentially based on previous tokens only. **Autoregressive** models are inherently directional, as they rely on past context without "looking ahead." This is also known as **causal masking,** where the model is trained in predicting the next most likely word (token). The core module of an LLM is (almost always) the "Decoder" module of the Transformer architecture, which engages with generating new (textual) data. Therefore, the characterization of LLMs as "**generative**" models was chosen. The masking approach enhances the model's capture of semantic nuances and contextual relationships. For instance, if a model is trained on a vast corpus of Shakespeare's poems, it will eventually internalize the dynamics, semantics, and stylistic patterns of this form of language.

Following this initial pre-training, the model's capabilities are largely limited to text completion, driven by its training on next-token prediction within the framework of masked attention. If the model is prompted with a fragment of text, it will only be able to generate continuations that adhere to the same poetic style, linguistic structures, and contextual cues, continuing the generation while imitating the original dynamics. However, achieving the communicative proficiency and contextual awareness characteristic of advanced chatbots requires further stages of training and refinement (see below).

3.2.3 Insights from Attention Blocks across Textual Fields

The attention blocks within transformer models, particularly in the context of MHA, provide intricate insights across various textual fields. In each transformer layer, attention blocks iteratively refine the model's understanding of the text by integrating information from different parts of the sequence. Each layer captures increasingly abstract representations, ranging from shallow syntactic structures in lower layers to high-level semantics in upper layers. Attention weights within each block reveal patterns regarding how a model perceives textual elements, which can be leveraged to improve interpretability in CH tasks. For example, attention layers in lower blocks may focus on word order and local context, while higher blocks capture broader relationships such as themes or narrative arcs, which are invaluable for tasks involving document contextualization or topic modelling.

These layered representations make transformers particularly well-suited for multilingual contexts within CH, where relationships across languages and cultural contexts can vary widely. One of the most transformative aspects of attention in LLMs is the emergence of "contextual embeddings" or the model's ability to generate token embeddings that vary depending on context. This contrasts with traditional static word embeddings, such as Word2Vec (Mikolov, 2013) or GloVe (Pennington, 2014), where a word has a single embedding, regardless of its meaning in different contexts. In transformers, attention blocks enable tokens to develop context-sensitive embeddings, which are critical for capturing the nuances of language and specific terms in domains such as CH. Contextual embeddings have proven especially effective for tasks such as translation and sentiment analysis, where word meanings shift depending on other surrounding context. In multilingual settings, transformers can align semantically similar concepts across languages, aiding in tasks such as cross-linguistic document alignment, multilingual retrieval, and automated translation of cultural texts.

3.2.4 An Example Walkthrough: A Culturally and Linguistically Adaptive Chatbot System

Imagine a chatbot system[1] designed for a virtual museum exhibit (see Figure 3.2). This system leverages a decoder-only transformer model equipped with MHA and masked token prediction to process user queries, retrieve relevant content, and generate culturally sensitive responses. We present a step-by-step breakdown of how the LLM (the core module of the chatbot) processes a sentence from the user ("Can you tell me about the traditional art of the Hmong people?") and adapts it for multilingual and culturally nuanced responses.

Phase 1: Input Parsing and Tokenization. Upon receiving the user's input, the chatbot first tokenizes the sentence into sub-word units or tokens, which are processed individually in a way that allows the model to consider each unit's contextual significance within the sentence. Tokenization ensures that the model can handle complex words and phrases that may have specific meanings in a CH context (e.g., "Hmong" as an ethnonym). Since transformers process sequences non-sequentially, the model can immediately attend to tokens like "Hmong" and "traditional art" without waiting to process all previous words. This ability to access any token's context through the entire sequence allows the model to "understand" the user's intention, even if the phrasing or word order differs across languages.

Phase 2: Contextual Representations with MHA. In this phase, the model leverages MHA to create a rich, context-aware representation of each token. Each attention *head* in the MHA attends to various aspects of the sentence, such as the cultural specificity of "Hmong," the concept of "traditional art," and how these elements interact within the user query. For instance:

Attention Head 1: may focus on the syntactic structure, identifying "Hmong people" as a noun phrase.

Attention Head 2: could highlight culturally relevant terms like "traditional art" and relate them to Hmong culture.

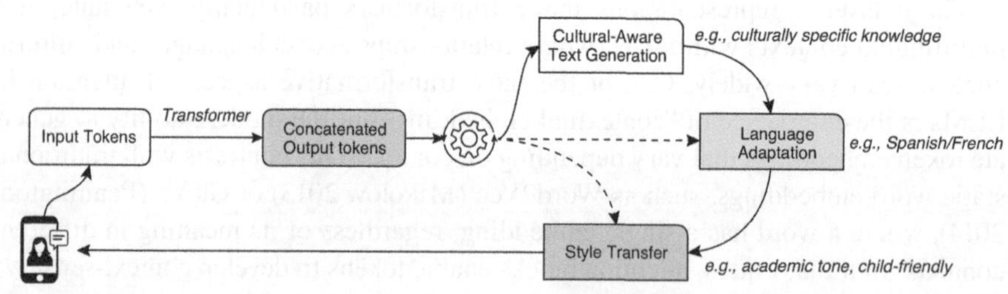

FIGURE 3.2 Adaptive response pipeline in a culturally aware chatbot system powered by a large language model (LLM). The model processes user input into tokens, which are contextualized through MHA within the transformer block. The resulting representations feed into downstream selectively activated modules, based on user needs or task requirements before providing the response to the user.

Attention Head 3: might identify the user's intent by recognizing query patterns associated with information requests (e.g., starting with "Can you tell me about").

These attention heads work in parallel, capturing the interrelationships between words, phrases, and cultural elements, resulting in contextual embeddings that encode both the linguistic structure and the cultural significance of the input. These contextual representations are then concatenated resulting in a rich representation of the input.

Phase 3: Masked MHA and Cultural Generation. As a decoder-only model, the LLM uses masked MHA to predict tokens in sequence while considering previous ones, refining its output based on context derived from the user's input. Here, the model's masked token prediction enables it to "guess" and generate tokens that complete the response meaningfully while accounting for cultural subtleties.

Assume that the chatbot generates the response: "The Hmong people have a rich tradition of textile art, known for its vibrant colors and symbolic patterns that often represent aspects of their cultural heritage." In generating this response, the model considers:

- **Relevant Cultural Facts**: Due to pre-training on vast and related datasets, the model has acquired cultural knowledge (encoded in its parameters), and the LLM now draws on information about the Hmong people's artistic traditions and symbolism, adapting its response based on the specific cultural context requested by the user.
- **Language Style Adaptation**: If the user's preferred language is French, the model may adjust to output: "*Le peuple Hmong possède une riche tradition d'art textile, réputée pour ses couleurs vibrantes et ses motifs symboliques, représentant souvent des aspects de leur patrimoine culturel.*" This translation and style adaptation happen through parallel attention heads that learn and apply language-specific rules, idioms, and nuances.
- **Engagement Optimization**: The model may also use stylistic elements to make the response more engaging, incorporating conversational phrases or local expressions to make the information more relatable and accessible to a lay audience. For example, it might add, "*It's a beautiful way of storytelling without words*" based on learned language generation patterns for engaging responses.

Phase 4: Cross-Language Semantic Alignment and Retrieval Enhancement. Given the user query's cultural context, the LLM retrieves and aligns semantic content across multiple languages. By using multilingual embeddings—representations aligned across languages, the model can understand culturally specific terms even if they do not have direct equivalents in other languages. For example, in translating "Hmong textile art," the model retains the core concept but adapts descriptors that might hold different connotations in other cultures or languages. Through cross-language semantic alignment, the model can respond to similar queries in other languages with culturally appropriate phrasing. For instance, if a Spanish-speaking user asked, "*¿Puedes contarme sobre el arte tradicional del pueblo Hmong*?" the model would produce an equally rich response in Spanish, bridging cultural differences while retaining factual accuracy.

Phase 5: Style Transfer and Cultural Preservation. As part of its curator role, the chatbot can perform style transfer to tailor descriptions to specific audience needs. If a younger audience interacts with the chatbot, the model might rephrase complex historical descriptions in simpler, more vivid language. For instance, it might transform "The traditional Hmong textile art embodies intricate symbolism" into "Hmong textile art uses patterns and colors to tell stories from their culture." Conversely, for a more academic audience, the model may elaborate with terminology relevant to ethnography or textile anthropology.

In cases where cultural preservation is essential, it is necessary to ensure that culturally sensitive terms remain unaltered or are contextualized carefully to respect their significance. The model must avoid over-simplifying complex cultural information that could misrepresent traditions or customs. This aligns with ethical NLP practices, as the model balances accessibility with respect for cultural accuracy.

Phase 6: Response Output and User Engagement. Finally, the generated response is outputted to the user in the preferred language and style. This output showcases the LLM's ability to integrate complex language understanding with culturally attuned generation, bridging linguistic and cultural divides. For example, the chatbot's output might be as follows:

- **User Query (English)**: "Can you tell me about the traditional art of the Hmong people?" + "Answer in Spanish"
- **Chatbot Response (Spanish)**: "El pueblo Hmong tiene una rica tradición de arte textil, conocido por sus colores vivos y patrones detallados. Estos diseños suelen tener significados profundos, reflejando la conexión de los Hmong con la naturaleza y sus historias ancestrales. Cada pieza es como un tapiz de historia, preservado y transmitido a través de generaciones."

3.3 THE ROLE OF LARGE LANGUAGE MODELS IN CULTURAL HERITAGE

3.3.1 Capabilities of LLMs

LLMs are trained on vast amounts of linguistic data using deep learning techniques, particularly transformer architectures (Zhao, 2023). These models, such as GPT-4, are characterized by their extensive parameter count, which has been proven to enable them to store a wide variety of learned patterns (Muennighoff, 2023). While the sheer number of parameters does not inherently grant the ability to understand language, it still allows LLMs to encode and represent more complex relationships between words, phrases, and contexts during training (Bhatia, 2024). This depth of representation improves their capacity to capture subtle nuances in language, handle ambiguous or context-dependent inputs, and provide more accurate predictions and outputs across multiple languages. This makes them particularly effective in producing coherent and contextually appropriate responses in various domains, including historical and cultural narratives. The capabilities of LLMs extend beyond mere text generation. They can:

Perform Language Translation: LLMs can perform language translation tasks accurately, improving over traditional rule-based or statistical machine translation methods. This is due to their ability to model context and semantic relationships within and across sentences using transformer architectures. Their utility extends to many languages, even with limited available resources (also known as *low-resource*) (Lankford, 2023), as transfer learning allows LLMs to generalize patterns learned from high-resource languages to improve translation in lesser studied languages. For historical or ancient languages, while direct training data may be limited, LLMs can incorporate relevant linguistic datasets (e.g., digitized texts or inscriptions) and context from related languages, improving their ability to assist in translating or interpreting these texts (Volk, 2024). This capability is particularly important in CH research, where accurate translation of historical documents can provide critical insights into past societies. Furthermore, approaches such as fine-tuning on specialized corpora and leveraging model understanding of cross-linguistic patterns have demonstrated their effectiveness in tasks where traditional methods struggled (Hauser, 2024).

Text Recognition and Transcription: Integrating multimodal LLMs, such as GPT-4 with vision or LLAVA, presents another approach to optical character recognition (OCR) systems by utilizing visual and contextual cues to enhance transcription accuracy. The physical state of the document often constrains traditional OCR systems—its shape, condition, or quality—which can lead to incomplete or inaccurate transcriptions, especially for historical or damaged texts. For example, an old, degraded manuscript may suffer from smudged ink, irregular handwriting, or partial text loss. While a traditional OCR system might struggle in such cases, multimodal LLMs can leverage additional contextual information, such as textual metadata, document type, or inferred historical context, to fill in the gaps where OCR alone would fail. For instance, when transcribing an ancient document, a multimodal LLM can be provided with prompts that include the estimated date of creation, the document's likely origin, or linguistic markers specific to that time period. This enables the model to use its semantic understanding to make more accurate guesses about missing or corrupted characters, reconstructing text in a way standard OCR system cannot. The multimodal capabilities of these models allow them to cross-reference visual and linguistic cues, providing a richer understanding of the text beyond simple character recognition. Another notable implementation of a pure VLM principle is found in the eScriptorium platform,[2] where a GPT-based plugin assists in deciphering complex historical scripts such as Gothic and Carolingian. Rather than merely transcribing visible characters, the system leverages pre-trained models on medieval Latin and Old French corpora to infer and restore missing or ambiguous words. It dynamically ranks multiple completions based on contextual coherence within the surrounding text, enabling editors to interactively select the most plausible restoration. This setup exemplifies how multimodal LLMs can augment traditional transcription workflows with historically and linguistically informed reasoning—not only reading but effectively interpreting heritage documents. However, LLMs can serve as a post-OCR refinement layer (Bourne, 2024). Once an OCR system has extracted text, the LLM can further analyze and refine the transcription, correcting errors or filling in missing snippets by comparing the output with similar historical documents or previously annotated texts. For example, if certain characters or words are missing due to damage, the LLM can infer the likely content by drawing on its knowledge of similar texts or linguistic patterns from the same historical period, effectively "guessing" missing information in a more informed and context-aware manner. This dual approach—OCR followed by LLM-based refinement—offers a powerful solution for digitizing and transcribing challenging CH materials that traditional systems struggle to handle. For example, Liu (2024b) demonstrates how GPT-4 Vision was used to restore degraded Qing Dynasty manuscripts by leveraging historical dictionaries and character structure priors, successfully reconstructing missing or smudged characters in classical Chinese.

Content Generation and Summarization: LLMs can generate concise summaries and explanations of cultural data, enhancing accessibility for specific user groups. Through prompt engineering techniques (Sahoo, 2024), LLMs can be directed to modify the output's tone, style, and complexity, tailoring it to different audiences. For example, descriptions of historical artifacts can be simplified for educational purposes aimed at children, with language adjusted for age-appropriate comprehension, and for educational purposes (Nayeem, 2024). Using style transfer, LLMs can further adapt the same content for various accessibility needs, such as creating auditory-friendly descriptions for visually impaired users or straightforward summaries for non-experts; therefore, presenting artifacts in multiple formats, broadening accessibility across different demographic groups. Moreover, in cases where multimodal inputs are available (such as images accompanying text), VLMs (e.g., GPT-4 with vision capabilities or LLAVA) (Zhang, 2024a) can incorporate text and visual data to generate more detailed descriptions. For example, when analyzing a historical document, these models can cross-reference visual elements (e.g., artifact images) with text to infer additional contextual information, particularly when visual context plays a crucial role in the object's significance.

A practical example of how multimodal LLMs cross-reference historical documents comes from the "Time Travel" benchmark project (Ghaboura, 2025). In this work, GPT-4V and similar models were evaluated on tasks involving the analysis of historical artifacts and manuscripts across time periods. One compelling use case involved presenting the model with a degraded museum record and asking it to generate a contextual description by pulling related information from linked archival databases, academic literature, and image collections. The model used retrieval methods to anchor its generated content in real, referenced documents—such as prior exhibition catalogs and digitized colonial archives—rather than hallucinating outputs. This process enabled the model to surface lost metadata and re-contextualize orphaned objects with provenance insights that had previously been dispersed across sources. Once again, this is an additional advantage compared with text-based LLMs alone, which we will analyze later.

Contextual Analysis and Interpretation: LLMs can facilitate natural language interactions with CH databases, enabling users to retrieve and analyze historical texts or artifacts more efficiently. Instead of relying on rigid keyword searches, scholars can pose detailed questions and receive relevant information, allowing for a more flexible exploration of archival materials (Xu, 2023). This approach enhances the accessibility of complex data, particularly in the digital humanities, where it supports more dynamic research workflows. Additionally, LLMs are able to assist in cross-linguistic exploration by processing multilingual collections, making it easier to access and understand materials from different cultures. This capability also extends to educational platforms (Almusaed, 2023), where users can interact with historical content in a more engaging and personalized way through natural language queries.

To further illustrate these capabilities in a CH context, the following case studies (see Section 3.4) provide grounded examples of LLMs in action. These include applications such as transcribing ancient manuscripts, powering interactive museum guides, and supporting collaborative research. While the focus in this section is primarily on the technical potential of LLMs, these real-world scenarios provide a clearer picture of how such capabilities translate into CH outcomes.

3.3.2 Relevance of LLMs to CH Preservation

Traditionally, preserving and interpreting cultural artifacts have required extensive manual labor and specialized knowledge, making it a time-consuming process. In this context, ArchGPT (Zhang, 2024b), for instance, autonomously retrieves and interprets heritage-specific data, replacing hours of manual archival work with efficient digital processes. This technology integrates seamlessly with cultural data, making it feasible to analyze, expand, and update archives in real-time—dramatically reducing the effort and expertise previously required. Studies like (Stopponi, 2023) illustrate the potential of distributional models for examining semantic relationships in low-resource ancient languages, such as Ancient Greek. By dynamically capturing word meanings in context, LLMs could facilitate automated annotation and contextualization for ancient languages, streamlining processes requiring manual expert validation. In a similar vein, authors in (Ryberg Smidt, 2024) explore how LLMs, such as the fine-tuned Multilingual BERT, contribute to understanding the grammar and syntax of ancient languages like Old Babylonian. They show that these models effectively handle part-of-speech tagging, creating preliminary syntactic structures that allow linguists to shift their focus from routine annotation to more complex interpretive tasks.

Additionally, (Dereza, 2024) reveals specific challenges and breakthroughs in using word embeddings for low-resource historical languages, particularly in tasks such as Part-of-Speech (POS)-tagging,[3] lemmatization,[4] and morphological prediction.[5] Notably, the task highlighted that even with limited datasets, LLMs like DeBERTa (He, 2020) and RoBERTa (Liu, 2019c) reached up to 96% accuracy in POS-tagging for languages like Gothic and Medieval Latin. These findings show how embedding-based models, while powerful, require fine-tuning to effectively capture the syntactic and morphological nuances of ancient languages—laying a foundation for future LLM applications in underrepresented and ancient language processing. The authors in (Spennemann, 2024) examine how generative AI models, including ChatGPT and Google Bard, may redefine CH management through targeted applications like digital preservation, virtual reconstructions, and the automation of artifact classification. The study identifies practical advancements, such as using AI for transcription and restoration of fragile manuscripts and creating interactive, immersive experiences of historical sites. However, it also highlights the risks of AI in heritage contexts—chief among them is the potential for biased digital reproductions and the erosion of human interpretative skills. The authors argue that while generative AI can increase accessibility to complex cultural archives, careful oversight is needed to ensure the authenticity, accuracy, and ethical use of AI-generated cultural representations.

Furthermore, LLMs can foster accessibility and engagement with CH by serving as interactive guides and storytellers, making digital content more appealing for global audiences. Museums and cultural institutions, increasingly reliant on digital platforms to reach wider audiences, now offer virtual tours and online exhibitions tailored to diverse visitor interests and backgrounds (De Carolis, 2023). Leveraging LLMs, these platforms can generate personalized content, adapting to visitor preferences, language, and cultural background to enrich engagement in real-time. As digital storytellers, LLMs excel in creating immersive narratives that blend historical accuracy with imaginative elements (Beguš, 2024). By doing so, they bring exhibits to life, transforming static artifacts into dynamic experiences, providing a more profound appreciation of cultural histories. Moreover, LLMs can also function as sophisticated recommender systems, offering personalized navigation through cultural spaces. They can suggest viewing paths, select artifacts based on a visitor's expressed interests, and even simulate interactive dialogue to deepen visitor engagement (Trichopoulos, 2023).

AI-powered platforms can profit from LLMs, integrating data from various institutions, simplifying collaboration and enabling the construction of unified, accessible digital repositories of CH artifacts (Bernasconi, 2024). Unlike traditional search methods, which typically rely on keyword-based retrieval, LLM-powered platforms empower semantic search capabilities. This allows them to understand and match concepts rather than just words offering more accurate, contextually relevant results. For researchers, this means faster, more precise access to a broader range of datasets, even across languages and specialized fields.

Finally, LLMs can contribute to the preservation of endangered languages and cultures (Pinhanez, 2024). By training LLMs on texts from minority languages and oral traditions, AI can help document and maintain linguistic and cultural diversity that might otherwise be lost. This aspect is critical for supporting communities striving to revive or maintain their linguistic heritage and for researchers studying language evolution and cultural continuity.

In summary, LLMs offer a powerful suite of tools for CH preservation, from digitization and accessibility enhancements to supporting interdisciplinary collaboration and safeguarding endangered cultural elements. Their integration into CH projects marks a significant shift toward a more efficient, inclusive, and globally accessible approach to understanding and preserving our shared past. These practical capabilities will be further contextualized in the next sections, where illustrative examples highlight how such techniques support tasks like digitization, interpretation, and public engagement in CH settings.

3.4 CASE STUDIES

To display some practical applications of LLMs in CH, this section presents three indicative case studies: digitization of ancient texts, interactive museum exhibits, and collaborative research platforms. These examples, along with the capabilities outlined in Section 3.3 and the accessibility-focused applications described in Section 3.5, can also be viewed through broader thematic lenses. These include efforts to **archive** (e.g., digitizing artifacts and texts), **understand** (e.g., interpreting ancient languages or historical records), **educate** (e.g., enhancing public engagement through museum exhibits), and **preserve** (e.g., safeguarding endangered languages or traditions). While the chapter is not organized explicitly around these categories, this interpretive lens may assist readers—particularly those new to the CH domain—in contextualizing the diverse ways LLMs can serve human-centered objectives (Figure 3.3).

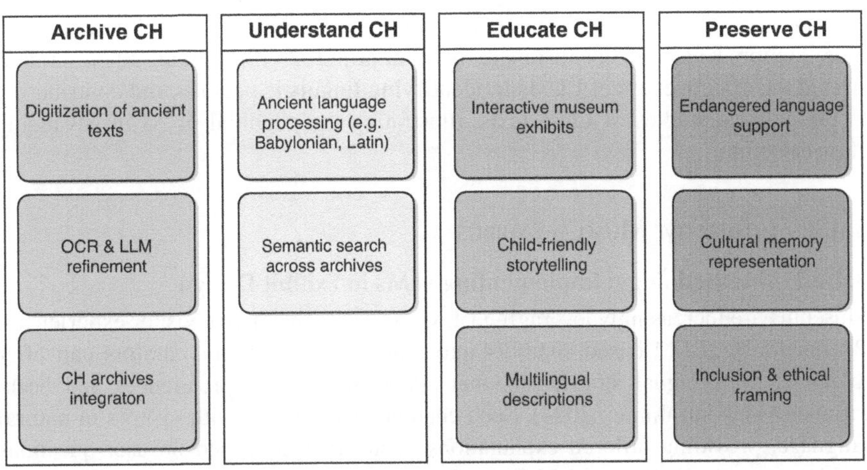

FIGURE 3.3 An interpretive grouping of the LLM applications discussed across Sections 3.3–3.5. These thematic areas help clarify the multifaceted contributions of LLMs to CH in terms of archiving, understanding, education, and preservation.

3.4.1 Digitization of Ancient Texts

3.4.1.1 Methodology: Using LLMs for Text Recognition and Translation

The digitization of ancient texts is a critical task in CH, as it enables the preservation and accessibility of invaluable historical documents. Traditional methods of transcription and translation require intensive manual labor and expertise, especially for languages like Ancient Greek, Old Babylonian, or classical Latin, which have complex scripts and limited contemporary fluency (Sommerschield, 2023). LLMs, in combination with OCR systems, could offer an efficient solution to this challenge. However, the current performance of LLMs in this domain highlights significant limitations (Boros, 2024). Challenges such as handling diverse languages, time periods, and transcription types remain significant, underscoring the need for further fine-tuning in historical text digitization tasks.

Despite these limitations, it is reasonable to speculate that with targeted fine-tuning and domain-specific training, LLMs could achieve significant improvements in future research. This aligns with findings from recent studies (Thomas, 2024), demonstrating the efficacy of instruction-tuned models.

3.4.1.2 Outcomes and Benefits

The use of LLMs in this context has dramatically accelerated the digitization process (Mandvikar, 2023), allowing cultural institutions to process and preserve large volumes of documents in a fraction of the time previously required. For example, applying LLMs to ancient Greek manuscripts has yielded preliminary success, as models have managed to transcribe, translate, and annotate these texts (Liu, 2024b). This not only improves academic accessibility but also makes these texts available to the general public, often through open-access digital libraries. The outcomes extend beyond efficiency; LLMs enable the interpretation of texts that might otherwise remain inaccessible due to their fragmentary nature. By filling in gaps and providing context, they can assist in reconstructing historical narratives and cultural contexts more comprehensively. For instance, the EvaLatin 2024 (Sprugnoli, 2024) demonstrated the effectiveness of LLMs in identifying linguistic patterns and contributing to philological studies of Latin texts, illustrating the potential of AI in advancing classical studies.

3.4.2 Interactive Museum Exhibits

3.4.2.1 Methodology: Implementing LLMs in Exhibit Design

Museums are increasingly leveraging LLMs to create interactive visitor experiences. By integrating LLM-based chatbots into museum applications, curators can offer dynamic, multilingual storytelling that adapts to visitors' preferences and backgrounds (Alabdulrahman, 2024). Such chatbots can interact with visitors in natural language, providing tailored explanations about artifacts based on user questions and interests. To implement LLM-based chatbots practically, museums might opt to utilize pre-trained models like the closed sourced GPT variants or the open-sourced LLaMA variants and fine-tune them with their own curated content. This fine-tuning process involves *instruction fine-tuning* (Zhang, 2023), where the model is trained to follow specific instructions and adopt particular roles or personas relevant to the museum's exhibits. For example, to create a chatbot that embodies the persona of Pablo Picasso, the training data would include prompts and responses formatted to guide the model's behavior. The training dataset might consist of pairs where the model is instructed to answer questions as Picasso, using first-person narrative and reflecting his known style and perspectives. Additionally, an advanced layer of fine-tuning incorporates the selection of preferred responses by professionals, refining the model's output using human feedback to adjust accuracy and relevance. An example of how the training data might look:

Instruction: "You are Pablo Picasso. Answer the following question in the first person, expressing your thoughts and feelings."
User Prompt: "What inspired you to paint 'Guernica'?"
Model Response: "The horrors of the Spanish Civil War deeply affected me. 'Guernica' was my way of expressing the anguish and suffering inflicted upon innocent people. I wanted to capture the brutality of war and make a powerful political statement."

By providing numerous such examples, the model learns to generate responses that reflect Picasso's persona, allowing visitors to engage in conversations that feel authentic and informative. To support such compelling user experiences, several underlying technical methods are employed to tailor LLM behavior to museum-specific needs. These approaches form a critical part of model alignment in real-world CH applications, where narrative fidelity, educational sensitivity, and inclusivity are essential.

In the following, we briefly highlight some of the most commonly adopted fine-tuning and interaction techniques—among others—that continue to shape the evolving field of cultural AI design. While the broader landscape of LLM engineering is rapidly expanding, the methods discussed here remain central to configuring models for domain-specific roles, particularly in heritage-centered environments that demand both factual precision and user-centered adaptability.

- **Instruction Fine-Tuning**: is a specialized training phase where the model is adjusted to perform tasks based on explicit instructions. During this process, the model is exposed to curated datasets containing task-specific annotations. These annotations specify the desired behaviors, styles, and tones the model should adopt, enabling it to personalize responses in alignment with the museum's goals. For example, fine-tuning could involve training the model to embody personas of historical figures, deliver exhibit-specific narratives, or use professional and educational tones. This phase further modifies the weights of the model to internalize these instructions for a more context-aware interaction.
- **Reinforcement Learning with Human Feedback (RLHF)**: By applying RLHF, museums can further refine the chatbot's responses. This method involves training the model by rewarding outputs that align with expert feedback, enabling it to better capture the desired tones, styles, and accuracy in responses. The refined data from RLHF can then be used to enhance instruction fine-tuning, leading to a more polished conversational

experience. Additionally, this technique can be applied online, leveraging visitor reactions to further optimize their models during their deployment lifecycle.

- **Prompt Engineering**: Prompt engineering involves designing precise prompts with contextual cues or constraints to guide the model's understanding of intent, such as: "Explain the significance of [Exhibit X] in layman's terms for a visitor unfamiliar with the subject." or "Adopt the tone of a historian when discussing the cultural context of [Artifact Y]." Unlike instruction fine-tuning, prompt engineering does not modify the model's internal parameters. Instead, it leverages the model's existing capabilities, making it ideal for quick adjustments and iterative testing during deployment.
- **Interactive Chatbot Development**: Combining instruction fine-tuning and RLHF results in an interactive chatbot that feels both authentic and informative. This process culminates in a chatbot that provides dynamic, personalized interactions with visitors.

Implementing LLM-based chatbots in museum exhibits can significantly enhance visitor engagement and satisfaction (Gaia, 2019). By offering conversational and customized digital guides, visitors experience a deeper connection to the content and show greater interest in exploring further. For example, "The Living Museum" project[6] allows visitors to chat with over 1.2 million objects from the British Museum's collection. This LLM-based chatbot provides real-time, tailored information about exhibits and aligns with individual learning styles, effectively communicating the collections' cultural and historical significance.

Moreover, LLM-based chatbots play a crucial role in improving accessibility. By adhering to Universal Design (UD) principles,[7] they can offer personalized audio guides, visual descriptions, and interactive touchpoints for visitors with disabilities. Museums utilizing these technologies can increase visits from diverse demographic groups, including visually impaired individuals who benefit from detailed artifact descriptions powered by LLMs. For instance, chatbots developed by Livdeo[8] enable museums to provide inclusive digital solutions, allowing visitors of all backgrounds to engage with exhibits through natural language conversations.

3.4.3 Collaborative Research Platforms

Collaborative research platforms serve as centralized systems for storing, annotating, and analyzing data collaboratively. Through them, researchers from diverse disciplines can access shared datasets, engage in comparative analyses, and generate insights that span geographical, linguistic, and temporal boundaries. By embedding LLM techniques, these platforms allow the integration of data-driven insights with domain expertise, establishing a unified infrastructure for research in fields like anthropology, archaeology, linguistics, and history (Garcia, 2023).

3.4.3.1 Methodology: LLMs in Facilitating Cross-Institutional Research

The success of cross-institutional research within collaborative platforms relies on methodologies that enable precise information retrieval, contextual understanding, and knowledge sharing. Retrieval-Augmented Generation (RAG) (Fan, 2024) and Retrieval-Augmented Fine-Tuning (RAFT) (Zhang, 2024c) are foundational techniques. RAG, a method that enhances model responses with information retrieved from databases or archives, enables LLMs to directly incorporate verified, up-to-date data into outputs, reducing the likelihood of hallucination and enhancing reliability. Conversely, RAFT allows models to prioritize specific data repositories or subject matter, improving the model's domain specificity and ensuring outputs remain tightly relevant to the research context. Collaborative research platforms often employ ontology-based annotation to enhance the quality and relevance of outputs. LLMs trained on these platforms gain improved semantic accuracy by aligning datasets with standardized ontologies recognizing relationships and classifications within each domain. Another critical technique, hierarchical modeling, incorporates multi-layered contextual information—such as time periods, cultural settings, and linguistic shifts—into the LLM framework, refining its ability to interpret data within specific historical and cultural narratives. These methods allow for highly specialized, accurate, and context-sensitive analyses across varied datasets. Semantic embedding models are also essential within collaborative platforms, enabling advanced clustering and retrieval of related documents by transforming text into vectorized semantic representations. This approach facilitates searches based on conceptual relevance rather than simple keywords, allowing researchers to locate datasets connected by thematic or contextual relationships. Additionally, dynamic, domain-adaptive fine-tuning with transfer learning is used to calibrate LLMs for specific research areas, minimizing re-training needs and allowing the models to maintain adaptability to new institutional requirements and updates to existing datasets. Knowledge Graph-Augmented Language Models (KG-LMs) (Pan, 2024) further support research by integrating structured data and relationships among entities, such as historical figures, locations, and events. This gives LLMs a robust relational context that enhances interpretative capabilities.

3.4.3.2 Outcomes and Benefits

Ontology-based annotation and hierarchical modeling enable consistent data interpretation across institutions, a critical factor in ensuring compatibility of insights derived from disparate datasets. By supporting LLMs with layered contextual information, these techniques ensure outputs remain sensitive to the historical and cultural nuances of the data, greatly benefiting interdisciplinary studies that require careful handling of contextualized knowledge. Embedding models contribute to efficient data retrieval, facilitating the discovery of thematic connections across datasets housed in different institutions or recorded in diverse languages and formats. This semantic clustering capability promotes interdisciplinary collaboration and knowledge-sharing, making it possible to uncover connections that would otherwise remain siloed. The ability of these platforms to support dynamic, domain-specific fine-tuning means that institutions can continuously update and refine model outputs without extensive re-training efforts, allowing collaborative research efforts to remain current and relevant across changing data landscapes. Knowledge graphs within LLMs hold the promise to facilitate the cross referencing of interconnected historical entities, allowing platforms to dynamically link artifacts, texts, and cultural records into cohesive networks (Chen, 2023). This relational understanding greatly enriches the scope and depth of research by providing a web of interconnected knowledge that spans disciplines and time periods (Abate, 2023).

Recent work demonstrates how the methodologies described above are actively employed in real-world cross-institutional CH research. For instance, (Maree, 2025) developed a neuro-symbolic system combining LLM-generated narrative explanations with a knowledge graph built on Cultural Heritage Linked Open Data.[9] The system ranks and contextualizes semantic relations between historical entities—such as artworks, figures, and institutions—providing interpretable insights that scale across archives.

Sartini (2024) offers a complementary case with IICONGRAPH, a semantically enriched iconographic knowledge graph aligned with FAIR data principles. Integrated into LLM-based platforms, this resource enables cultural question answering and inferential reasoning with KG across museum and datasets like ArCo,[10] demonstrating ontology-aware alignment as a practical tool for interinstitutional interoperability. He (2024) shows how participatory users employ visual generative models to narrate heritage site experiences. Their work argues for grounding generative outputs in culturally authentic data, advocating integration with RAG to ensure that multimodal narratives remain historically faithful and context-sensitive.

Together, these cases illustrate the tangible impact of techniques such as domain-adaptive RAG architectures in fostering collaborative, scalable, and cultural research infrastructures.

3.5 ENHANCING ACCESSIBILITY WITH LLMs

Ensuring accessibility in CH is paramount, as it allows individuals from all backgrounds and abilities to engage with and appreciate cultural knowledge. LLMs provide innovative ways to enhance accessibility, making cultural experiences inclusive, dynamic, and tailored to the needs of diverse groups. This section explores various applications and benefits of LLMs in improving accessibility, with examples and suggestions for visual aids to further engage readers.

3.5.1 Text In, Text Out: LLMs for Enhanced Descriptions

LLMs have the ability to transform dense, technical text into simplified, accessible language, which is especially useful in museums and archives. For instance, LLMs can take an exhibit label or a scholarly article and convert it into a concise, easy-to-read summary or an interactive story suitable for a broad audience.

How It Works: "Text In, Text Out" architectures (such as Llama 3.1[11]) use a sequence of layers within a large language model to process and generate text. At a basic level, the model first breaks down the input text into tokens—small units like words or parts of words—and maps them to numerical vectors. These vectors pass through multiple transformer layers, where the model captures context and meaning through attention mechanisms. Finally, the processed vectors are used to generate new text, ensuring that the output is coherent and relevant to the input. This way, the model can effectively translate complex language into simpler, more engaging content.

Example: A museum exhibit on ancient Egyptian artifacts could have an LLM-generated summary that simplifies the historical background and importance of hieroglyphics for children or non-experts. Meanwhile, the LLM can offer deeper linguistic or archaeological insights for advanced/professional visitors.

While "Text In, Text Out" systems typically refer to simplification or summarization, recent research shows that LLMs can also operate in reverse—translating modern language *into* ancient or non-standard scripts. For example, *HieroLM* (Cai, 2025) treats hieroglyph recovery as a next-word prediction task, enabling an LLM to generate plausible sequences of Egyptian hieroglyphs from partial or modern prompts. Similarly, Shih et al. (2025) demonstrate that GPT-4o and other advanced LLMs can reason over rare glyphs and simulate their linguistic logic, using prompting methods that connect text with non-Unicode visual languages. Even in the visual domain, work on *Maya glyph segmentation* (Fnu, 2024) highlights a pipeline where LLMs and image models jointly facilitate the interpretation and eventual production of ancient writing. These cases suggest that "Text In" systems in CH could soon support "Ancient Script Out" as a real output modality, broadening both engagement and restoration efforts.

3.5.2 Text, Image In—Text Out: Enriching Descriptions with Visual Context

LLMs can also process both text and images simultaneously to generate descriptive, accessible content that considers visual context. Such an indicative open model is LLAVA (Liu, 2024a). This capability is particularly useful in settings like museums, where a combination of images and text can provide a richer and more inclusive experience for visitors given the understanding of visual elements alongside written information.

How It Works: In "Text, Image In - Text Out" models (such as Llama 3.2,[12] which combines the language capabilities of the Llama 3.1 backbone with an image encoder like a Vision Transformer (ViT) (Radford, 2021), the architecture is designed to handle multiple types of input. First, the text input is tokenized into numerical vectors, as in traditional language models. Simultaneously, the image is processed through an image encoder, which breaks the image into smaller patches and translates them into vectors that capture essential visual features. These text and image vectors are then integrated in a shared space using either concatenation in the input or fusion with attention mechanisms, allowing the model to understand the relationship between the visual and textual content. Finally, the model generates text that incorporates insights from both the image and the accompanying text, making the description richer and more contextually aware.

Enhanced Example: Consider a museum exhibit featuring a statue of the Greek goddess Athena. Using a "Text, Image In - Text Out" model, the LLM could analyze a photograph of the statue along with a description of its historical significance. For children or non-experts, the LLM might generate a simplified explanation, highlighting Athena's role as the goddess of wisdom and her importance in Greek mythology. For experts, it could offer a more detailed analysis, such as insights into the statue's artistic style and historical context, referencing features like the craftsmanship of the shield or the symbolism of the owl depicted alongside Athena.

3.5.3 Personalized Descriptions for Disabled People

The ability of LLMs to dynamically adapt descriptions and narratives based on individual needs represents a significant advancement in accessibility for disabled people, enhancing their engagement with CH. Unlike traditional static descriptions, LLM-powered personalized interactions enable detailed, contextually relevant information that caters specifically to the sensory and cognitive preferences of each user. For visually impaired visitors, LLMs can integrate with screen readers or voice-over applications to provide nuanced audio descriptions, which go beyond standard exhibit labels. These descriptions can be tailored to offer layered details, such as the physical attributes of artifacts, their textures, dimensions, and color variations, alongside contextual information like the artifact's historical or cultural significance. This multi-level approach ensures that visually impaired users can build a rich mental image of the exhibit and feel equally engaged in the experience as sighted visitors. To further improve accessibility, LLMs can deliver real-time, interactive guidance to visually impaired users as they navigate cultural spaces. For example, an LLM could respond to questions about nearby exhibits, providing directions or detailed descriptions of specific items based on the visitor's proximity to those objects. Through integrating geolocation or Bluetooth beacon technology, the LLM could deliver precise, location-based information, helping visually impaired visitors explore exhibitions independently and confidently. In addition to visual accessibility, LLMs can adapt content for individuals with cognitive disabilities. Many cultural descriptions are inherently complex, containing layers of historical, social, and linguistic details. LLMs can simplify these elements, presenting information in manageable, easy-to-understand pieces. For instance, rather than offering one comprehensive paragraph about an artifact's historical context, the LLM can break down the information into short, digestible sections. Each section can be presented one at a time, allowing the user to process the information at their own pace. LLMs can enhance accessibility in museum settings by supporting the application of UD principles (Design, 1997), which aim to create environments usable by the widest range of individuals without requiring specialized adaptations. The seven principles of UD include equitable use, flexibility in use, simple and intuitive operation, perceptible information, tolerance for error, low physical effort, and appropriately sized spaces for interaction. In this context, LLMs can generate textual descriptions of exhibits, which can be integrated with audio guides, tactile resources, or visual displays. While LLMs primarily operate within the textual domain, their outputs can be paired with multimodal resources (e.g., audio or visual aids) by museum systems to create a more inclusive visitor experience. Finally, personalized, disability-friendly descriptions have another dimension: supporting neurodiverse individuals, such as those on the autism spectrum, who may find social interaction challenging in traditional museum settings. LLMs can be fine-tuned to recognize cues or preferences set by these users, offering structured, predictable interactions without overwhelming them with too much information at once. For instance, rather than an open-ended exploration, the

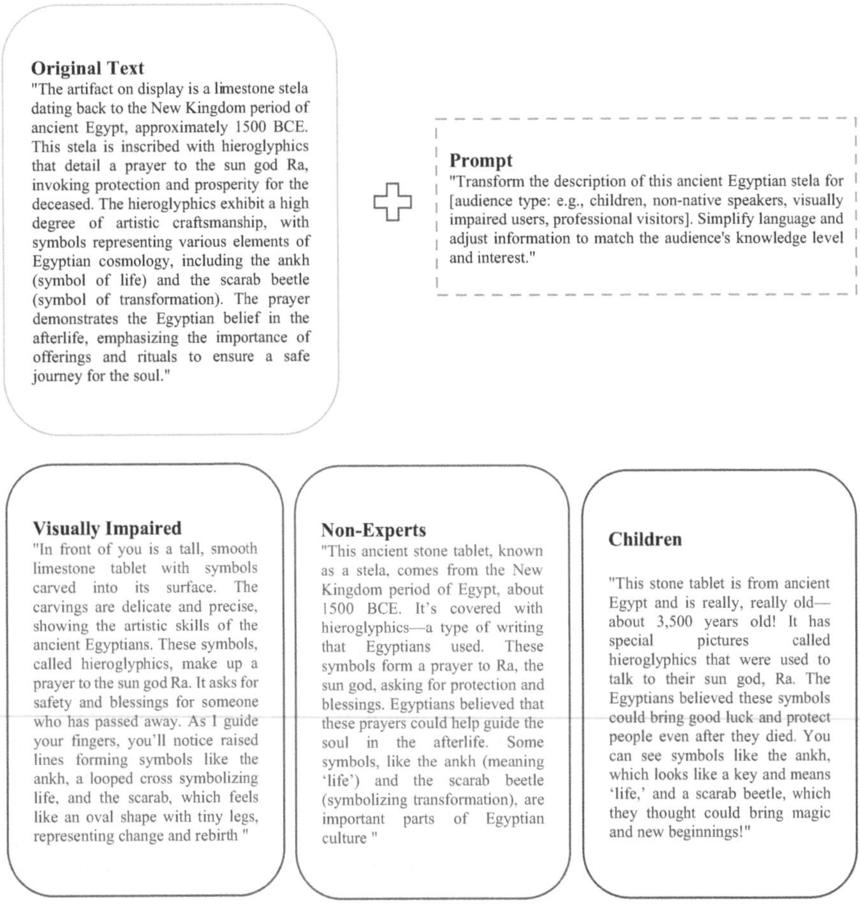

FIGURE 3.4 An illustrative example demonstrating how the description of an ancient Egyptian stela can be tailored for different audiences—visually impaired, non-experts, and children—created to showcase text adaptation techniques. Generated using Meta-Llama/Meta-Llama-3-70B-Instruct.

LLM could guide neurodiverse visitors through exhibits in a step-by-step manner, adjusting its responses based on their pace and feedback. In summary, LLMs provide a highly inclusive cultural experience through personalized descriptions (see Figure 3.4), transforming how museums, archives, and cultural sites serve visitors of all abilities. By fostering environments that accommodate diverse sensory, cognitive, and social needs, LLMs enable a broader segment of society to appreciate and interact with CH, advancing the mission of universal accessibility in the cultural sector.

3.5.4 Applications in Museums: Leveraging Metadata and User Cues

Museums increasingly use LLMs to enhance accessibility through the intelligent use of metadata and user cues (Rachabatuni, 2024). As already discussed, this is particularly effective when combined with interactive exhibits or digital kiosks, where visitors can engage directly with LLM-powered systems. LLMs can also adapt their outputs based on user behavior and preferences. For example, if a visitor expresses interest in a particular historical period or type of artifact, the LLM can use this information to provide related content or guide the visitor to other exhibits that match their interests. Additionally, by analyzing user engagement patterns, these systems can continuously learn and refine the information they offer, ensuring it remains relevant and engaging. This application provides dynamic, responsive information tailored to diverse needs, including language preferences and varying levels of detail.

3.5.5 Broad Accessibility: Benefits for Children and Other Groups

LLMs make cultural content accessible to diverse groups, including children and non-native language speakers. Museums and cultural institutions can leverage LLMs to create age-appropriate content that engages children by simplifying and presenting complex information through storytelling techniques. For instance, interactive narratives designed specifically for young audiences can transform exhibits into engaging adventures, enhancing the learning experience while making it fun and memorable.

LLMs achieve this simplification by utilizing fine-tuning techniques on specialized datasets, such as text corpora aimed at young readers. The model learns to adjust vocabulary, sentence structure, and narrative tone to suit a younger audience, making the content both accessible and educational. These models can even incorporate playful language patterns or storytelling formats to keep children engaged. Additionally, for audiences who speak different languages or are learning new ones, LLMs offer multilingual capabilities that allow the seamless translation of descriptions and narratives into various languages. This broadens cultural content's reach, ensuring that linguistic barriers do not limit it, enabling a global audience to engage with CH. For example, models such as the "No Language Left Behind" model (Costa-jussà, 2022) employ advanced tokenization and multilingual pre-training techniques. These methods ensure high-quality translations across numerous languages, addressing even low-resource languages efficiently.

Recent developments further demonstrate how LLMs are being adapted for broad accessibility across age groups and language boundaries. FairyLandAI (Makridis, 2024) exemplifies this trend by generating culturally contextualized fairy tales tailored to children's age and background, using GPT-based models fine-tuned on pedagogical corpora and integrated with visual storytelling tools. Similarly, Mathemyths (Zhang, 2024d) introduces a co-creative educational platform where children help construct narratives that incorporate foundational concepts, such as early mathematics, through guided interaction with an LLM. For multilingual access, the Pangea model (Yue, 2024) offers a fully open-source LLM trained across 39 languages and validated with a culturally diverse benchmark, demonstrating robust performance even in low-resource linguistic settings. These initiatives confirm the potential of LLMs to support inclusive engagement with CH through age-appropriate, linguistically adaptive, and culturally sensitive design.

For the significant use case of visually impaired users, LLMs and multimodal LLMs (such as VLMs) offer particularly transformative potential. One such system is *ViAssist* (Yang, 2024), a real-time platform built on the LLAVA architecture that helps visually impaired users interpret complex environments such as museums. By fusing visual perception with natural language generation, ViAssist enables users to query their surroundings conversationally—for example, by asking "What is in front of me?"—and receive scene-aware descriptions that dynamically adapt to their preferences. This offers an immediate path for museums to deploy AI-based guides that assist with both exhibit interpretation and spatial navigation.

Similarly, *VisionGPT* (Wang, 2024) presents a framework that combines anomaly detection with LLM-based contextual feedback. While originally designed for general navigation safety, its ability to provide human-readable, real-time commentary on visual changes or hazards makes it highly transferable to museum settings. For instance, the model could detect obstructions in exhibit pathways or highlight objects of interest based on their relevance, thereby supporting both safety and engagement for visually impaired visitors.

Complementing these tools, *VIALM* (Zhao, 2024) introduces a benchmark suite that evaluates LLM-based assistance across tasks like object recognition, and scene understanding. Tested on a diverse set of VLMs, VIALM offers a robust framework for assessing and improving how assistive technologies can support the visually impaired in everyday environments with insights that can inform applications in cultural settings. It also surfaces key challenges such as maintaining cultural sensitivity, supporting multilingual contexts, and handling ambiguous visual scenes—issues especially pertinent to heritage spaces.

Together, these systems demonstrate how visually grounded LLMs are evolving into interactive accessibility agents holding promise to establish a new frontier in inclusive CH engagement.

3.5.6 Integration with Chatbots: Fine-Tuned for Task-Specific Objectives

Integrating LLMs into chatbots is one of the most effective ways to enhance accessibility in cultural settings. LLM-powered chatbots can provide visitors with personalized, on-demand information, simulating the experience of a human guide (Spadoni, 2023). These chatbots are trained to respond to various inquiries, from basic exhibit information to deeper historical context, adjusting their responses based on the visitor's knowledge level and interest.

It is important to clarify that base LLMs are typically pre-trained on vast textual corpora to perform tasks like next-word prediction based on previous word sequences. Widely known platforms such as ChatGPT or Gemini utilize these pre-trained models in a more interactive format by applying additional fine-tuning steps. For the purposes of CH applications, it is useful to briefly explain this transformation process to distinguish general-purpose language models from chatbots specialized in tasks like museum engagement or artifact explanation.

3.5.6.1 How We Go from LLM to Chatbot

Transforming an LLM into an interactive chatbot involves several key steps. Base models, such as Meta-Llama's Llama-3.2–3B,[13] are designed primarily for text completion tasks. To create a conversational agent, organizations or research groups often release a corresponding "chat model," like Llama-3.2–3B-Instruct,[14] which has been fine-tuned on conversational datasets to align the base model with interactive dialogue tasks. From this foundation, tools such as LangChain[15] can be used to integrate the chat model with essential components, including text processing pipelines, embedding models, and vector databases. Additionally, this chat model can be further fine-tuned on task-specific objectives, enabling it to specialize in domains such as museum exhibits or cultural artifacts.

When a user then asks a question, the system employs an embedding model to convert the query into a high-dimensional vector. This vector is compared with stored content in the vector database to retrieve the most relevant context. The chatbot then uses a structured prompt to generate a human-like response, ensuring that the information provided is accurate and contextually appropriate.

3.5.6.1.1 From Chatbots to AI Agents

While traditional chatbots rely on prompt-response paradigms, recent research highlights the emergence of *agentic systems*—LLM-driven frameworks capable of orchestrating complex, multi-step reasoning workflows. Ke (2025) describes this shift across three dimensions: inference scaling, learning to reason, and agentic workflows. These agentic systems integrate planning modules, memory components, and tool-use abilities, enabling them to go beyond static exchanges and operate as autonomous cultural mediators. In a heritage context, such systems could dynamically retrieve archival materials, adapt content across sessions, or even verify historical claims using integrated search APIs. Unlike conventional chatbots, which are constrained to pre-defined interactions, agentic LLMs represent a move toward proactive, context-aware digital curators capable of evolving with user intent and institutional goals.

3.5.7 Utilizing RAG for Dynamic Content Generation

RAG combines the strengths of LLMs with real-time retrieval capabilities, making it possible to generate dynamic and up-to-date content. In CH, this is particularly useful for creating interactive educational platforms or updating exhibit information as new discoveries emerge (Garcia, 2023). By connecting LLMs with extensive cultural databases, RAG enables the generation of responses that are not only accurate but also contextually enriched with the most relevant information available.

How It Works: RAG systems operate by first embedding user queries into a vector space and retrieving relevant documents from a database based on semantic similarity. However, to improve accuracy, advanced techniques like query expansion and cross-encoder re-ranking are employed (Zhao, 2024). Query expansion refines the original question, broadening the search scope and hypothesizing potential answers to surface more precise information. Cross-encoder models further re-rank retrieved documents, scoring them based on direct relevance to the query, ensuring the most contextually relevant results are used to inform the LLM's response. This multi-step approach minimizes distractors—irrelevant but related documents—allowing the LLM to generate focused and accurate outputs.

Example: An RAG system in a museum could provide visitors with the latest research findings about an ongoing archaeological excavation. By dynamically retrieving new academic publications or recent field reports, the chatbot ensures exhibit information stays current and informative. Similarly, if the museum acquires a new artifact, the RAG system can quickly generate updated descriptions and contextual information, enriching the visitor experience in real time.

3.5.8 Making Entertainment More Accessible to Excluded Target Groups

The implementation of LLMs for cultural entertainment, such as historical games, films, and educational apps, also holds significant potential for enhancing representation and accessibility for marginalized groups. By leveraging models specifically trained with high-quality, demographically relevant data, developers can ensure that narrative content is both engaging and culturally sensitive. For instance, LLMs can generate stories that incorporate indigenous languages or reflect underrepresented cultural experiences in a manner that feels authentic and inclusive.

A critical prerequisite is minimizing stereotype propagation and harmful content biases by making use of advanced techniques that prioritize safety and contextual appropriateness. Furthermore, for audiences with disabilities, these models can facilitate adaptive and personalized content, dynamically adjusting language complexity and providing alternative formats like text-to-speech or visual aids. By applying these methods thoughtfully (as demonstrated by models like KidLM) (Nayeem, 2024), cultural and educational experiences can be tailored to the diverse needs of all users, fostering a more inclusive digital environment.

3.6 CHALLENGES AND FUTURE DIRECTIONS

In recent years, the adoption and evolution of LLMs have presented a range of challenges spanning technical limitations, ethical and privacy concerns, and substantial resource requirements. This section explores these fundamental challenges and proposes areas for future advancements.

3.6.1 Technical Challenges, Ethical and Privacy Considerations, and Resource and Training Requirements

Applying LLMs to CH presents technical, ethical, and resource challenges. Technically, LLMs often struggle with preserving the cultural specificity and linguistic nuance, which are critical for representing diverse heritage contexts. Standard benchmarking metrics (e.g., BLEU, ROUGE) fail to capture cultural sensitivity, making it challenging to accurately assess model performance in these areas. Additionally, fine-tuning models for heritage-specific tasks can lead to catastrophic forgetting, where adaptations to niche knowledge diminish the model's general capabilities. Finally, hallucination (cases in which LLMs generate coherent and grammatically correct outputs but factually incorrect or nonsensical) is a long-standing problem.

Ethical and privacy concerns further complicate LLMs' use in CH. Models trained on vast, generalized datasets often contain cultural biases that risk reinforcing stereotypes or overlooking underrepresented groups, particularly Indigenous and minority communities. These models also risk producing historically inaccurate or misleading content, which poses a significant ethical risk in the heritage sector, where cultural narratives must be respected and preserved accurately.[16] Privacy issues arise when training on datasets that include sensitive historical documents or proprietary heritage data, raising concerns about data security and misuse. Advanced techniques such as differential privacy and federated learning, even the RLHF method, which has been used to improve alignment with ethical standards, can offer solutions. Yet, they are computationally intensive and may affect model performance. Lastly, the substantial resource demands of LLMs—such as the extensive data, GPU requirements, and energy consumption involved in training models like GPT-3 (trained with 175 billion parameters and over 300 billion tokens) and PaLM (Chowdhery, 2023) (540 billion parameters)—pose a barrier for many cultural institutions with limited resources. Even methods like RLHF, which can enhance cultural sensitivity, add significant computational overhead. Thus, addressing these challenges will require innovations in model efficiency and ethical safeguards, ensuring that LLM scan responsibly and effectively serve the CH domain.

3.6.2 Potential Advancements in LLM Technology

Advancements in LLMs technology are directed toward refining model performance, contextual understanding, and ethical alignment, all of which hold significant implications for CH applications. Bias mitigation remains a focal point, with efforts to reduce data-driven biases through improved data preprocessing and advanced debiasing techniques, ensuring balanced and sensitive representations of diverse cultural content. Efficiency optimization, particularly through model compression techniques like knowledge distillation, addresses the high computational demands of LLMs, making them more accessible and sustainable, especially for resource-intensive heritage preservation projects. Improved dynamic context handling will allow LLMs to process and interpret longer, complex narratives—a critical need for analyzing historical documents and layered cultural records. Continuous learning approaches further enable these models to adapt to evolving knowledge, ensuring they remain relevant and valuable in heritage applications as new insights emerge. Multimodal learning capabilities are also crucial, enabling LLMs to interpret diverse cultural artifacts, from text and images to audio and video, which enriches digital archiving efforts and enhances understanding of cultural materials. Finally, improved human–AI collaboration frameworks and dynamic evaluation metrics ensure that LLMs augment rather than replace expert insight, fostering more effective and context-sensitive interactions that maintain alignment with cultural preservation standards.

3.6.3 Multimodal Language Models (VLM, ALM)

Recent advancements in multimodal LLMs like GPT-4, GPT-o, Gemini,[17] and LLaMA 3.2, and the newer LLaMA 3.3 have significantly expanded the possibilities for CH preservation and accessibility by enabling models to process and synthesize data across multiple modalities—text (Zhao, 2023), images (Zhang, 2024a), and audio (Triantafyllopoulos, 2024). OpenAI's GPT models, with their enhanced multimodal capabilities, integrate both visual and textual data processing, which can assist in tasks like interpreting historical documents alongside their visual contexts, such as artifacts or manuscripts. This functionality is pivotal for CH applications, where text-based metadata are often paired with visual artifacts. Similarly, Google's Gemini models incorporate capabilities across text, audio, and visual inputs, with variants that support complex audio reasoning and video input, pushing forward the frontiers of CH engagement through multimodal interaction. Gemini's functionality allows for detailed scene understanding and object recognition within videos, which could revolutionize how historical events or restoration projects are documented and analyzed over time. For example, a CH archive could leverage Gemini's abilities to automatically annotate and categorize video footage of archaeological digs, making these resources accessible for educational or research purposes. Meta's LLaMA 3.2 advances multimodal language model efficiency by integrating quantization aware training with LoRA adaptors and SpinQuant post-training quantization, reducing memory usage and model size by half and boosting processing speed. These optimizations make LLaMA 3.2 with its 1B and 3B sizes, ideal for on-device CH applications, supporting tasks like real-time transcription and artifact analysis on mobile devices.

Such multimodal models extend AI's utility in CH by supporting preservation and educational efforts tailored to the complexity of historical and cultural artifacts.

3.6.4 Emerging Applications in CH

Like various domains, the applications of LLMs in CH will continue to evolve. Emerging areas include dynamic virtual restoration of artifacts, where LLMs, especially when combined with VLMs and ALMs, can simulate lost or damaged elements of historical objects and environments, offering more immersive experiences in museums and digital exhibits. Another promising direction is personalized interactive storytelling, allowing LLMs to generate historically accurate narratives based on a visitor's interests or cultural background, creating deeper connections with the past. Furthermore, multimodal models could serve as comprehensive digital curators, integrating language, image, and audio data to provide a holistic understanding of artifacts—supporting not only traditional preservation efforts but also democratizing access to cultural knowledge.

3.6.5 Long-Term Implications for Cultural Preservation

The integration of LLMs into CH initiatives presents significant long-term implications for cultural preservation. LLMs can democratize access to cultural knowledge by transcending language barriers, enabling global audiences to engage with diverse artifacts and narratives. They can play a vital role in preserving endangered languages through content generation and processing, supporting revitalization efforts. Additionally, LLMs can transform educational approaches by providing personalized learning experiences, fostering deeper connections with history and culture. Ethical considerations are crucial; addressing inherent biases in AI models ensures fair representation of all cultures, especially marginalized ones. Sustainability is another concern, as the high computational demands of LLMs necessitate advancements in energy-efficient technologies to mitigate environmental impact. Finally, as AI-generated content becomes part of the cultural record, questions about authorship and authenticity arise, emphasizing the need for collaborative efforts among technologists, cultural experts, and communities to navigate these complexities responsibly.

3.7 CONCLUSION

Our chapter introduced the key technical underpinnings of contemporary LLMs and provided a detailed discussion of potential applications in the realm of CH understanding, preservation, and curation. To that end, we also compiled a list of recent use-cases reflecting the increasing use of LLMs by CH specialists. We expect the rapidly advancing model capabilities, as well as the push for more optimization and transition toward efficient models that have reduced computational complexity, to further accelerate their uptake. Nevertheless, additional involvement of the different stakeholders—including CH specialists, curators, and local communities—is needed to ensure the fair and responsible development and deployment of this exciting technology.

NOTES

1 While the term *chatbot* is used throughout this example for simplicity, it refers to a broader system where the core language processing component is an LLM. This system may also include retrieval mechanisms, user interface layers, and additional logic tailored to cultural contexts.
2 https://www.sofer.info/.
3 *POS-tagging* refers to assigning parts of speech (e.g., noun, verb, adjective) to each word in a sentence.
4 *Lemmatization* is the process of reducing words to their base or dictionary form (e.g., "running" → "run").
5 *Morphological prediction* involves predicting grammatical features of words such as tense, number, case, or gender.
6 https://www.livingmuseum.app/.
7 https://universaldesign.ie/about-universal-design/the-7-principles.
8 https://www.livdeo.com/en/products-and-services/vocal-chatbots.
9 https://query.wikidata.org/.
10 https://github.com/ICCD-MiBACT/ArCo.
11 https://ai.meta.com/blog/meta-llama-3-1/.
12 https://ai.meta.com/blog/llama-3-2-connect-2024-vision-edge-mobile-devices/.
13 https://huggingface.co/meta-llama/Llama-3.2-3B.
14 https://huggingface.co/meta-llama/Llama-3.2-3B-Instruct.
15 https://www.langchain.com/.
16 https://www.nytimes.com/2024/02/22/technology/google-gemini-german-uniforms.html.
17 https://gemini.google.com/app.

REFERENCES

Abate, N. V. (2023). Potential impact of using ChatGPT-3.5 in the theoretical and practical multi-level approach to open-source remote sensing archaeology. *Archaeologies*, 6(12), 402.

Alabdulrahman, M. K. (2024). Sarid: Arabic storyteller using a fine-tuned LLM and text-to-image generation. In: *2024 16th International Conference on Computer and Automation Engineering (ICCAE)*, Tokyo, Japan, IEEE, pp. 1–5.

Almusaed, A. A. (2023). Enhancing student engagement: Harnessing "AIED"'s power in hybrid education: A review analysis. *Education Sciences*, 13(7), 632.

Bahdanau, D. (2014). Neural Machine Translation by Jointly Learning to Align and Translate. arXiv preprint arXiv:1409.0473.

Beck, M. P. (2024). xLSTM: Extended Long Short-Term Memory. arXiv preprint arXiv:2405.04517.

Beguš, N. (2024). Experimental narratives: A comparison of human crowdsourced storytelling and AI storytelling. *Humanities and Social Sciences Communications*, 11(1), 1–22.

Bernasconi, E. (2024). Mining literary trends: A tool for digital library analysis. In: A. Antonacopoulos, A. Hinze, B. Piwowarski, M. Coustaty, G. M. Di Nunzio, F. Gelati, & N. Vanderschantz (Eds.), *International Conference on Theory and Practice of Digital Libraries*, Springer Nature, Switzerland, pp. 342–359.

Bhatia, S. (2024). Transformer networks of human conceptual knowledge. *Psychological Review*, 131(1), 271.

Boros, E. E.-M. (2024). Post-correction of historical text transcripts with large language models: An exploratory study. In: *Proceedings of the 8th Joint SIGHUM Workshop on Computational Linguistics for Cultural Heritage, Social Sciences, Humanities and Literature (LaTeCH-CLfL 2024)*, St. Julians, Malta, pp. 133–159. Association for Computational Linguistics.

Bourne, J. (2024). CLOCR-C: Context leveraging OCR correction with pre-trained language models. *arXiv preprint* arXiv:2408.17428.

Cai, X., & Zhang, E. (2025). HieroLM: Egyptian hieroglyph recovery with next word prediction language model. In A. Kazantseva, S. Szpakowicz, S. Degaetano-Ortlieb, Y. Bizzoni, & J. Pagel (Eds.), Proceedings of the 9th Joint SIGHUM Workshop on Computational Linguistics for Cultural Heritage, Social Sciences, Humanities and Literature (LaTeCH-CLfL 2025) (pp. 25–31). Association for Computational Linguistics. https://doi.org/10.18653/v1/2025.latechclfl-1.4.

Chen, J. D.-R. (2023). Knowledge Graphs for the Life Sciences: Recent Developments, Challenges and Opportunities. arXiv preprint arXiv:2309.17255.

Chowdhery, A. N. (2023). Palm: Scaling language modeling with pathways. *Journal of Machine Learning Research*, 24(240), 1–113.

Costa-jussà, M. R. (2022). No Language Left Behind: Scaling Human-Centered Machine Translation. arXiv preprint arXiv:2207.04672.

Cui, C. D. (2024). AI-powered built heritage: Enhancing interpretation and recovery processes with generative AI models. In: *ReUSO 2024 Documentazione, Restauro e Rigenerazione Sostenibile del Patrimonio Costruito*, Rome, Italy, pp. 117–126.

De Carolis, B. M. (2023). AI-assisted virtual reality for cultural heritage education: The experience of the MARTA museum. In: *6th Experiment@ International Conference (exp. at'23)*, Évora, Portugal, IEEE, pp. 152–156.

Dereza, O. D. (2024). Findings of the SIGTYP 2024 shared task on word embedding evaluation for ancient and historical languages. In: *Proceedings of the 6th Workshop on Research in Computational Linguistic Typology and Multilingual NLP*, St. Julian's, Malta, pp. 160–172.

Design, U. (1997). *The Principles of Universal Design: Version 2.0.* North Carolina State University, Raleigh, NC. https://design.ncsu.edu/wp-content/uploads/2022/11/principles-of-universal-design.pdf.

Fan, W. D. (2024). A survey on rag meeting llms: Towards retrieval-augmented large language models. In: *Proceedings of the 30th ACM SIGKDD Conference on Knowledge Discovery and Data Mining*, Barcelona, Spain, pp. 6491–6501.

Fiorucci, M. K. (2020). Machine learning for cultural heritage: A survey. *Pattern Recognition Letters*, 133, 102–108.

Fnu, Shivam., Leight, M., Kelly, M. K., Davis, C., Clodfelter, K., Thrasher, J., Reddy, Y., & Gyawali, P. (2024). Segmentation of Maya hieroglyphs through fine-tuned foundation models. arXiv. https://doi.org/10.48550/arXiv.2405.16426

Gaia, G. B. (2019). Engaging museum visitors with AI: The case of chatbots. In: T. Giannini & J. P. Bowen (Eds.), *Museums and Digital Culture: New Perspectives and Research*, Springer, Cham, pp. 309–329.

Garcia, G. G. (2023). If the Sources Could Talk: Evaluating Large Language Models for Research Assistance in History. arXiv preprint arXiv:2310.10808.

Gîrbacia, F. (2024). An analysis of research trends for using artificial intelligence in cultural heritage. *Electronics*, 13(18), 3738.

Ghaboura, S. (2025). Time Travel: A comprehensive benchmark to evaluate LMMs on historical and cultural artifacts. *arXiv preprint* arXiv:2502.14865.

Gu, A. (2023). Mamba: Linear-Time Sequence Modeling with Selective State Spaces. arXiv preprint arXiv:2312.00752.

Hauser, J. K.-C. (2024). Large language models' expert-level global history knowledge benchmark (HiST-LLM). In: *The Thirty-Eight Conference on Neural Information Processing Systems Datasets and Benchmarks Track*, Vancouver, Canada.

He, P. L. (2020). Deberta: Decoding-Enhanced Bert with Disentangled Attention. arXiv preprint arXiv:2006.03654.

He, Z., Su, J., Chen, L., Wang, T., & LC, R. (2024). "I recall the past": Exploring how people collaborate with generative AI to create cultural heritage narratives. arXiv. https://doi.org/10.48550/arXiv.2501.00359.

Kalyan, K. S. (2023). A survey of GPT-3 family large language models including ChatGPT and GPT-4. *Natural Language Processing Journal*, 6, 100048.

Ke, Z., Jiao, F., Ming, Y., Nguyen, X.-P., Xu, A., Long, D. X., Li, M., Qin, C., Wang, P., Savarese, S., Xiong, C., & Joty, S. (2025). A survey of frontiers in LLM reasoning: Inference scaling, learning to reason, and agentic systems. arXiv. https://doi.org/10.48550/arXiv.2504.09037

Lankford, S. A. (2023). Adaptmllm: Fine-tuning multilingual language models on low-resource languages with integrated llm playgrounds. *Information*, 14(12), 638.

Lee, B. C. (2023). Human-AI Interaction for Exploratory Search & Recommender Systems with Application to Cultural Heritage. M.S. Thesis, University of Washington, Seattle, WA, USA.

Liu, H. L. (2024a). Visual instruction tuning. *Advances in Neural Information Processing Systems*, 36, 34892–34916.

Liu, J. M. (2024b). How can generative artificial intelligence techniques facilitate intelligent research into ancient books? *ACM Journal on Computing and Cultural Heritage*, *17*(4), 3690391.

Liu, Y. O. (2019c). Roberta: A Robustly Optimized Bert Pretraining Approach. arXiv preprint arXiv:1907.11692, 364.

Makridis, G. (2024). FairyLandAI: Personalized fairy tales utilizing ChatGPT and DALL·E-3. *arXiv preprint* arXiv:2407.09467.

Mandvikar, S. (2023). Augmenting intelligent document processing (IDP) workflows with contemporary large language models (llms). *International Journal of Computer Trends and Technology*, 71(10), 80–91.

Maree, M. (2025). Quantifying relational exploration in cultural heritage knowledge graphs with LLMs: A neuro-symbolic approach for enhanced knowledge discovery. Data, 10(4), 52. https://doi.org/10.3390/data10040052.

Mikolov, T. (2013). Efficient Estimation of Word Representations in Vector Space. arXiv preprint arXiv:1301.3781, 3781.

Muennighoff, N. R. (2023). Scaling data-constrained language models. *Advances in Neural Information Processing Systems*, 36, 50358–50376.

Nayeem, M. T. (2024). KidLM: Advancing Language Models for Children: Early Insights and Future Directions. arXiv preprint arXiv:2410.03884.

Pan, S. L. (2024). Unifying large language models and knowledge graphs: A roadmap. *IEEE Transactions on Knowledge and Data Engineering*, *36*(7), 3580–3599

Pennington, J. S. (2014). Glove: Global vectors for word representation. In: *Proceedings of the 2014 Conference on Empirical Methods in Natural Language Processing (EMNLP)*, Doha, Qatar.

Pinhanez, C. C. (2024). Harnessing the Power of Artificial Intelligence to Vitalize Endangered Indigenous Languages: Technologies and Experiences. arXiv preprint arXiv:2407.12620.

Rachabatuni, P. K. (2024). Context-aware chatbot using MLLMs for cultural heritage. In: *Proceedings of the 15th ACM Multimedia Systems Conference*, Bari, Italy, pp. 459–463.

Radford, A. K. (2021). Learning transferable visual models from natural language supervision. In: *International Conference on Machine Learning*, Virtual Conference, pp. 8748–8763.

Ryberg Smidt, G. D. (2024). Keep me PoS-ted: Experimenting with Part-of-Speech prediction on Old Babylonian letters. *IT-Information Technology*, 65(6), 264–274.

Sahoo, P. S. (2024). A Systematic Survey of Prompt Engineering in Large Language Models: Techniques and Applications. arXiv preprint arXiv:2402.07927.

Sartini, B. (2024). IICONGRAPH: Improved iconographic and iconological statements in knowledge graphs. In: A. Meroño Peñuela et al. (Eds.), *The Semantic Web – ESWC 2024* (Lecture Notes in Computer Science, Vol. 14665, pp. 57–74). Springer.

Sharma, A. R. (2017). Literature survey of statistical, deep and reinforcement learning in natural language processing. In: *International Conference On Computing, Communication and Automation (ICCCA),* Noida, India, IEEE, pp. 350–354.

Shih, Y.-F., Lin, Z.-L., & Hsieh, S.-K. (2025). Reasoning over the glyphs: Evaluation of LLM's decipherment of rare scripts. arXiv. https://doi.org/10.48550/arXiv.2501.17785.

Sommerschield, T. A. (2023). Machine learning for ancient languages: A survey. *Computational Linguistics*, 49(3), 703–747.

Spadoni, E. G. (2023). A personalized expert guide for the hybrid museums of the future. In: A. Antonacopoulos, E. Bernasconi, & F. Fioravanti (Eds.), *International Conference of the Italian Association of Design Methods and Tools for Industrial Engineering*, Springer Nature, Cham, pp. 261–272.

Spennemann, D. H. (2024). Will artificial intelligence affect how cultural heritage will be managed in the future? Responses generated by four genai models. *Heritage*, 7(3), 1453–1471.

Spina, S. (2023). Artificial Intelligence in Archival and Historical Scholarship Workflow: HTS and ChatGPT. arXiv preprint arXiv:2308.02044.

Sprugnoli, R. I. (2024). Overview of the EvaLatin 2024 evaluation campaign. In: *Proceedings of the Third Workshop on Language Technologies for Historical and Ancient Languages (LT4HALA)@ LREC-COLING-2024*, Marseille, France, pp. 190–197.

Stopponi, S. P.-M. (2023). Evaluation of distributional semantic models of ancient Greek: Preliminary results and a road map for future work. In: *Proceedings of the Ancient Language Processing Workshop (ALP) Associated with RANLP-2023*, Varna, Bulgaria, pp. 49–58. Association for Computational Linguistics (ACL).

Su, Y. (2022). Recurrent neural networks and their memory behavior: A survey. *APSIPA Transactions on Signal and Information Processing*, 11(1), e26.

Thomas, A. G. (2024). Leveraging LLMs for post-OCR correction of historical newspapers. In: *Proceedings of the Third Workshop on Language Technologies for Historical and Ancient Languages (LT4HALA)@ LREC-COLING-2024*, Marseille, France, pp. 116–121.

Touvron, H. L. (2023). Llama: Open and Efficient Foundation Language Models. arXiv preprint arXiv:2302.13971.

Triantafyllopoulos, A. T. (2024). Computer Audition: From Task-Specific Machine Learning to Foundation Models. arXiv preprint arXiv:2407.15672.

Trichopoulos, G. (2023). Large language models for cultural heritage. In: *Conference of the ACM Greek SIGCHI Chapter*, Athens, Greece, pp. 1–5.

Vaswani, A. (2017). Attention is all you need. In: I. Guyon, U. V. Luxburg, S. Bengio, H. Wallach, R. Fergus, S. Vishwanathan, & R. Garnett (Eds.), *Advances in Neural Information Processing Systems 30 (NeurIPS 2017)*. Curran Associates, Inc, Red Hook, New York, pp. 5998–6008.

Volk, M. F. (2024). LLM-based Translation Across 500 Years. The Case for Early New High German. M.S. Thesis, University of Zurich, Zurich.

Wagner, A. (2023). Safeguarding cultural heritage in the digital era: A critical challenge. *International Journal for the Semiotics of Law-Revue Internationale de Sémiotique Juridique*, 36(5), 1915–1923.

Wang, H. (2024). VisionGPT: LLM-assisted real-time anomaly detection for safe visual navigation. *arXiv preprint* arXiv:2403.12415.

Xu, R. F. (2023). Chatgpt vs. Google: A Comparative Study of Search Performance and User Experience. arXiv preprint arXiv:2307.01135.

Yang, B. (2024). VIAssist: Adapting multi-modal large language models for users with visual impairments. *arXiv preprint* arXiv:2404.02508.

Yue, X. (2024). Pangea: A fully open multilingual multimodal LLM for 39 languages. *arXiv preprint* arXiv:2410.16153.

Zhang, J. H. (2024a). Vision-language models for vision tasks: A survey. *EEE Transactions on Pattern Analysis and Machine Intelligence*, 46(8), 5625–5644.

Zhang, J. X. (2024b). ArchGPT: Harnessing large language models for supporting renovation and conservation of traditional architectural heritage. *Heritage Science*, 12(1), 220.

Zhang, S. D. (2023). Instruction Tuning for Large Language Models: A survey. arXiv preprint arXiv:2308.10792.

Zhang, T. P. (2024c). Raft: Adapting Language Model to Domain Specific Rag. arXiv preprint arXiv:2403.10131.

Zhang, C., (2024d). Mathemyths: Leveraging large language models to teach mathematical language through child–AI co-creative storytelling. In *Proceedings of the CHI Conference on Human Factors in Computing Systems (CHI '24)*. ACM, Article 274, 1–23. https://doi.org/10.1145/3613904.3642647

Zhao, S. Y. (2024). Retrieval Augmented Generation (RAG) and beyond: A Comprehensive Survey on How to Make Your LLMs Use External Data More Wisely. arXiv preprint arXiv:2409.14924.

Zhao, W. X. (2023). A Survey of Large Language Models. arXiv preprint arXiv:2303.18223.

4 Large Language Models in Education

Applications, Challenges, and Use Cases

Mahnaz Moallem, Hoda Harati, and Suzhen Duan

4.1 INTRODUCTION

The rapid progress of generative artificial intelligence (GenAI), which was built on large language models (LLMs), drives immense applications and tools combining human-like text, images, video, audio, and other data types (e.g., ChatGPT, Llama, Claude, Gemini). The new large-scale LLMs can perform an increasingly broad range of tasks, from text manipulation and analysis to image generation and music composition to speech recognition (Nestor Maslej, 2024). The improved capabilities of LLMs to solve diverse tasks with human-level performance (e.g., GenAI) could transform education, presenting both opportunities and challenges.

Since its launch in November 2022, GenAI has infused various domains of our lives. New large-scale AI models have been released every month, reshaping knowledge acquisition and learning, disrupting educational institutions' traditional role, necessitating changes in curricula and learning environments, prompting a reevaluation of teaching and learning processes and delving into practical applications and implications of GenAI across diverse educational contexts. GenAI's rapidly maturing capabilities offer the opportunity to positively shift education systems and create new teaching and learning models. It provides opportunities to improve the adaptability of learning resources to students' strengths and needs. This includes customizing curricular resources to address local needs, personalizing learning content according to the knowledge and experience students bring to their learning, and providing support to teachers through automated assistants or other tools. Additionally, it enables teachers to extend their support to individual students beyond classroom time (U.S. Department of Education, 2023).

GenAI is also redefining and reshaping the learning design (LD) and technology field by assisting learning designers in exploring initial research ideas on specific subjects/content, developing data-driven customized instruction, automating formative assessment, providing personalized feedback, and enhancing alignment of learning strategies with learning outcomes, assessment strategies, and the needs of the learners and their capabilities.

DOI: 10.1201/9781003492252-4

The chapter will analyze and explore the impact of GenAI on higher education (HE), PK-12 education, and, more specifically, the field of LD and technology, as well as educational research. It will further review the adoption and use cases of GenAI tools in HE, PK-12 education, and LD fields in creating high-quality learning experiences, enhancing personalized and adaptive learning, making data-informed decisions, predicting performance with embedded learning analytics, and providing automated assessment and immediate feedback to learners. Furthermore, this chapter examines the challenges of GenAI despite its potential to drastically change teaching and learning approaches and impose new ways of thinking. It will discuss why the responsible use of GenAI tools is required to address challenges, such as privacy concerns, the risk of perpetuating biases, and ethical complexity.

4.2 THE IMPACT OF GENAI ON HE: AN ANALYSIS

HE claims that it has always embraced innovation and attempted to provide challenging learning environments where students are encouraged to think deeply and critically about complex issues with no simple solutions (Sarofian-Butin, 2024). Before the advent of LLM models, traditional teaching and learning in HE classrooms underwent a transformation grounded in several decades of pedagogical evolution (Moraes et al., 2023). Influenced by constructivist epistemologies, social learning theories, and evidence-based LD principles, HE faculty increasingly emphasized active, learner-centered approaches over didactic, lecture-based strategies in their courses. Flip-classroom, problem-based learning, or case-based reasoning have become more prominent, encouraging learners to engage in complex, authentic tasks and challenging them to pause to think differently and reflect on themselves and their world (Sarofian-Butin, 2024). Furthermore, with the support of innovative pedagogical approaches, HE institutions integrated various technologies to improve access, facilitate communication, and enrich learning experiences to meet the demands and challenges of digital and global society (Miranda et al., 2021). The increase in databases and online scholarly collections further improved exponentially, allowing students access to extensive materials. Nonetheless, while the shift from more traditional, teacher-centered instruction marked significant changes in HE's learning environments, faculty and students are still constrained by the time-intensive, tailored feedback and the cognitive load associated with finding, filtering, and synthesizing large volumes of information. Faculty wishing to provide more individualized scaffolding to support students' success or offer more authentic assessment opportunities are still limited by their class size, time constraints, and the challenges of curating high-quality, discipline-specific resources. Thus, within the emerging learner-centered model of teaching and learning, traditional assessment strategies (e.g., assign papers, quizzes, and end-of-semester presentations) continue to survive (Webber & Tschepikow, 2011). Students also face the intellectual demands of both engaging deeply with the content and developing complex skills, such as critical evaluation of sources, argumentation, and creative problem-solving.

The introduction of LLMs into HE, a gatekeeper of knowledge, providers of credentials and research centers, incubators of innovation, and drivers of social change, abruptly disrupted various aspects of academia (Diab Idris et al., 2024) and further offered opportunities to address constraints. As LLMs became increasingly more sophisticated and more accessible, they transformed the landscape of information acquisition and learning (Biri et al., 2023; Walczak & Cellary, 2023) and accelerated the introduction of digital technologies that were capable of emulating human intelligence, simulating human reasoning, offering personalization, improving learning through teaching assistance, generating new content close to human production, increasing work efficiency, developing statistical algorithms, and understanding complex data distributions to make prediction and improve performance. Although colleges and universities generally have been cautious in adopting technology tools (Hodges & Ocak, 2023), the evolution of LLMs and their broad capabilities, accessibility, and integration into many other tools that HE faculty and students use profoundly impacted how faculty teach, learn, assess, and research and how students learn and perform (see Table 4.1).

To understand how HE began to pass through the stages of grief about the evolution of LLMs and admit rethinking how teaching and learning are done with LLMs and GenAI as a catalyst, we should examine how students, faculty, and HE institutions responded to this disruptive change and navigated the associated technological considerations. This analysis will draw a meaningful connection between the valued pedagogy of the learning-centered model and the age of AI in HE, providing a framework for better integrating these technologies to improve teaching, learning, and research.

4.2.1 Navigating the Generative AI Revolution: HE's Adaptive Journey with LLMs

The systematic review of current empirical studies (e.g., Batista, Mesquita, & Carnaz, 2024; Pozdniakov et al., 2024; Wilson, Olivier, & Morkel, 2024) suggests that after the launch of ChatGPT, HE students rapidly adopted GenAI to perform many of their academic tasks. The studies also showed that students are increasingly familiar with GenAI tools and are willing to use them for learning (e.g., Batista et al., 2024; Chan & Hu, 2023; Tala et al., 2024). Students' perceptions appear to be influenced by effort expectancy, ease of use, and social influence (e.g., peers and instructors) (e.g., Batista et al., 2024; Bernabei, Colabianchi, Falegnami, & Costantino, 2023; Lemke, Kirchner, Anandarajah & Herfurth, 2023; Strzelecki & ElArabawy, 2024; Yilmaz, Yilmaz, & Ceylan, 2023). Students tend to use LLMs and GenAI in their course assignments (both formally cited or unattributed), taking quizzes or finding passable responses to assessment questions, writing/editing computer programs, and preparing for advanced placement tests. Furthermore, students use AI chatbots as tutors to collaborate on generating ideas, obtaining explanations for complex topics, writing assignments, summaries, and reviews, and conducting research. However, despite students' positive perceptions of LLMs' usefulness and assistance and their intention to use them, they show moral discomfort when using GenAI (Chan et al., 2024) and continue to fear violating academic integrity and engaging in unethical practices.

In contrast to studies of students' favorable views and perceptions on the use of GenAI, the research on HE faculty highlights concerns and apprehension regarding the use of GenAI (Chan & Tsi, 2024; Lee et al., 2024). Faculty's early reaction to

TABLE 4.1
Higher Education before and after the Advent of LLMs

	HE before the Advent of LLMs		HE after the Advent of LLMs
Pedagogical Transformation	**Improved Teaching and Learning**	**Continued Constraints**	**LLMs and Opportunities**
• Traditional teaching and learning was influenced by constructivist epistemology, social learning theories, and evidence-based design. • Integration of various learning technologies to improve access, facilitate communication, and enrich learning.	• Learner-centered approach over didactic, lecture-based strategies (flip classroom, problem-based learning, case-based reasoning, use of real-world complex and authentic tasks). • Students' access to extensive materials (databases, online access to scholarly collections), yet the intellectual demands of engaging deeply with the content and developing complex skills.	• Time-intensive tailored feedback and cognitive load associated with finding, filtering, and synthesizing large volumes of information. • Large class sizes to offer individualized scaffolding to support students' success. • Time constraints to offer authentic assessment opportunities. • Within the emerging learner-centered model of teaching and learning, traditional assessment strategies (e.g., assign papers, quizzes, and end-of-semester presentations) continued to survive.	• Accelerated adoption of digital technologies. • Offered personalization and teaching assistance. • Generated content close to human production quickly. • Increased work efficiency (time-saving). • Offered statistical algorithm and understanding of complex data distribution. • Made predictions and improved performance. • Increased accessibility.

GenAI was banning its use in fear of student cheating or a breach of academic integrity and AI-induced plagiarism. The concerns related to ethical issues (e.g., academic integrity and bias), along with the lack of institutional policies and frameworks and, in many cases, technical knowledge gaps, have continued to hinder faculty adoption of GenAI. However, research suggests that the faculty in some disciplines, such as STEM fields and health and medical sciences, demonstrate higher levels of willingness to adopt and integrate GenAI tools into teaching and learning (Khlaif et al., 2024; Kovari, 2024; Wang et al., 2024) compared with faculties of social sciences, humanities, and business and management fields. Overall, regarding integrating GenAI in their teaching and learning, faculties seem to recognize that ignoring GenAI or banning its use is naïve, and it is more ethical to guide students in using AI tools properly rather than pretending they do not exist and are not used by students.

Nevertheless, studies across various nations report that educators generally have a negative attitude toward the use of GenAI and express their concerns about the adverse impact of GenAI on the quality of instruction and learning outcomes for students; thus are cautious about adopting GenAI in their fields (e.g., Iqbal et al., 2022; Khlaif et al., 2024; Kiryakova & Angelova, 2023; Yusuf, Pervin, & Román-González, 2024). Faculty, however, acknowledge the opportunities GenAI offers and utilize them in their own work to brainstorm, summarize ideas, gather information, generate content, and design more engaging courses, ultimately improving their teaching practice. This duality of faculty perceptions points to many issues that current research has documented regarding the lack of institutional policies, frameworks, and guidelines, amplifying the gaps between recognizing the benefits of GenAI and hesitating to integrate it fully into the practice.

Before the emergence of GenAI technologies, reviews of the AI literature in HE (e.g., Bond et al., 2024; Sikström et al., 2022; Zawacki-Richer et al., 2019; Zawacki-Richter, 2023) offer evidence that colleges and universities utilize various aspects of AI such as built-in spelling, grammar and punctuation, adaptive quizzes and exercises, personalized tutoring in disciplines such as mathematics, science, and engineering, automated responses to frequently asked questions, AI-generated feedback for closed-ended quiz/test questions and grading, advanced plagiarism detection, personalized advice based on predictive analytics, simulating real-world training scenarios (e.g., Bond et al., 2024; Desmarais & Baker, 2012; Luckin et al., 2022) as well as institutional operation and efficiency improvements such as assisting students with administrative queries, real-time translation and text-to-speech functionalities to support special needs students, admission process, campus navigation, and many others (e.g., Bond et al., 2024). However, competing views regarding accommodating AI in academic practices have always been contested in HE. For example, some question whether AI, as part of a dynamic and modern data system, reduces and alters the human teacher's role, pedagogy, and professional judgment and whether human and non-human (AI) might work together in a teaching and learning environment and if so, how does it look like and would key human capabilities be lost as a result of employing AI (Bearman, Ryan & Ajjawi, 2023)? The proponents of this view are further concerned that AI technologies could hinder students' potential for new and original discoveries and negatively impact the acquisition of knowledge and skills (e.g., Chan & Tsi, 2024; Civil, 2023; Kiryakova & Angelova, 2023; Warschauer et al., 2023). Other views center on the vital role of HE in adopting innovations with AI components emphasizing, enhancing, and improving teaching and learning, and student success in areas such as intelligent tutoring systems, assessment and evaluation (automated feedback and personalization), adaptive systems, and learning analytics (Zawacki-Richter et al., 2019). Yet, another perspective highlights ongoing challenges related to the ethical use of AI (Holmes et al., 2021), the development of transparency and explainability in systems (Khosravi et al., 2022), and the need to address the pedagogical implications of increasingly autonomous AI systems in educational environments (Han et al., 2023).

Reflecting the ongoing debate regarding a diverse range of viewpoints about the utilization and integration of AI and a broader discussion on AI ethics, it is not surprising that as a preventive and vigilant measure, many HE institutions globally spent time responding to the launch of LLMs with a cautious approach (with many banning the use of GenAI or altering assessment strategies at the early stages) to develop policy and guidelines on the ethical use of GenAI, in both academic and non-academic activities. Between 2022 and 2023, in an attempt to balance the vital role of HE for innovation (e.g., the GenAI's potential to enhance assessment methods and the improvement of student learning outcomes) and acknowledging the potential and benefits of GenAI with academic integrity and ethical and bias concerns, HE institutions engaged in three interrelated initiatives to respond to the inevitable impact of GenAI: *policy development*, *professional development and adequate support for faculty*, and *resource allocation*.

4.2.1.1 Policy Development

HE leading institutions, as well as other global institutions such as the Quality Assurance Agency for Higher Education), Organization for Economic Co-Operation and Development (2023), RUSSEL GROUP (24 leading UK universities, 2023), and United Nations Educational, Scientific, and Cultural Organization (UNESCO, 2023), to name a few, began developing GenAI use policy in teaching, learning, and administration after the public release of GenAI tools. In HE institutions, stakeholders, including faculty committees, student representatives, and educational technologists, collaborated with AI researchers and ethics faculty to frame policies addressing GenAI risk, plagiarism, and bias. Chan and Colloton (2024, p. 6) identify several principles included in the HE policy guidelines to address GenAI's concerns and weaknesses.

- **Transparency**: To increase transparency, the policy provides information about how the AI system was trained and what data were used.
- **Explainability and Interpretability**: To provide clear and understandable explanations, the policy explains how the AI system makes decisions or recommendations and how the internal logic of AI models works.
- **Fairness and Bias**: To enhance fairness and avoid perpetuating biases and stereotypes, the policy explains how to recognize and address AI system biases and foster an inclusive digital space, ensuring fairness and reducing discrimination of GenAI in certain groups.
- **Accountability**: To establish ownership and ensure proper attribution of content generated by the AI system, the policy determines who is responsible

for incorrect information generated by GenAI for fact-checking and verifying the content and to develop a comprehensive framework to educate users and developers about the ethical implications of AI and develop guidelines for the development and use of GenAI.

- **Safety/Security and Robustness**: To ensure that GenAI is safe, the policy recommends careful plans to test and monitor GenAI's inputs and outputs to fix any issues.
- **Privacy and Data Protection**: To address the significant concerns about user privacy and data protection, the policy proposes a robust legal framework, transparent practices, and ongoing scrutiny to ensure these technologies are used responsibly and follow human rights principles and that the tools comply with the General Data Protection Regulations.
- **Autonomy and Human Oversight**: Autonomous GenAI systems can make mistakes, and humans need to identify and correct those mistakes. To address the challenge of balancing AI autonomy and human oversight, the policy defines domains and situations where complete autonomy for AI is acceptable and where it is not. It also proposes regulatory frameworks that offer clear guidelines when human oversight is mandatory for specific applications of GenAI.
- **AI alignment for Humanity**: As GenAI systems become more intelligent, the policy recommends ensuring that they are used for the good of humanity. In other words, GenAI technologies should be incorporated as supplementary tools to assist teachers and students rather than replacements for human interaction.

While leading universities and organizations spearheaded policy development and implementation after the advent of GenAI, by 2024, most HE institutions globally have engaged in developing their institutions' policies and how to implement them. However, institutions continued to face difficulties in addressing potential biases and plagiarism risks associated with GenAI. Furthermore, developing policies requires expertise and significant investment in training and technology. Thus, ethical dilemmas and resource inequalities remain, making integrating GenAI fully into teaching, learning, and administrative activities challenging.

4.2.1.2 Professional Development and Support

The Technology Acceptance Model (TAM), as a research framework, predicts when and how people use a new technology (Davis, 1989). According to this model, the two factors of perceived usefulness and perceived ease of use influence the user's acceptance of new technology. Studies using TAM in assessing faculty and students' acceptance of GenAI demonstrate that if the faculty, students, and staff think that new technology is valuable and easy to use, then they are more likely to adopt it (e.g., Iqbal et al., 2022; Ivanov et al., 2024; Khlaif et al., 2024; Kiryakova & Angelova, 2023; Rahiman & Kodikal, 2024). Additionally, studies suggest other factors such as AI-readiness/awareness (the deeper level of understanding of GenAI technologies and its many components) (e.g., Chergarova et al., 2023; Luckin et al., 2022), the perceived risks associated with the GenAI adoption (instructional and institutional), performance expectancy, and social issues (e.g., ethics, privacy and inequality) (e.g., Strzelecki & ElArabawy, 2024) have significant impact on the two factors of usefulness and ease of use and on the HE faculty's overall decisions to adopt GenAI and integrate the technologies in the curriculum—furthermore, research shows that the faculty's perception and view of the usefulness of GenAI impact student perception and proper use.

Recognizing the importance of faculty training alongside the development of GenAI usage policy and guidelines (Chan et al., 2023), colleges and universities began dedicating time and effort to equip and support faculty to use GenAI tools effectively and appropriately in their teaching and learning practices. Given that the appropriate uses of GenAI tools are likely to differ between academic disciplines and tend to be formed by policies and guidance from discipline-specific fields and organizations, institutions addressed faculty development using several institution-wide strategies such as developing workshops and online modules to train faculty, establishing dedicated AI support teams and helplines, incentivizing GenAI adoption through grants and rewards as well as encouraging academic departments or specific areas within the institutions to determine the proper training opportunities for faculty and students and how to implement the institution-wide GenAI policies and guidelines.

4.2.1.3 Resource Allocation

The rapid growth of LLMs and GenAI demanded that HE institutions consider substantial investments in infrastructure, training, and equitable access to ensure effective and inclusive implementation. AI technologies, particularly LLMs, require robust computing power and infrastructure such as scalable cloud services, high-performing computing systems, data management systems to ensure secure and scalable data pipelines, specialized AI and GenAI applications, AI experts, technologists, and ethics and policy advisors. Simultaneously, universities had to allocate significant resources to faculty development in AI literacy, ethical practices, and pedagogical integration to effectively incorporate GenAI into teaching, learning, and administrative processes, as well as provide equal access to GenAI tools for students to avoid disparities.

In summary, HE institutions are still challenged to fund and invest significantly in technical infrastructure, faculty training, and equitable access to integrate GenAI effectively. Colleges and universities will continue addressing these challenges or initiatives in the coming years.

4.2.2 Adopting GenAI in HE: Use Cases in Teaching, Learning, and Research

In the rapidly evolving field of research, current analyses and practical use cases of GenAI in HE institutions indicate that GenAI is promising and is actively being adopted as:

1. **Teaching/Learning Assistant**: GenAI is used to generate informative content, create presentations or learning materials, suggest templates, images, and graphics, including recommendations to enhance learner engagement, generate quizzes, assist in evaluating and providing feedback, integrate

simulations and virtual laboratory to enhance practical learning experiences, and leveraging data analytics to examine student performance and as a result, adjusting teaching strategies (e.g., Jacobs-Basadien & Shaun, 2024; Lee & Low, 2024; Rodriguez-Donaire, 2024; Yuk Chan, & Colloton, 2024);
2. **Research Assistant**: GenAI chatbots streamline research activities by assisting in research ideation, reviewing and summarizing the literature, suggesting research methodology, data collection, and data generation, and research execution stages (e.g., Chergarova et al., 2024; Naeem & Thomas, 2024; Wan, Kitzie, & Alsaid, 2024);
3. **Writing Assistant**: GenAI is used to help improve writing skills by providing feedback on the structure, grammatical issues, and punctuation, or generating content and offering organization ideas (e.g., Beck & Levine, 2023; Kaharuddin, 2021; Kovari, 2024) as well as critical thinking and reasoning skills (Lee & Low, 2024); and
4. **Personalization Assistant/Expert**: GenAI aids faculty in customizing exams or tasks to the background and experiences of each student (personalization), enhancing student engagement by aligning the content to students' interests and needs and assisting students by helping them in completing homework or assignments (e.g., Barry & Walkden, 2024).

Recommended or applied use cases show how rapidly GenAI can transform HE and offer opportunities to enhance teaching, learning, research, and administrative efficiency (see Figure 4.1). However, it is unclear how and in what ways these instances

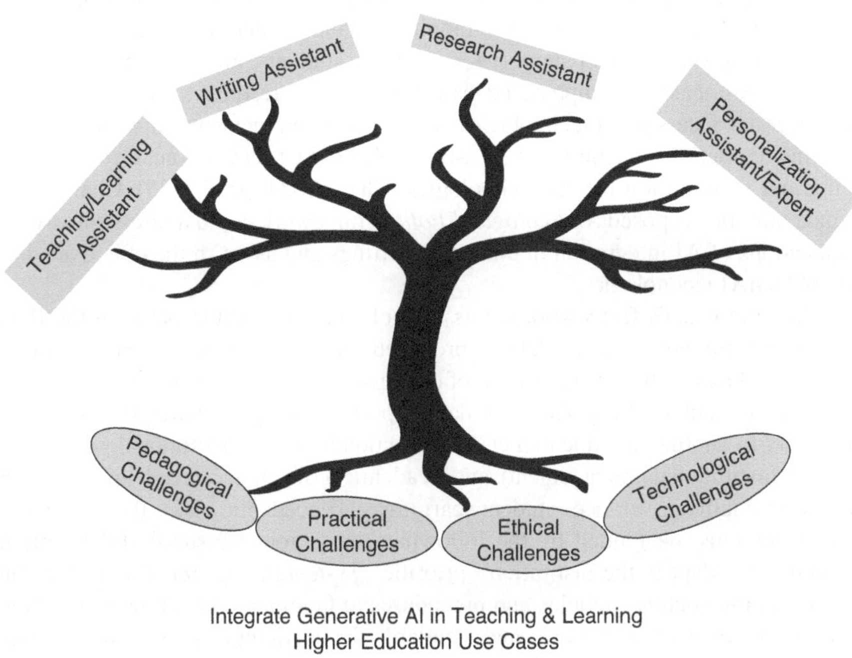

FIGURE 4.1 Higher education and generative artificial intelligence (GenAI) use cases.

of use in HE incorporate the primary challenges of GenAI's pedagogical (accuracy, hallucination, transparency, biases), ethical (academic integrity, privacy, clear guidelines), technological (accessibility, equity, diversity, inclusion), and practical components (monitoring and evaluation of GenAI implementation, training, support of stakeholders). Therefore, a review of the institutional-level frameworks proposed by researchers to address the adoption challenges can guide the optimal adoption and integration of GenAI in HE and provide possible roadmaps for the successful acceptance and implementation of GenAI.

4.2.3 Frameworks for Adoption and Integration of GenAI

As a result of a comprehensive study and attempt to develop an institution-level framework for integrating GenAI in HE, Chan (2023) proposed an *AI Ecological Policy* framework to address the primary challenges of GenAI. Chan organized the challenges into three broad categories: *pedagogical*, *governance*, and *operational.*[1] The *pedagogical* dimension addressed the teaching and learning aspects of GenAI integration, where teachers are the initiators and manage the design and implementation. It included five key areas: rethinking assessments, developing students' holistic competency skills, preparing students for an AI-driven workforce, using a balanced approach to AI adoption, and developing students' holistic competency skills. The *governance* dimension included five key areas: academic integrity, ethical issues, AI governance, AI technologies, and equity in assessing AI technologies. The institution's senior management is expected to manage the *governance* dimension. The *operational* dimension concentrated on the practical implementation of AI in university settings. It comprised two areas: monitoring and evaluating AI implementation and providing training and support for teachers, students, and staff in AI literacy. The *operational* aspects emphasized providing training, resources, and support, promoting equal access to AI technologies, and fostering an inclusive environment. The teaching and learning team and the institution's IT staff play a crucial role in managing and maintaining the AI technologies (Chan, 2023, pp. 20–25). Chan (2023) argued that the proposed *AI-Ecological Policy* framework could address the diverse implications of AI integration in university settings and assist in the effective adoption of GenAI technologies.

Critiquing Chan's framework for its lack of clear interaction between the three dimensions, Shailendra et al. (2024) proposed an institution-level, holistic framework that addresses the entire process of the university education system, curriculum development, and teaching. Arguing that the policies and procedures vary from one university to another and due to a clear distinction between institutional governance (overall institutional management) and academic governance (academic decision, academic quality assurance, student learning, and academic risk), the conceptual framework must be guided by the following key factors for successful adoption: *flexibility*, to adapt to the institution's priorities, *systematic approach* to organizing and managing content (policies and procedures to facilitate content generation) and defining parameters, directions, and standards for curriculum policy and practice, ensuring that content development is aligned with specific requirements of the institutions, and *engagement of all stakeholders* for consultation.

Basing their framework on these key factors, Shailendra et al. (2024) proposed the *4E (Embrace, Enable, Experiment, and Exploit*[2]*)* model to facilitate the incorporation of necessary updates during the adoption process.

- The *Embrace* phase of the framework focuses on articulating the intent of adopting GenAI in the education system. In this phase, the institution's stakeholders (management, leadership, and academic committee representatives) document agreed upon intention within their policy framework. This phase results in redefining the vision and mission of education if needed, as well as crafting new policies that empower academic units to incorporate GenAI as an integral component of the curriculum design and teaching strategies.
- The *Enable* phase, the core of GenAI adoption within an academic ecosystem, involves identifying the requirements and developing plans for AI-friendly curriculum design, including academic evaluation matrix, professional development, and ethical academic integrity guidelines. The *Experiment* phase (as an iterative process with the *Enable* phase) focuses on using and experimenting with the developed curriculum and modified evaluation approach to provide feedback on the effectiveness and challenges faced during the adoption process, the usefulness of the tools and training, and the missing gaps.
- The *Exploit* phase marks the scaling-up phase in which academic units are exposed to the updated curriculum and the tools and techniques necessary to deliver quality education using GenAI (Shailendra et al., 2024, pp. 780–781).

While there are other proposed models such as PAIR (Problem formulation, AI tool selection, Interaction, Reflection) (Acar, 2023), a five-stage "AI Literacy Model" (enable AI, understand AI, apply AI, evaluate AI, and AI ethics) (Southworth et al., 2023), IDEE (Identify the desired outcome, Determine the appropriate level of automation, Ensure ethical consideration, and Evaluate the effectiveness) (Su & Yang, 2023), and TPACK (Technological, Pedagogical, and Content Knowledge) (Mishra, Warr & Islam, 2023), they are more focused on the curriculum-level adoption of GenAI.

In summary, the analysis of the impact of GenAI in HE indicates that, despite numerous challenges and lack of research, colleges and universities face significant internal and societal demands to integrate GenAI to ensure that graduates are well-prepared and equipped with the skills necessary to effectively utilize these tools upon entering the job market. To achieve this objective, institutions must consider various influencing factors, including institutional priorities, ethical considerations, regulatory frameworks, and the societal acceptance of AI-driven educational practices. Simultaneously, systematic research and evaluation of the outcomes of the adoption of GenAI must inform the responsible integration of GenAI into HE.

4.3 ANALYZING THE PK-12 EDUCATION SYSTEM'S RESPONSES TO GENAI

Before analyzing the reviews and reports regarding the emergence of GenAI in PK-12 education after November 2022, it is important to understand the context of PK-12 education. Reviews and reports suggest that the introduction of LLMs and GenAI in PK-12 education was met with cautious optimism. While educators and the education systems quickly recognized technology's potential to transform learning, they became concerned about equitable access to these tools. This heightened awareness of a potential new digital divide led to questions about who will have the opportunity to learn with and about AI. Moreover, educators and the education system viewed LLMs and GenAI within the context of computer science (CS) and computational thinking (CT) education, which was increasingly included in the school curriculum, yet still facing equitable participation of minoritized populations and attending to the needs of neurodiverse learners and learners with disabilities. Thus, in contrast with HE, in which students were the first to adopt LLMs and GenAI quickly, in addition to the fact that LLMs were only available to age 18 and above, PK-12 education was alert due to ethical and equity concerns and primarily viewed the use of LLMs and GenAI within CS and CT and AI literacy education and envisioned LLMs and GenAI as tools alongside human teachers (human-centered) to improve instruction (Grover, 2024; Klopfer, Reich, Abelson, & Breazeal, 2024).

Considering this unique PK-12 education environment, the analysis of available reviews and reports suggests that, similar to HE, the rise of LLMs sparked varied and evolving responses across the PK-12 education system. In the United States, at the national level, the U.S. Department of Education, in its 2023 report, explored how AI, including GenAI, can be used responsibly in classrooms, emphasizing data privacy and ethical and equity access. Later, in its 2024 report, the U.S. Department of Education offered designing with AI guidelines for developers, emphasizing human-in-the-loop AI, AI as a tool to augment human capabilities, and placing teachers at the center of the design (Grover, 2024). The International Society for Technology in Education (ISTE[3]) offered best practices for teachers to incorporate AI tools into lesson planning and student engagement activities. Teachai.org, a non-profit group, emphasized GenAI's societal implications and provided *TeachAI guidance for school toolkits.*[4] At the state level, state departments in technology (e.g., New York, California, Texas, Massachusetts, and North Carolina) developed their own policies and guidelines. They also began pilot programs integrating GenAI into teacher professional development and personalized learning platforms.

At the local level, school districts' earlier responses were diverse, from collaborating with educational technology companies to testing AI-powered tools such as adaptive learning platforms to those that remained cautious and waited for state and national guidelines. In mid-2023 and early 2024, many local school boards began creating district-level AI policies, addressing data privacy, plagiarism, and ethical use concerns. Pilot programs were expanded, with some districts focusing more on integrating AI tools into STEM education, particularly mathematics problems and creative arts classrooms. The increasing advances in GenAI in 2024 further resulted in more collaboration across the states in policymaking, enhancement of efforts on equity, and experimentation with the integration of GenAI. Companies (e.g., Google, OpenAI, Microsoft) and not-for-profit groups and organizations also developed online resources and AI literacy workshops to prepare teachers and students.

At the global level, the responses of the K-12 education system also varied across nations. In early 2023, UNESCO (2023) called for regulations on AI use in schools and provided policy guidelines for integrating AI technologies into K-12 education systems, focusing on ethical challenges and equitable access. Later, UNESCO led the efforts to define AI-literacy competency for teachers and students (2024). In Europe, the European Commission (2022) released *ethical guidelines on using artificial intelligence (AI) and data in teaching and learning for educators.*[5] European nations like Finland initiated programs to integrate AI literacy into school curricula, emphasizing foundational AI concepts. The Australian government released guidelines for safe and responsible AI use in education, aligning with its *AI Ethics Principles.*[6] In Asia, South Korea began a significant initiative, evaluating AI's impact on education and framing AI integration as part of their national digital education strategy (Education for Global Development, 2024). In their proposed transformational model, teachers as facilitators focused more on mentoring, coaching, and fostering problem-solving and critical thinking skills, leaving routine tasks to AI-driven tools.

Overall, in the United States and globally, GenAI has been integrated into PK-12 classrooms in a limited way. In addition to the AI literacy for teachers and students to support learning about AI, AI in support of learning (GenAI integration) is student-, teacher-, and institution-focused. Student-focused AI integration includes adaptive tutoring and assisted writing systems or chatbots. Teacher-focused AI integration is done by tailoring lesson plans, providing assessment support, and curating resources. The institution-focused AI integration helps school administration and operations handle scheduling, address safety concerns, and identify at-risk students (NEA Report, 2024). Like the United States, by mid-2023, many nations acknowledged the lack of guidelines for teachers on using AI effectively in their teaching and the concern about ethics, thus developing policies and frameworks to guide GenAI integration in PK-12 education. Over time, countries expanded on testing GenAI tools for tailoring content to students' needs, using the tools for automated grading and providing real-time feedback, and recognizing the need for professional development to equip teachers with the skills to use GenAI effectively.

4.3.1 Adapting to GenAI: Use Cases in PK-12 Education

Between 2022 and 2024, PK-12 teachers across the nations experimented with GenAI in various ways (Klopfer et al., 2024). Early use cases centered on creative writing, generating essay prompts, brainstorming ideas with students (Mills et al., 2024), and STEM applications (Lee et al., 2023). Mills et al. (2023) observed a case in which a middle school teacher used GenAI to guide students in reading and writing skills by internalizing key elements of a story (e.g., character, setting, plot, genre, tone, etc.). This GenAI use case allowed for scaffolded writing instruction support, especially given to students who struggled with the stamina to construct passages. As an example of a STEM use case, Lee et al. (2023) employed a rule-based AI chatbot in elementary school science classes to examine the impact of rule-based

AI chatbot use on Grade 6 Korean elementary students' science achievement and attitudes toward science. The results revealed a positive effect on students' science achievement and interest, with a particularly significant impact on those with lower achievement levels. Notably, the chatbot played a crucial role in elevating female students' interest in science.

Later implementation centered on teachers using GenAI in their lesson planning (e.g., creating homework prompts, instruction for lab experiments, visual aids, etc.) and tailoring learning materials for students at different proficiency levels (personalization) and enhancing their lessons' quality (e.g., revising assignments, improving assessment strategies, etc.) while saving time (Bahroun et al., 2023; Celik et al., 2022; Klopfer et al., 2024; Swindell et al., 2024; Zhang & Tur, 2024). GenAI has also been utilized to help students solve mathematical problems and explain solutions in accessible language (tutoring buddy) (Sawyer, 2024). The results of using GenAI as a math tutor suggest increased confidence among students when ChatGPT assists in breaking down complex equations. GenAI has been used for project-based learning, allowing students to co-create visual projects with tools like DALL-E (Lai & Tu, 2024). The results demonstrated improved collaboration skills when students used GenAI for group projects.

Overall, the framework of "AI with the human in the loop" (U.S. Department of Education, 2023) or "putting teachers at the center of AI design (human-centered approach)" (UNESCO, 2023) has guided the adoption of the LLMs and GenAI in PK-12 education (Figure 4.2). Within this humanistic framework, the use cases have focused on the application of AI in support of teaching and learning (GenAI

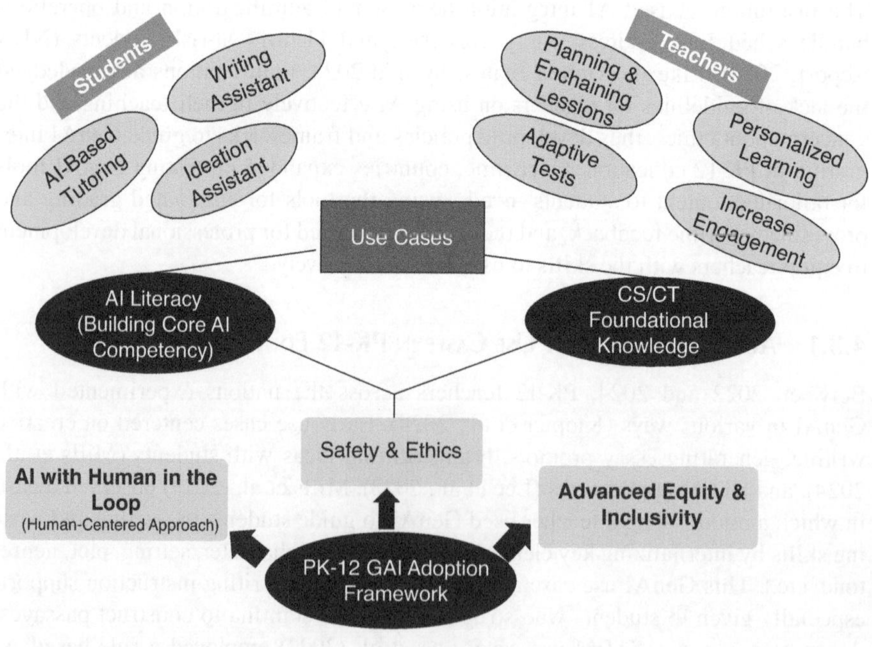

FIGURE 4.2 PK-12 GenAI use cases.

integration) and support of learning about AI (AI literacy). Early PK-12 adopters seem to focus on creative writing and STEM applications (experiment with GenAI integration), while later implementations explored differentiated instruction and ethical AI education. PK-12 teachers in the United States and Canada led initiatives in using AI for personalized learning, whereas Singapore, for instance, pioneered curriculum designs embedding AI ethics.

However, despite promising outcomes of use cases for PK-12 education, challenges like accessibility, professional training for teachers, and ethical concerns still limit applications of GenAI. Equitable access to AI resources remains critical in the United States and across nations, particularly in rural areas of Brazil and South Africa. The future of AI in PK-12 education depends on robust teacher training programs, innovative curriculum integration, and policy frameworks that address both opportunities and risks.

4.4 LLMs AND GENAI IN THE CONTEXT OF LD

4.4.1 Overview of the Field of LD and Technology

LD and technology is a multidisciplinary field that analyzes, designs, develops, implements, and evaluates instructional systems mediated by learning technologies to enhance teaching and learning processes (see Figure 4.3). It encompasses principles of LD and user-experience design, cognitive and learning sciences, instructional design (ID) pedagogies, learning analytics, and elements of design thinking, aiming to create creative, effective, and engaging learning environments across diverse contexts (West, 2018) (Figure 4.3).

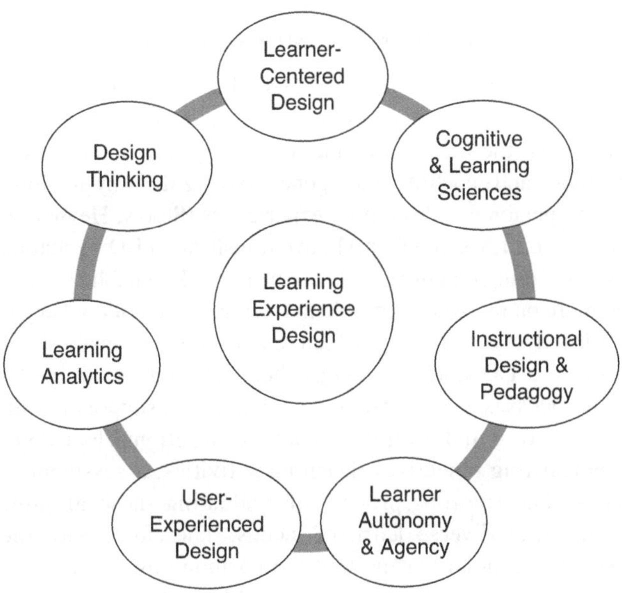

FIGURE 4.3 Principles of learning design.

As education systems grapple worldwide with digital transformation, the LD and technology field has evolved to highlight iterative, creative, user-centric, and equity-oriented participatory design framework that integrates advanced technology tools like learning management systems, adaptive technologies, virtual reality, and data analytics to adapt to diverse learner needs and improve educational outcomes (Gachago, Bali & Pallitt, 2023; Tawfik et al., 2024). Traditionally, learning designers often work as facilitators of pedagogical innovation, emphasizing learner-centered approaches, accessibility, and evidence-based learning practices. They engage in:

- Using the learner-centered approach to create dynamic learning environments (computer-mediated or instructor-guided).
- Incorporating user-experienced design and ID principles to engage learners at deeper and more personal levels.
- Using cognitive and learning sciences principles to create a learning environment that integrates the cognitive, emotional, behavioral, and social dimensions of learning.
- Designing learning experiences that empower learners to find their own way, define their goals, and discover what learning strategies work for them.

The emergence of LLMs and GenAI as co-designers enhancing human–AI partnerships challenges the learning designers to improve their collaborative skills and work with AI systems to co-create innovative instructional strategies. This shift also requires developing new skills, including proficiency in prompt engineering, data interpretation, and ethical AI usage.

4.4.2 The Impact of GenAI on LD and Technology

The integration of AI and, more recently, GenAI tools like LLMs indicates a paradigm shift in how learning is conceptualized, delivered, and assessed. With the advent of LLMs and GenAI, the field of LD and technology now stands at the intersection of human creativity and machine intelligence, paving the way for innovative learning strategies and personalized learning experiences (Storey, Hevner & Yoon, 2024; Shneiderman, 2020). LLMs and GenAI have transformed LD by automating content creation and streamlining repetitive and data-heavy tasks, enabling learning designers to concentrate more on innovation, problem-solving, and emotional engagement while prioritizing ethical decision-making, which remains the responsibility of humans.

Furthermore, the LD and technology field has long examined the value of automated LD processes as a promising mechanism to support learning designers in carefully connecting and aligning various instructional/learning systems elements, such as learning objectives, learning activities, assessments, and instructional materials. The rigorous process of evaluating these alignments impacts learning and meeting diverse learners' needs. Therefore, with the emergence of GenAI tools, LD scholars argue that the harmonious interaction between the human cognitive process and the power of GenAI capabilities amplifies the impact and value of LD. Also, putting human designers at the center of design activities to capture the nuances of human cognition and learning context resolves GenAI's oversimplification of complex human learning processes.

4.4.3 Generative AI in LD and Technology: Use Cases

Integrating LLMs and GenAI into LD and technology has opened transformative possibilities, offering a wide array of use cases that span instructional development, learner engagement, and optimizing educational processes. While research is still limited, by leveraging the capabilities of LLMs and other generative tools, learning designers can now enhance LD activities in unprecedented ways. This section outlines the key dimensions of GenAI integration in LD, some of which are still under investigation, followed by two specific use cases implemented by the co-authors.

- **Content Development and Curation**: GenAI can be incorporated into LD processes to assist designers in performing initial research on specific subjects/content and information on critical facts, barriers, concepts, principles, processes, and procedures. GenAI tools can help learning designers summarize lengthy materials found through other traditional searches for a much broader set of inputs. It assists in creating and curating the content quickly and efficiently, saving time and resources (before or after collaborating with stakeholders and subject matter experts). It also allows for adapting the content to diverse learners and their needs. The learning designers can learn about a particular topic using GenAI-generated content, then fact-check that content, add to it, and explore other sources of information to create an extended learning experience (e.g., Chai et al., 2024; Ullmann, 2024).
- **Developing Data-Driven Insights**: Incorporating GenAI into LD processes allows learning designers to analyze learner performance and preferences to create customized instruction (adjust the difficulty and the order of the learning materials, etc.). It assists in rapidly analyzing vast amounts of data from various sources (e.g., learner surveys, performance metrics, learner feedback, social media, interactions, etc.) to GenAI insights into learning patterns, the effectiveness of previous and proposed instructional strategies, and areas where target students struggle. The emphasis on data-driven information can help designers generate or refine learning objectives and align them with learners' needs (e.g., Caporusso et al., 2024).
- **Generating Customized Learning Objectives**: GenAI can be incorporated into LD processes to help learning designers create learning objectives catering to each learner's needs and abilities. With GenAI assistance in analyzing and finding patterns in target learners' prior performance and preferences, LDs can create learning objectives tailored to students' preferences. GenAI can also assist in aligning the learning objectives to curriculum and standards. It can ensure that learning objectives are measurable and achievable. However, novice learning designers must master the knowledge and skills of writing measurable objectives before seeking GenAI assistance.
- **Generating Customized Learning Strategies**: Incorporating GenAI in LD processes makes it easy for designers to define problems and generate

the best-fit solutions. Because defining a learning/instructional strategy requires designers to take a lot of data and analyze it, GenAI can assist effectively. Furthermore, GenAI can assist in customizing strategies to learners' needs. However, novice designers must develop expertise in knowing what data to feed GenAI and what questions to ask first to define the problem they are trying to solve and, second, to co-create the best-fit solution to the problem.

- **Automated Feedback and Assessment**: GenAI tools can provide automated feedback and formative assessments, which are essential components of effective LD. By providing immediate, personalized feedback on student work, LLMs can improve learning outcomes while reducing the workload of educators (Ankerstein, 2024; Ruwe & Mayweg-Paus, 2024). Furthermore, GenAI tools enhance the timeliness and quality of formative assessment and can be incorporated into LD processes to assist designers in developing assignments, quizzes, and other assessment strategies. GenAI can generate a set of questions that are more closely aligned with the learning objectives and provide a more accurate measure of learner progress, aligning them with the learning objectives.
- **Accessibility, Inclusion, and Development**: GenAI can assist learning designers in producing initial drafts of the learning materials, such as lesson plans, scripts, and explanations, based on the learning objectives and content structure defined in the earlier design processes. It could assist in creating multimedia resources. AI image-creator tools and illustrations can be used to convert textual descriptions to visual aids tailored to the specific content and learning objectives. However, novice learning designers must ensure they have opportunities to master their development skills before seeking GenAI assistance. GenAI also improves accessibility by offering real-time language translation, text-to-speech conversion, and simplified explanations of complex concepts to accommodate diverse learners, including those with disabilities or non-native speakers of the instructional language, ensuring that educational resources are inclusive and equitable (see Figure 4.4).

4.4.4 Integrating GenAI into Graduate Design and Development Course

Two sections of a foundational course titled "Instructional Design and Development," offered in the Instructional/Learning Design and Development graduate program at a university in the Mid-Atlantic region of the United States, were selected to integrate GenAI. The course is offered as a 7-week asynchronous online course, which serves as an overview of a systems approach to LD (analysis, design, development, implementation, and evaluation) with a focus on the application of LD principles, theories, and practices to solve instructional problems using learning technologies. The course was redesigned and offered in the spring of 2024 to explore novice instructional designers' perceptions of feedback regarding the integration of GenAI to assist in the analysis phase of the LD process (see Figure 4.5). The total number of students

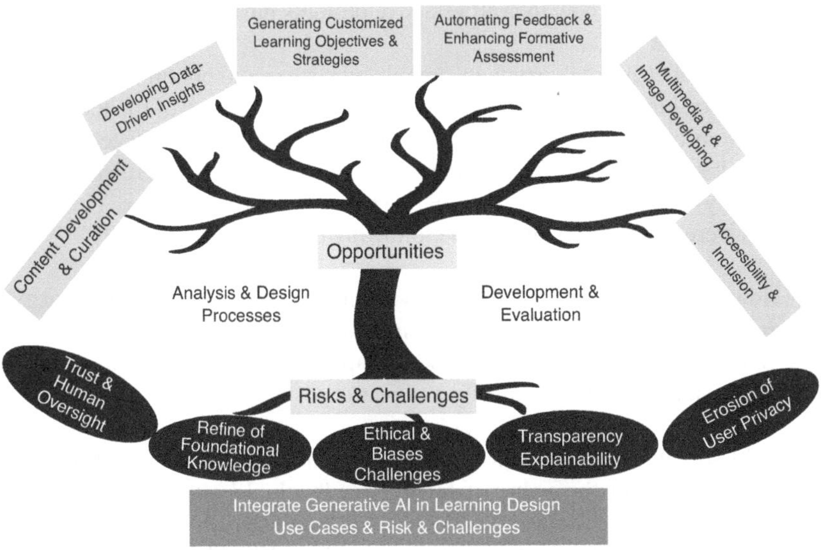

FIGURE 4.4 GenAI in learning design use cases.

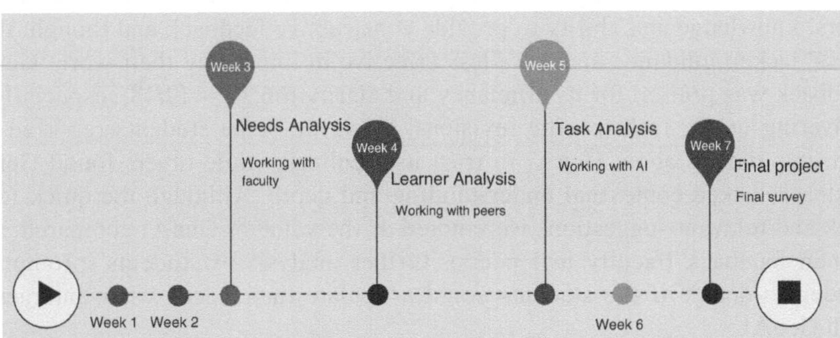

FIGURE 4.5 Integration of GenAI in course redesign.

in both sections was 20 ($N = 20$). Typically, to enrich students' learning experiences when designing and developing their learning environments, the course instructors require students to submit their work in each stage of analysis, design, and development for review and feedback from the instructor and sometimes from their peers. In the course redesign, the analysis module was selected to integrate GenAI. For the three stages of the analysis phase (needs analysis, learner analysis, and task analysis), students were required to submit their (1) learner analysis report to the instructor for critical review and feedback, (2) needs analysis report to their peers for critical review and feedback, and (3) task analysis work to a GenAI tool of their choice using prompts to obtain feedback for improvement. To ensure students were familiar with AI tools and prompt engineering, they were offered training resources and webinars with guest speakers on how to craft effective prompts during the first 2 weeks of the course (see Figure 4.5).

At the end of each week, students completed a survey to obtain their perceptions of the trustworthiness, clarity, constructiveness, and usefulness of the feedback they received from faculty, peers, and GenAI. A final survey also evaluated the impact of the feedback on students' learning experience and the LD project outcomes. The majority of students used ChatGPT, while a few used Gemini. The students' responses were analyzed to answer the following questions:

1. How did students perceive the trustworthiness and usefulness of faculty, AI, and peer feedback in improving their instructional design skills?
2. What were AI-generated feedback's comparative strengths and weaknesses versus human feedback (faculty and peer)?

Not all 20 students responded to the survey. Nine students out of 20 responded to the question about the value of faculty feedback, 18 out of 20 responded to the value of peers' feedback, and 12 out of 18 responded to the value of GenAI feedback. The results suggested that students found faculty feedback clear and in-depth, offering technical guidance (100%—9/9). Peer feedback was also received positively (100%—18/18), with some students indicating the value they saw in their peers' shared practical experiences, while others hesitated about their peers' knowledge and ability to provide constructive feedback and thought their peers' lacked guidance and were less effective in improving their work. GenAI feedback was praised for its efficiency and clarity (66.7%—12/18), especially in delivering quick, task-specific revisions. However, some students reported not knowing enough about GenAI to trust its feedback, while others found GenAI feedback lacked contextual understanding and depth. Although the quick feedback and relevant suggestions were noted as the value of GenAI compared with human feedback (faculty and peers), further analysis of students' prompts is needed to decide if the students could articulate their needs when interacting with GenAI.

Furthermore, to appreciate the feedback, learners have to manage their affective reaction to the feedback provider and prevent negative influences from getting in the way of engaging with the feedback information. Thus, it is not surprising that students either preferred not to complete the survey after faculty provided feedback or those who responded found faculty feedback offering more guidance. Students likely felt an emotional connection with the instructors and peers and were cautious in their evaluation. Furthermore, they did not need to explain what feedback they needed from the instructors compared with GenAI, which required prompt engineering skills and establishing relationships and trust (Ruwe & Mayweg-Paus, 2024). Although this was a quick case study of how LD students perceive GenAI as a trusted tutor and guide, the results confirm the development of new skills, including proficiency in prompt engineering, data interpretation, and ethical AI usage for learning designers.

4.4.5 Challenges and Risks

Despite the transformative opportunities, adopting GenAI in LD raises several challenges.

- If the content generated by GenAI is not continuously evaluated and checked by experts/designers, it could increase and amplify misinformation ("LLMs can generate persuasive text, even if it is not true"). LLMs do not use pedagogy and cannot reliably suggest strategies supported by learning sciences. These issues raise the challenges of preparing novice learning designers in the age of AI. Learning designers' nuanced understanding of the domain knowledge is critical for successfully evaluating, adapting, and contextualizing the content. Should the field redefine the foundational knowledge for learning designers? When and how can novice designers benefit from GenAI and LLM models? What strategies should be used to prepare learning designers or other LD professionals? Should training LD professionals be changed?
- GenAI models can inadvertently perpetuate biases present in their training data. Biased training data and outputs could impact the analysis, evaluation, and content GenAI provides ("AI models are only as good as the data they are trained on"). The challenge is how learning designers can mitigate biases and ensure fairness in AI outputs. Given the current constraint of GenAI, learning designers should consciously reflect on their LD practices and design decisions with ethical considerations to avoid misuse and biases of AI and harmful impacts on learners.
- While designers can benefit from AI assistance and support, erosion of user privacy can happen ("Training AI models require massive amounts of data. This data is often collected without user consent"). Many AI applications require access to sensitive learner information, such as performance data and interaction history, which creates challenges in ensuring data security and compliance with privacy regulations. Furthermore, there is a risk that personal data collected by GenAI systems could be misused. Privacy challenges and risks require LD programs to consider ethical AI usage training to prepare future designers.
- The emergence of GenAI raises the issue of redefining the foundational knowledge for learning designers. In other words, effective integration of GenAI requires new skills, such as AI literacy, prompt engineering, and ethical AI use. However, learning designers' lack of technical expertise to leverage these tools effectively creates a skills gap that hinders widespread adoption. The LD field should develop strategies to prepare learning designers or other professionals for the age of AI.
- The "black-box" nature of GenAI systems poses challenges in understanding how AI models generate specific outputs. This lack of transparency and explainability limits designers' ability to trust AI-generated recommendations, affecting the adoption of GenAI tools.

To address these challenges, LD and technology programs must adopt a multifaceted approach that includes clear ethical and privacy training and guidelines, identified knowledge and skills, improved understanding of GenAI literacy, removal of technical barriers and infrastructure limitations, and equitable access to GenAI tools. By navigating these constraints thoughtfully, the LD and technology field can leverage GenAI to enhance learning outcomes while preserving its core commitment to equitable, human-centered LD.

4.5 GenAI AND ITS ROLE IN ADVANCING EDUCATIONAL RESEARCH

The advent of GenAI represents a groundbreaking development in educational research. Research employing GenAI is increasing across fields and is anticipated to accelerate and transform scientific knowledge. Several reviews (some focused specifically on the potential of GenAI on quantitative and qualitative data analysis) of the potential applications of GenAI in academic research suggest the following GenAI use cases (e.g., Anderson et al., 2024; Gierhart et al., 2025; Laajini & Tadjousti, 2024; Peres et al., 2023; Perkins & Roe, 2024c).

- **Systematic Review Methodology**: GenAI can assist with literature review, helping scholars efficiently process vast amounts of relevant literature. GenAI tools can support searching databases, extracting the key information from studies, and summarizing the key findings of large volumes of literature. However, researchers have been alerted to fact-check and be aware of incomplete, biased, or even false GenAI outputs in drafting their literature reviews. Furthermore, the reviews highlight the risk of using GenAI when summarizing key findings of specialized research areas (van Dis et al. 2023). Future research should investigate the reliability and validity of using LLMs to assist human researchers in conducting literature reviews. Research should also perform a comparative analysis using various new GenAI research tools.
- **Data Collcction and Data Generation**: Several researchers have explored the application of GenAI for quantitative and qualitative data. For qualitative studies, GenAI tools can assist in transcription and text processing to significantly improve the process of converting audio recordings to textual data (Perkins & Roe, 2024b). Coding qualitative/narrative data is time-intensive. GenAI can enhance the process by analyzing large volumes of textual data and identifying patterns, recurring themes, and anomalies that might be missed by human observation. Perkins and Roe (2024b) used a new approach to inductive thematic analysis using ChatGPT. They employed a dual-analysis approach: one researcher conducted a traditional manual analysis, whereas the other utilized ChatGPT to assist in the analysis process. The researchers discovered that ChatGPT rapidly generated an initial set of codes from the dataset and demonstrated proficiency in theme identification based on the relationships between codes, offering new

perspectives on data. GenAI can analyze large, complex datasets for empirical research and uncover patterns, relationships, and anomalies that may not be readily noticeable using traditional statistical methods. However, there are currently fewer use cases for the quantitative analysis, possibly due to concerns about the GenAI operational transparency and explainability (the issue of the "black box"). The increase in the development of new and more advanced GenAI tools and the integration of GenAI with statistical software applications offer promises in handling increasingly more complex datasets.

- **Research Ideation and Research Execution**: GenAI can assist researchers in generating research ideas and brainstorming solutions for research problems. It can assist novice researchers in generating research questions and, based on research questions, help write research plans. With the proper prompts to provide context, examples, and advice, GenAI tools, like ChatGPT and Microsoft Copilot, can craft helpful research questions, tasks, and interview questions.
- **Visual, Predictive, and Prescriptive Analytics**: Data visualization is critical for interpreting and communicating insights from data. GenAI-powered tools enhance this process by generating customized, interactive visualizations based on researchers' queries or datasets. The GenAI tools can assist researchers by writing code for tasks like merging datasets, filtering observations, cleaning, organizing, and transforming raw data into formats ready for analysis, and creating statistical summaries. Through machine learning algorithms, these tools identify missing values, detect outliers, and normalize datasets, ensuring the data are suitable for visualization and analysis. GenAI further enhances data-driven decision-making by incorporating predictive and prescriptive analytics.

The use cases show that GenAI enables new research methodologies and efficiently enhances traditional research practices. Researchers suggest that GenAI offers significant potential for data analysis (Dennard, 2024; Gierhart et al., 2025; Laajini & Tadjousti, 2024), generating new research ideas, creating research intervention and learning materials (Gupta et al., 2024; Oye & Rhoda, 2024), and exploring innovative research methodologies (Perkins & Roe, 2024a). Studies have also shown that GenAI tools like ChatGPT and other LLMs could enhance the speed and scalability of research tasks, enabling researchers to analyze vast datasets, synthesize literature, and even generate hypotheses with unprecedented efficiency (e.g., Al-Zahrani, 2023; Laajini & Tadjousti, 2024; Perkins & Roe, 2024b). Nevertheless, due to inherent ethical and privacy uncertainty, data quality, and the possibility of fabricated information associated with GenAI, there are intense debates on the potential benefits, risks, and challenges of GenAI for research purposes (e.g., Anderson et al., 2024; Al-Zahrani, 2023; Peres et al., 2023).

Therefore, like other areas, there are diverse opinions about whether academics should use GenAI in research and how they assess research integrity across various use cases. Anderson et al. (2024) conducted a large-scale survey study of 2,534

Danish researchers to explore the use of GenAI and research integrity assessments of 32 GenAI use cases by researchers. The respondents evaluated the use cases across five research phases: idea generation, research design, data collection, data analysis, and writing/reporting. They also assessed whether the practices in the use cases were considered good research practices. The findings showed varied opinions on GenAI's research integrity implications. Language editing and data analysis were generally viewed positively, whereas experiment design and peer review tasks faced more criticism. Controversial areas included image creation/modification and synthetic data, with comments highlighting the need for critical and reflexive use of GenAI. Usage differed by main research area, with technical and quantitative sciences reporting slightly higher usage and more positive assessments. Junior researchers used GenAI more than senior colleagues, while no significant gender differences were observed (Anderson et al., 2024, p. 34). A few researchers (e.g., Laajini & Tadjousti, 2024; Perkins & Roe, 2024b) also explored the opportunities of using GenAI for advanced data analysis in qualitative methodologies. They highlighted the significant role of GenAI in improving the quality, validity, and creativity of qualitative research (Laajini & Tadjousti, 2024). They anticipated that integrating GenAI tools in qualitative research methodologies could revolutionize how researchers approach data analysis, interpretation, and visualization (Perkins & Roe, 2024c).

With the advancement of LLM models and the development of new GenAI tools, the impact of GenAI on academic research will continue to evolve, and future tools will likely offer improved accuracy and consistency in the outputs, addressing some of the current concerns (Perkins & Roe, 2024c). However, it is evident that to harness GenAI's full potential, researchers must remain cautious about data quality, ethical considerations, and the interpretability of AI-driven outputs.

AI USAGE DISCLAIMER

While preparing this work, the authors used ChatGPT and Grammarly to develop content, proofread, and improve the readability of the text throughout the production of the manuscript. After using these tools/services, the authors reviewed and edited the content as needed and they take full responsibility for the publication's content.

NOTES

1 See "AI ecological education policy framework" in Chan (2023, https://educationaltechnologyjournal.springeropen.com/articles/10.1186/s41239-023-00408-3).
2 See "4E-An adoption framework of GenAI in education" in Shailendra et al. (2024, https://ieeexplore.ieee.org/document/10629211).
3 ISTE (https://iste.org/).
4 TeachAI guidance for school toolkits (https://www.teachai.org/toolkit).
5 Ethical guidelines on the use of artificial intelligence (AI) and data in teaching and learning for educators (https://op.europa.eu/en/publication-detail/-/publication/d81a0d54-5348-11ed-92ed-01aa75ed71a1/language-en).
6 AI Ethics Principles (https://www.industry.gov.au/publications/australias-artificial-intelligence-ethics-principles).

REFERENCES

Acar, O. A. (2023). *Are Your Students Ready for AI? A 4-Step Framework to Prepare Learners for a ChatGPT World.* Harvard Business Publishing Education. https://hbsp.harvard.edu/inspiring-minds/are-your-students-ready-for-ai?.

Al-Zahrani, A. M. (2023). The impact of generative AI tools on researchers and research: Implications for academia in higher education. *Innovations in Education and Teaching International*, 61(5), 1029–1043. doi:10.1080/14703297.2023.2271445.

Anderson, J. P., Degn, L, Fishberg, R., Graversen, E. K., Horbach, S. P. J. M., Schmidt, E. K., Schneider, J. W., & Sørensen, M. P. (2024). *Generative Artificial Intelligence (GenAI) in the Research Process: A Survey of Researchers' Practices and Perceptions.* OSF. doi:10.31235/osf.io/83whe.

Ankerstein, C. A. (2024). ChatGPT and Me: Implementing and evaluating a custom GPT for written corrective feedback. *Ubiquity Proceedings*, 4(1), 32. doi:10.5334/uproc.154.

Bahroun, Z., Anane, C., Ahmed, V., & Zacca, A. (2023). Transforming education: A comprehensive review of generative artificial intelligence in educational settings through bibliometric and content analysis. *Sustainability*, 15, 12983. doi:10.3390/su151712983.

Barry, M. L., & Walkden, G. (2024). Reflection on using generative AI in tertiary education: The case of a project management course. *ICERI Proceedings.* https://library.iated.org/view/BARRY2024REF.

Batista, J., Mesquita, A., & Carnaz, G. (2024). Generative AI and higher education: Trends, challenges, and future directions from a systematic literature review. *Information*, 15, 676. doi:10.3390/info15110676.

Bearman, M., Ryan, J., & Ajjawi, R. (2023). Discourses of artificial intelligence in higher education: A critical literature review. *Higher Education,* 86(2), 369–385. doi:10.1007/s10734-022-00937-2.

Beck, S. W., & Levine, S. R. (2023). Backtalk: ChatGPT: A powerful technology tool for writing instruction. *Phi Delta.* doi:10.1177/00317217231197487.

Bernabei, M., Colabianchi, S., Falegnami, A., & Costantino, F. (2023). Students' use of large language models in engineering education: A case study on technology acceptance, perceptions, efficacy, and detection chances. *Computers, and Education: Artificial Intelligence*, 5, 100172, doi:10.1016/j.caeai.2023.100172.

Biri, S. K., Kumar, S., Panigrahi, M., Mondal, S., Behera, J. K., & Mondal, H. (2023). Assessing the utilization of large language models in medical education: Insights from undergraduate medical students. *Cureus,* 15, e47468. doi:10.7759/cureus.47468.

Bond, M., Khosravi, H., De Laat, M. et al. (2024). A meta systematic review of artificial intelligence in higher education: a call for increased ethics, collaboration, and rigor. *International Journal of Educational Technology in Higher Education,* 21, 1–41. doi:10.1186/s41239-023-00436-z

Caporusso, N., Doan, M. H., Acharya, B., Pandit, P., Shrestha, S., Le, N., Khatri, R., & Pond, W. (2024). From angry reviews to classroom success: Using LLMs to synthesize RateMyProfessors.com Data. In: *2024 Proceedings of the ISCAP Conference*, Baltimore, MD. 10(62), 2473–49. https://iscap.us/proceedings/.

Celik, I., Dindar, M., Muukkonen, H., & Järvelä, S. (2022). The promises and challenges of artificial intelligence for teachers: A systematic review of research. *TechTrends,* 66, 616–630. doi:10.1007/s11528-022-00715-y.

Chai, G. W., Kim, S. H., Lee, D., & Moon, J. (2024). Utilizing generative AI for instructional design: Exploring strengths, weaknesses, opportunities, and threats, *TechTrends*, 68, 832–844. doi:10.1007/s11528-024-00967-w.

Chan, C. K. Y. (2023). A comprehensive AI policy education framework for university teaching and learning. *International Journal of Educational Technology in Higher Education*, 20(1), pp. 1–25, 2023.

Chan, C. K. Y., & Hu, W. (2023). Students' voices on generative AI: Perceptions, benefits, and challenges in higher education. *International Journal of Educational Technology in Higher Education*. doi:10.1186/s41239-023-00411-8.

Chan, C. K. Y., & Tsi, L. H. (2024). Will generative AI replace teachers in higher education? A study of teacher and student perceptions. *Studies in Educational Evaluation*, 83, 101395. doi:10.1016/j.stueduc.2024.101395.

Chergarova, V., Tomeo, M., Provost, L., De la Peña, G., Ulloa, A., & Miranda, D. (2023). Case study: Exploring the role of current and potential usage of generative artificial intelligence tools in higher education. *Issues in Information Systems*, 24, 282–292.

Civil, B. (2023). ChatGPT can hinder students' critical thinking skills: Artificial intelligence is changing how students learn to write. *The Queen's Journal*. https://www.queensjournal.ca/story/2023-03-16/opinions/chatgpt-can-hinder-stude nts-critical-thinking-skills/.

Davis, F. D. (1989). Perceived usefulness, perceived ease of use, and user acceptance of information technology. *MIS Quarterly*, 13(3), 319–340. doi:10.2307/249008.

Dennard, A. (2024). AI enhanced pedagogy: Examining how teachers elevate their practice through innovativee lesson design. Dissertations & Theses, Lamar University-Beaumont, ProQuest No. 31635857. 2024.

Desmarais, M. C., & Baker, R. S. D. (2012). A review of recent advances in learner and skill modeling in intelligent learning environments. *User Modeling and User-Adapted Interaction*, 22, 9–38.

Diab Idris, M., Feng, X., & Dyo, V. (2024). Revolutionizing higher education: Unleashing the potential of large language models for strategic transformation. *IEEE Access,* 12, 67738–67757.

Education for Global Development (Oct. 2024). Teachers are leading an AI revolution in Korean classrooms. https://blogs.worldbank.org/en/education/teachers-are-leading-an-ai-revolution-in-korean-classrooms.

European Commission, Directorate-General for Education, Youth, Sport and Culture. (2022). *Ethical Guidelines on the Use of Artificial Intelligence (AI) and Data in Teaching and Learning for Educators*. Publications Office of the European Union. https://data.europa.eu/doi/10.2766/153756.

Gachago, D., Bali, M. & Pallitt, N. (2023). Equity-oriented learning design: An entangled future. *Postdigital Science and Education*, 6, 173–193. doi:10.1007/s42438-023-00420-w.

Gierhart, A. R., Shefferly, N., Li, Y, & Speetzen, E. (2025). Generative artificial intelligence and postsecondary education: Rethining policy and course design. In: A. R. Gierhart (Ed.), *Cases on Enhancing P-16 Student Engagement with Digital Technologies* (pp. 31–54). IGI. https://doi.org/10.4018/979-8-3693-5633-3.ch002.

Grover, S. (2024, March). Teaching AI to K-12 learners: Lessons, issues, and guidance. In: *Proceedings of the 55th ACM Technical Symposium on Computer Science Education,* Minneapolis, MN, USA (Vol. 1, pp. 422–428). ACM.

Gupta, N., Khatri, K., Malik, Y., Lakhani, A., Kanwal, A., Aggarwal, S., & Dahuja, A. (2024). Exploring prospects, hurdles, and road ahead for generative artificial intelligence in orthopedic education and training. *BMC Medical Education*, 24(1), 1544. doi:10.1186/s12909-024-06592-8.

Han, B., Nawaz, S., Buchanan, G., & McKay, D. (2023). Ethical and pedagogical impacts of AI in education. In: *International Conference on Artificial Intelligence in Education* (pp. 667–673). Springer Nature.

Hodges, C. & Ocak, C. (August 30, 2023). Integrating generative AI into higher education: Considerations. *Educause Review*. https://er.educause.edu/articles/2023/8/integrating-generative-ai-into-higher-education-considerations.

Holmes, W., Porayska-Pomsta, K., Holstein, K., Sutherland, E., Baker, T., Shum, S. B., & Koedinger, K. R. (2021). Ethics of AI in education: Towards a community-wide framework. *International Journal of Artificial Intelligence in Education,* 31, 1–23.

Iqbal, N., Ahmed, H., & Abid Azhar, K. (2022). Exploring teachers' attitudes towards using ChatGPT. *Global Journal of Management and Administrative Sciences*, 3, 97–111.

Ivanov, S., Soliman, M., Tuomi, A., Alkathiri, N. A., & Al-Alawi, A. N. (2024). Drivers of generative AI adoption in higher education through the lens of the theory of planned behaviour. *Technology in Society*, 77, 102521. doi:10.1016/j.techsoc.2024.102521.

Jacobs-Basadien, M., & Shaun, P. (2024). Towards an ethical framework for artificial intelligence applications in higher education teaching and learning: A conceptual paper. In: *Digital Commons at Kennesaw State University, African Conference on Information Syste ms and Technology* (Vol. 28). Digital Commons at Kennesaw State University. https://digitalcommons.kennesaw.edu/acist/2024/presentations/28.

Kaharuddin, K. (2021). Assessing the effect of using artificial intelligence on the writing skill of Indonesian learners of English. *Linguistics and Culture Review*, 5(1), 288–304.

Khlaif, Z. N., Ayyoub, A., Hamamra, B., Bensalem, E., Mitwally, M. A. A., Ayyoub, A., Hattab, M. K., & Shadid, F. (2024). University teachers' views on the adoption and integration of generative AI tools for student assessment in higher education. *Education Sciences*, 14(10), 1090. doi:10.3390/educsci14101090.

Khosravi, H., Shum, S. B., Chen, G., Conati, C., Tsai, Y. S., Kay, J., & Gašević, D. (2022). Exicial intelligence in education. *Computers and Education: Artificial Intelligence,* 3, 100074.

Kiryakova, G., & Angelova, N. (2023). Chat GPT. A challenging tool for the university professors in their teaching practice. *Education Sciences*, 2023, 13, 1056. doi:10.3390/educsci13101056.

Klopfer, E., Justin Reich, J. Abelson, H., & Breazeal. C. (2024). Generative AI and K-12 education: An MIT perspective. *An MIT Exploration of Generative AI.* doi:10.21428/e4baedd9.81164b06.

Kovari, A. (2024). Ethical use of ChatGPT in education: Best practices to combat AI-induced plagiarism. *Frontiers in Education*, 9, 1465703. doi:10.3389/feduc.2024.1465703.

Laajini, T., & Tadjousti, H. (2024). Generative AI in qualitative research: A systematic review (2022-2024). In: B. Bentalha & L. Alla (Eds.), *Data Collection and Analysis in Scientific Qualitative Research* (pp. 119–134). IGI.

Lai, C. L., & Tu, Y. F. (2024). Roles, strategies, and research issues of generative AI in the mobile learning era. *International Journal of Mobile Learning and Organization*, 18(4), 516–537. doi:10.1504/IJMLO.2024.141836

Lee, C. C., & Low, M. Y. H. (2024). Using GenAI in education: The case for critical thinking. *Frontiers in Artificial Intelligence*, 7, 1452131. https://www.frontiersin.org/journals/artificial-intelligence/articles/10.3389/frai.2024.1452131/full.

Lee, D., Arnold, M., Srivastava, A., Plastow, K., Strelan, P., Ploeckl, F., Lekkas, D., & Palmer, E. (2024). The impact of generative AI on higher education learning and teaching: A study of educators' perspectives. *Computers and Education: Artificial Intelligence*, 6, 100221. doi:10.1016/j.caeai.2024.100221.

Lee, J., An, T., Chu, H., Hong, H., & Martin, S. N. (2023). Improving science conceptual understanding and attitudes in elementary science classes through the development and application of a rule-based AI chatbot. *Asia-Pacific Science Education*, 9(2), 365–412. https://doi.org/10.1163/23641177-bja10070.

Lemke, C., Kirchner, K., Anandarajah, L, & Herfurth, F. N. (2023). Exploring the student perspective: Assessing technology readiness and acceptance for adopting large language models in higher education. In: *Proceedings of the 22nd European Conference on e-Learning*, ECEL 2023, Pretoria, South Africa. doi:10.34190/ecel.22.1.1828.

Luckin, R., Cukurova, M., Kent, C., & Du Boulay, B. (2022). Empowering educators to be AI-ready. *Computers and Education: Artificial Intelligence*, 3, 100076. doi:10.1016/j.caeai.2022.100076.

Mills, K., Ruiz, P., Lee, K., Coenraad, M., Fusco, J., Roschelle, J., & Weisgrau, J. (2024, May). AI literacy: A framework to understand, evaluate, and use emerging technology.

Miranda, J., Navarrete, C., Noguez, J., Molina-Espinosa, J. M., Ramírez-Montoya, M. S., Navarro-Tuch, S. A., Bustamante-Bello, M. R., Rosas-Fernández, J. B., & Molina, A. (2021). The core components of education 4.0 in higher education: Three case studies in engineering education. *Computers & Electrical Engineering*, 93, 107278. doi:10.1016/j.compeleceng.2021.107278.

Mishra, P., Warr, M., & Islam, R (2023). TPACK in the age of ChatGPT and generative AI. *Journal of Digital Learning in Teacher Education*, 39(4), 1–17.

Moraes, E. B., Kipper, L. M., Hackenhaar Kellermann, A. C., Austria, L., Leivas, P., Moraes, J. A. R., & Witczak, M. (2023). Integration of industry 4.0 technologies with education 4.0: Advantages for improvements in learning. *Interactive Technology and Smart Education,* 20, 271–287. doi:10.1108/ITSE-11-2021-0201.

National Education Association Report (June 2024). Report of the NEA Task Force on artificial intelligence in education. https://www.nea.org/resource-library/artificial-intelligence-education.

Naeem, M., & Thomas, L. (2025). Case study research and artificial intelligence: A step-by-step process to using ChatGPT in the development of case studies in qualitative research. *International Journal of Qualitative Methods*, 24. https://doi.org/10.1177/16094069251371478

Nestor Maslej, L. F. (April 2024). *The AI Index 2024 Annual Report.* AI Index Steering Committee, Institute for Human-Centered AI, Stanford University.

OECD. (2023). Generative AI in the classroom: From hype to reality? *EDU/EDPC*(11). https://one.oecd.org/document/EDU/EDPC(2023)11/en/pdf.

Oye, E., & Rhoda, P. (2024). Evaluating the impact of generative ai on personalized learning outcomes. https://www.researchgate.net/publication/387504736_EVALUATING_THE_IMPACT_OF_GENERATIVE_AI_ON_PERSONALIZED_LEARNING_OUTCOMES.

Peres, R., Schreier, M., Schweidel, D., & Sorescu, A. (2023). On ChatGPT and beyond: How generative artificial intelligence may affect research, teaching, and practice. *International Journal of Research in Marketing*, 40(2), 269–275. doi:10.1016/j.ijresmar.2023.03.001.

Perkins, M., & Roe, J. (2024a). Academic publisher guidelines on AI usage: A ChatGPT supported thematic analysis [version 2; peer review: 3 approved, 1 approved with reservations]. In F1000Research (Vol. 12, Issue 1398). doi:10.12688/f1000research.142411.2.

Perkins, M., & Roe, J. (2024b). The use of Generative AI in qualitative analysis: Inductive thematic analysis with ChatGPT. *Journal of Applied Learning and Teaching*, 7(1), 1. doi:10.37074/jalt.2024.7.1.22.

Perkins, M., & Roe, J. (2024c). Generative AI tools in academic research: Applications and implications for qualitative and quantitative research methodologies. doi:10.48550/arXiv.2408.06872.

Pozdniakov, S., Brazil, J., Abdi, S., Bakharia, A., Sadiq, S., Gašević, D., Denny, P., & Khosravi, H. (2024). Large language models meet user interfaces: The case of provisioning feedback. *Computers and Education: Artificial Intelligence*, 7, 100289.

Rahiman, H. U., & Kodikal, R. (2024). Revolutionizing education: Artificial intelligence empowered learning in higher education. *Cogent Education*, 11(1), 2293431. doi:10.1080/2331186X.2023.2293431.

Rodriguez-Donaire, S. (2024). Influence of prompts structure on the perception and enhancement of learning through LLMs in online educational contexts. *IntechOpen*. doi:10.5772/intechopen.1006481.

Russel Group. (2023). Russell Group principles on the use of generative AI tools in education. https://www.russellgroup.ac.uk/policy/policy-briefings/principles-use-generative-ai-tools-education.

Ruwe, T., & Mayweg-Paus, E. (2024). Embracing LLM feedback: The role of feedback providers and provider information for feedback effectiveness. *Frontiers in Education,* 9, 1461362. doi:10.3389/feduc.2024.1461362.

Sarofian-Butin, D. (2024, July 17). At the crossroads of innovation: Embracing AI to foster deep learning in the college classroom. *Teaching and Learning Viewpoints, Educause Review.* https://er.educause.edu/articles/2024/7/at-the-crossroads-of-innovation-embracing-ai-to-foster-deep-learning-in-the-college-classroom.

Sawyer, A. G. (2024). Artificial intelligence chatbot as a mathematics curriculum developer: Discovering preservice teachers' overconfidence in ChatGPT. *International Journal on Responsibility*, 7(1), 1. doi:10.62365/2576-0955.1106. https://commons.lib.jmu.edu/ijr/vol7/iss1/1.

Shailendra, S., Kadel, R., & Sharma, A. (2024). Framework for adoption of generative artificial intelligence (GenAI) in education. *IEE Transactions on Education*, 67(5), 777.

Shneiderman, B. (2020). Human-centered artificial intelligence: Three fresh ideas. *AIS Transactions on Human-Computer Interaction,* 12(3), 109 124. doi:10.17705/1thci.00131.

Sikström, P., Valentini, C., Sivunen, A., & Kärkkäinen, T. (2022). How pedagogical agents communicate with students: A two-phase systematic review. *Computers & Education,* 188, 104564. doi:10.1016/j.compedu.2022.104564.

Southworth, J., Migliaccio, K., Glover, J., Glover, J. N., Reed, D., McCarty, C., Brendemuhl, J., & Thomas, A. (2023). Developing a model for AI across the curriculum: Transforming the higher education landscape via innovation in AI literacy. *Computers and Education: Artificial Intelligence,* 4, 100127.

Storey, V. C., Hevner, A. R., & Yoon, V. Y. (2024). The design of human-artificial intelligence systems in decision sciences: A look back and directions forward. *Decision Support Systems*, 182, doi:10.1016/j.dss.2024.114230.

Strzelecki, A., & ElArabawy, S. (2024). Investigation of the moderation effect of gender and study level on the acceptance and use of generative AI by higher education students: Comparative evidence from Poland and Egypt. *British Journal of Educational Technology*, 55, 1209–1230.

Su, J., & Yang, W. (2023). Unlocking the power of ChatGPT: A framework for applying generative AI in education. *ECNU Review of Education*, 6(3), 355–366.

Swindell, A., Greeley, L., Farag, A., & Verdone, V. (2024). Against artificial education: Towards an ethical framework for generative artificial intelligence (AI) use in education. *Online Learning Journal*, 28(2). doi:10.24059/olj.v28i2.4438.

Tala, M. L., Müller, C. N., Năstase, I. A., State, O., & Georghe, G. (2024). Exploring university students' perceptions of generative artificial intelligence in education. *Amfiteatru Economic Journal*, 26(65), 71–88. doi: 10.1007/s11423-024-10355-z.

Tawfik, A., Schmidt, M., Payne, L., & Huang, R. (2024). Advancing understanding of learning experience design: Refining and clarifying definitions using an eDelphi study approach. *Educational Technology Research and Development*, 72, 1539–1561 doi:10.1007/s11423-024-10355-z.

The EU Artificial Intelligence ACT. https://artificialintelligenceact.eu.

U.S. Department of Education, Office of Educational Technology (2023). *Artificial Intelligence and the Future of Teaching and Learning: Insights and Recommendations.* U.S. Department of Education.

U.S. Department of Education, Office of Educational Technology (2024). *Designing for Education with Artificial Intelligence: An Essential Guide for Developers.* U.S. Department of Education.

Ullmann, T. D., Bektik, D., Edwards, C., Herodotou, C., & Whitelock, D. (2024). Teaching with Generative AI: Moving forward with content creation. *Ubiquity Proceedings*, 4(1), 35. doi:doi:10.5334/uproc.157.

UNESCO. (2023, June 9). Artificial intelligence: UNESCO publishes policy paper on AI foundation models. https://www.unesco.org/en/articles/artificial-intelligence-unesco-publishes-policy-paper-ai-foundation-models.

van Dis, E. A. M., Bollen, J., Zuidema, W., van Rooij, R., & Bockting, C. L. (2023). ChatGPT: Five priorities for research. *Nature*, 614, 224–226. doi:10.1038/d41586-023-00288-7.

Walczak, K., & Cellary, W. (2023). Challenges for higher education in the era of widespread access to generative AI. *Economics and Business Review,* 9. doi:10.18559/ebr.2023.2.743.

Wan, Y., Kitzie, V., & Alsaid, M. (2024). The AI-empowered researcher: Using AI-based tools for success in Ph.D. programs. In: *ALISE Annual Conference Proceedings*, Portland, OR, USA. https://iopn.library.illinois.edu/journals/aliseacp/article/view/1710.

Wang, H., Dang, A., Wu, Z., & Mac, S. (2023). Generative AI in higher education: Seeing ChatGPT through universities' policies, resources, and Guidelines. *Computers and Education: Artificial Intelligence*. doi:10.1016/j.caeai.2024.100326.

Warschauer, M., Tseng, W., Yim, S., Webster, T., Jacob, S., Du, Q., & Tate, T. (2023). The affordances and contradictions of AI generated text for second language writers. *Journal of Second Language Writing,* 62. doi:10.2139/ssrn.4404380.

Webber, K. L., & Tschepikow, K. (2011). Learner-centered assessment: A comparison of faculty practices in US colleges and universities 1993 to 2004. In: Paper presented at the *Annual Forum of the Association for Institutional Research 51st*, Toronto, Ontario, May 21–25, 2011. https://eric.ed.gov/?id=ED531715.

West, R. E. (2018). Foundations of learning and instructional design technology. https://edtechbooks.org/lidtfoundations.

Wilson, O., Olivier, C., & Morkel, J. (2024). Chatting or cheating: Test of a first-rate intelligence? *Journal of Ethics in Higher Education*, (4), 61–89. https://doi.org/10.26034/fr.jehe.2024.5970.

Yilmaz, F. G. K., Yilmaz, R., & Ceylan, M. (2023). Generative artificial intelligence acceptance scale: A validity and reliability study. *International Journal of Human–Computer Interaction,* 39, 1–13.

Yuk Chan, C. K., & Colloton, T. (2024). *Generative AI in Higher Education: The ChatGPT Effect*. Routledge. doi:10.4324/9781003459026.

Yusuf, A., Pervin, N., & Román-Gonzále, M. (2024). Generative AI and the future of higher education: A threat to academic integrity or reformation? Evidence from multicultural perspectives. *International Journal of Educational Technology in Higher Education*, 21. doi:10.1186/s41239-024-00453-6.

Zawacki-Richter, O. (2023). Umbrella review in ODDE. In: *Herbsttagung der Sektion Medienpädagogik* (DGfE), FernUniversität in Hagen, Germany..

Zawacki-Richter, O., Marín, V. I., Bond, M., & Gouverneur, F. (2019). Systematic review of research on artificial intelligence applications in higher education—where are the educators? *International Journal of Educational Technology in Higher Education*. doi:10.1186/s41239-019-0171-0.

Zhang, P., & Tur, G. (2024). A systematic review of ChatGPT use in K-12 education. *European Journal of Education*, 59, e12599. doi:10.1111/ejed.12599.

5 Healthcare in the Age of LLMs

Perspectives on Radiology

Rikhiya Ghosh, Shima Asaadi, Sanjeev Kumar Karn, Ullas Ullaskrishnan, and Oladimeji Farri

ABBREVIATIONS

AI	Artificial Intelligence
BERT	Bidirectional Encoder Representations from Transformers
BLEU	Bilingual Evaluation Understudy
BLIP	Bootstrapping Language-Image Pre-training
CAPEC	Common Attack Pattern Enumeration and Classification
CDE	Common Data Element
CDSS	Clinical Decision Support Systems
CLIP	Contrastive Language-Image Pretraining
CNN	Convolutional Neural Network
CVE	Common Vulnerabilities and Exposures
CVSS	Common Vulnerability Scoring System
CWE	Common Weakness Enumeration
CXR	Chest X-Ray
DAPT	Domain-Adaptive Pretraining
EHR	Electronic Health Record
GPT	Generative Pretrained Transformer
IFT	Instruction Finetuning
LLM	Large Language Model
MAE	Mean Absolute Error
MDM	Medical Device Manufacturer
MIMIC	Medical Information Mart for Intensive Care
MPT	MosaicML Pretrained Transformer
MRI	Magnetic resonance imaging
NLP	Natural Language Processing
PACS	Picture Archiving and Communication System
PEFT	Parameter-Efficient Fine-Tuning
RAG	Retrieval-Augmented-Generation
RDE	Recommended Data Element
RIS	Radiology Information System
RNN	Recurrent Neural Network
ROUGE	Recall-Oriented Understudy for Gisting Evaluation

DOI: 10.1201/9781003492252-5

SLAKE	Semantically-Labeled Knowledge-Enhanced Dataset
TF-IDF	Term Frequency–Inverse Document Frequency
US	Ultrasound
VLM	Vision-Language Model
VQA	Visual Question Answering

5.1 INTRODUCTION

In a 2023 interview with Eric Topol (renown cardiologist, scientist and author), Geoffrey Hinton (renown computer scientist and 2024 Nobel Prize in Physics winner, regarded as the Godfather of AI) stated "in 2016, I made a daring and incorrect prediction that within 5 years, the neural nets were going to be better than radiologists at interpreting medical scans. It was sometimes taken out of context. I meant it for interpreting medical scans, not for doing everything a radiologist does, and I was wrong about that. But at the present time, they're comparable. This is like 7 years later" (Topol, 2023).

The "incorrectness" of Hinton's prediction is increasingly being reversed with rapid advancements in the development and applications of large language models (LLMs) in healthcare. Extensive experiments, evaluations, and scrutiny of LLMs relative to their potential to automate diagnosis, treatment, research, and administrative tasks in healthcare, remain central topics in medical publications and gatherings across the globe. Ethical and safety-related implications of errors and bias with using LLMs for critical processes and decisions in healthcare have triggered academic and regulatory bodies, together with their industry counterparts, to look closer at how appropriate guardrails and oversights can ensure that the use of LLMs in caring for patients does not violate one of the popularly accepted physician oaths: do no harm.

As part of the patient's journey toward identifying disease risk through image-based screening, unraveling etiology, and morphological manifestations in organs via diagnostic imaging, radiology can be considered pivotal to providing quality healthcare and achieving the best possible patient outcomes. Radiologists utilize diverse sources of patient information to ensure accurate image interpretation and informed clinical decision-making. Patient information is available in multiple formats, including textual data (clinical context), radiographic images, and tabular data (genomic information). Clinical context encompasses past medical history, presenting complaints and imaging findings, while imaging data are derived from various modalities such as X-ray, computer tomography (CT), magnetic resonance imaging (MRI), and ultrasonography (US).

Recent research on LLMs in radiology covers a wide variety of applications. These include driving auto-reporting by generating summaries, or *Impressions* based on the *Findings* in a medical image and standardizing the reports to ensure interoperability across healthcare systems. In addition, the combination of imaging and text in radiology provide opportunities to train, evaluate, and apply multimodal LLMs as assistants that can answer questions about the images and possibly execute repetitive clinical tasks while the human experts devote their cognitive capital to relatively more difficult tasks. Another interesting area of research involves integrating LLMs to ensure resilience of medical devices and thus patient safety (including those used in radiology) against cybersecurity attacks.

When developing LLMs for radiology, it's critical to acknowledge their generative nature, particularly the potential for hallucinations, where models may generate plausible but incorrect information. In clinical settings, unlike general AI applications, accuracy and reliability are paramount, as mistakes could directly affect patient outcomes. Studies (Agarwal et al., 2024) have indicated that LLMs are highly vulnerable to adversarial hallucination attacks in clinical decision support scenarios. A research study (Omar et al., 2025) using 300 physician-validated simulated vignettes observed hallucination rates between 50% and 82% across various models and prompting methods. The lowest hallucination rate achieved with even the most state-of-the-art post-training techniques is 23%, indicating substantial risk when integrating LLMs into clinical workflows. Such findings are consistent across other studies (Vishwanath et al., 2024) which find hallucinations in all aspects of generated answers, e.g., medical event (patient details, symptom, diagnosis etc.) consistency, chronological consistency, and incorrect reasoning. The adverse impact of these inconsistencies ranges from perpetuating violence, inaccurate associations, and false authority, leading to harmful AI, increased workload for medical practitioners, and ethical and legal issues (Koenecke et al., 2024). This necessitates developing LLMs with narrow, specific tasks where outputs can be rigorously validated by human experts. For example, rather than allowing an LLM to freely generate radiology reports, it should assist in tasks such as report summarization, text structuring, or suggesting differential diagnoses based on predefined imaging data, ensuring that the results remain interpretable and verifiable.

Overall, connecting LLMs to clinical and operational workflows beckons for careful investigation of the level of human expert supervision needed, expected efficiency gains, and the balance of risks versus benefits if (more like when) the LLMs hallucinate and prove that "to err is not just human, it is AI too."

A comprehensive discussion about the need and applicability of LLMs to radiology domain raises several important questions that require addressing:

a. **Analysis**: Which LLMs are suited for specific radiology tasks and which mechanisms of training/prompting gives the best results? A thorough analysis should examine various types of LLMs—including open-source versus proprietary models and general-purpose versus domain-specific models—in terms of performance metrics, interpretability, and robustness.
b. **Limitations**: What is the feasibility of LLM-based solutions in radiology and what are the inherent limitations of LLMs in applications? Specifically, can an end-to-end trainable LLM effectively address complex clinical tasks, or is it necessary to decompose these tasks into multiple sub-tasks? This consideration raises questions about factors like integration, usability, cost-effectiveness, reliability, performance, and regulatory compliance of LLMs in radiology workflow, and search for practical and optimal solutions for radiology applications.
c. **Data**: Which datasets should be used to benchmark LLMs used in the radiology domain? Given that much of the available clinical data are unlabeled, what strategies can be implemented to train LLMs without incurring the extensive time and costs associated with data annotation? This inquiry emphasizes the importance of identifying robust benchmarking datasets

and exploring semi-supervised or unsupervised learning techniques to maximize the utility of existing data.

d. **Improvement**: How can insights gained from the aforementioned research questions inform the development or adaptation of LLMs to achieve superior performance for radiology applications? By synthesizing findings from analyses of existing models, their limitations, and data considerations, we can identify pathways for enhancing LLM capabilities in clinical settings, ultimately leading to more effective and reliable tools for healthcare practitioners.

This chapter seeks to address several research questions by analyzing the application of various types of LLMs, exploring datasets relevant to different aspects of the radiology workflow, and suggesting potential improvements for LLM-powered applications, while also briefly discussing their limitations. While offering a glimpse into the future promises of LLMs in radiology, it humbly acknowledges that the rapid advancements in AI, and specifically LLMs, may necessitate revisions of the chapter sooner or later. Overall, while the future remains unknown, the expanse of innovation in radiology that this chapter portrays should hopefully make us intrigued and excited about what can be achieved by leveraging LLMs in this field and, by extension, in healthcare.

5.2 COMMON METHODOLOGIES FOR APPLICATION-SPECIFIC LLMs

The most commonly used paradigms of LLM integration in applications has been prompt engineering and retrieval-augmented generation (RAG) (Lewis et al., 2020). Though many applications benefit from using general-purpose LLMs, their application to the highly specialized medical domain is limited. To address this limitation, common methods for adapting pretrained LLMs to specific domains and tasks include:

a. **Domain-Adaptive Pretraining (DAPT)**: It involves an additional pretraining using domain-specific data.
b. **Instruction Finetuning (IFT)**: It involves fine-tuning on downstream tasks by reformulating the task using textual instructions and prompts, like the original pretraining objectives (Liu et al., 2023). Instructions for multi-task training are formatted consistently (Chung et al., 2024).

During DAPT and IFT, it is common to apply parameter-efficient finetuning methods (PEFT) (Xu et al., 2023) to train LLMs without fine-tuning all model's parameters. This reduces the computation and storage cost by almost 80%–90% while achieving competitive performance to the fully finetuned model. These methods train only a subset of the model's parameters, such as the last few layers of the LLM or newly introduced parameters, while other parameters remain frozen. One of the most effective PEFT methods is the Low-Rank Adaptation (LoRA) technique (Hu et al., 2021). LoRA preserves the model weights and introduces trainable rank decomposition matrices into different layers of the LLM, significantly reducing the number of trainable parameters in the LLM. In the following sections, different adaptation methods are applied to address the requirements for several radiology-focused applications.

In addition, ensemble learning (Lu et al., 2023), knowledge integration (Feng, Zhang and Fei, 2023), transfer learning (Chronopoulou, Baziotis and Potamianos, 2019), data augmentation (Whitehouse, Choudhury and Aji, 2023), and reinforcement learning (Lee et al., 2023) are other widely used methods that help in developing application-specific LLMs with high accuracy.

5.2.1 LLM Performance Metrics

The ubiquity of LLMs makes it difficult to define a universal set of metrics for the evaluation of LLM-based applications. On one hand, there are general natural language generation-based measures that compare LLM-generated text with annotated ground truth or reference text. On the other hand, there are application-based metrics, e.g., standard classification or regression metrics, or derived metrics that are domain-dependent.

Common metrics for evaluating the completeness of text generation in LLMs include ROUGE (Recall-Oriented Understudy for Gisting Evaluation) (Lin, 2004) and BLEU (Bilingual Evaluation Understudy) (Papineni et al., 2002). BLEU measures the precision of n-grams in generated text against reference texts, focusing on exact matches, while ROUGE evaluates the overlap of n-grams, emphasizing recall and capturing meaningful content. BLEU-4 is one of the most commonly used BLEU scores, and it calculates that fraction of 4-grams in generated text that overlap with the ground truth text. One of the most popular metrics in ROUGE is ROUGE-L that measures the longest common subsequence between the ground truth and generated output. Additionally, to evaluate factual correctness of the generated text, BertScore (Zhang et al., 2020) is used. It computes the cosine similarity score between tokens in the generated text and the ground truth text.

Application-based metrics can include classification metrics like F1, accuracy, precision/recall, and regression-based metrics like mean squared error or mean absolute error (MAE). Derived metrics for radiology domain, e.g., F1RadGraph (Delbrouck et al., 2022) and F1CheXbert (Xie et al., 2023), are also introduced specifically for radiology report generation. While F1CheXBERT measures how accurately a model predicts the presence or absence of 14 medical observations identified in chest X-rays, such as pneumonia, lung opacity, or other pathologies, F1RadGraph is based on the overlap in radiological entities and clinical relations between the generated report and the reference report (Sun et al., 2024).

The following sections describe some of the common applications in radiology workflow using a combination of the methodologies described in this section. We use the appropriate metrics discussed in this section to analyze performance of these methodologies in the specific applications and understand the potential improvements and limitations of LLM in radiology.

5.3 SYNOPTIC AI: STRUCTURING RADIOLOGY DATA USING LLMs

A radiologist's workflow typically begins with image acquisition, where medical imaging modalities like X-ray, CT, MRI, or ultrasound capture diagnostic images. The images are then routed to a Picture Archiving and Communication System (PACS), where the radiologist retrieves and organizes cases based on priority. Next, the radiologist performs image interpretation, analyzing the scans for abnormalities, making differential diagnoses, and correlating findings with clinical history. They then generate a radiology report, documenting key observations, impressions, and recommendations for referring physicians. Throughout the day, radiologists may also collaborate with clinicians, discuss urgent cases, and provide input for treatment planning. Additionally, they may be involved in image-guided procedures, quality assurance tasks, and administrative responsibilities such as protocol optimization and peer review. Automating or assisting portions of this workflow not only reduces workload and improves patient care but also becomes far more efficient if the reports and documents written are machine-readable. Structured reports that align with standardized data formats streamline data exchange across electronic health record (EHR) systems and clinical decision support tools, ultimately enhancing both efficiency and collaboration in patient care (Figure 5.1).

Radiology reports capture essential observations and diagnoses from imaging studies, forming a foundation for clinical decisions. However, these reports are usually free text, with significant variation in format and content across radiologists, institutions, and regions. The variability in clinical terminology used by radiologists, coupled with the contextual ambiguity of qualifiers in unstructured reports, poses significant challenges for automated information extraction. For example, a single finding might be described as "mild left pleural fluid," "small left-sided pleural effusion," or "trace fluid in the left lung base," yet all expressions refer to the same condition. These inconsistencies hinder interoperability with other EHR systems or clinical decision support tools, and hinder large-scale data analysis, often requiring labor-intensive manual work or complex natural language processing (NLP) solutions.

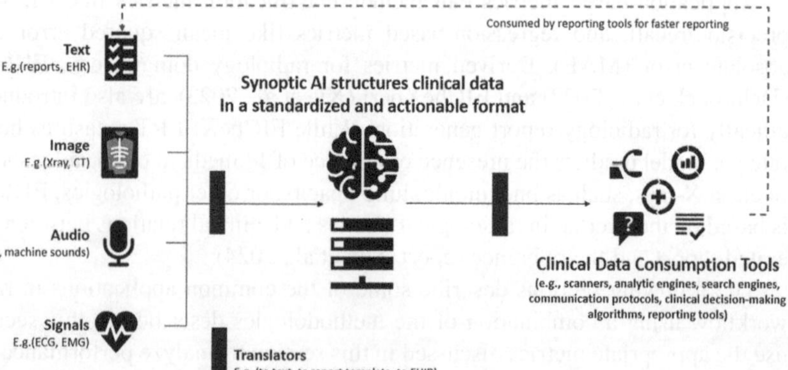

FIGURE 5.1 Synoptic artificial intelligence (AI) application in clinical workflow. A flowchart showing different types of clinical input data, translators to convert diverse clinical data into text or report template, and large language model (LLM)-based synoptic AI for structuring these data into standardized format. These comprehensive reports are machine-readable and clinically disambiguated for use by clinical data consumption tools such as clinical decision support tools, scanners, search and analytics engines, and reporting tools, and are integrated into clinical workflow.

Synoptic AI—so named because "synoptic" refers to producing a concise, standardized overview—directly addresses this issue by converting unstructured radiology reports into structured, machine-readable formats. Synoptic AI, as developed in this case study, focuses on converting unstructured radiology reports into structured, machine-readable formats using common data elements (CDEs) (Rubin and Kahn, 2017). These standardized data formats are maintained through coding standards like RadElement.[1] For instance, the LLM should structure "a small pleural effusion in the left lung," to "Presence of Pleural Effusion" (coded as RDE1652_Presence), with additional elements capturing "Laterality" (coded as RDE1304_Location with a value of Left) and "Size" (coded as RDE1372_Pleural_effusion_size with a value of Small). By structuring the data in this way, the AI ensures that all relevant clinical information is preserved in a consistent, interpretable format, ready for seamless integration into electronic health systems.

5.3.1 Challenges

Early approaches to structure radiology reports relied on keyword-based extraction or rule-based NLP (Woznicki et al., 2024). These systems struggled with challenges of a prose-based radiology report e.g., complexity and variability, different writing styles of radiologists, clinical ambiguity in reports, institutional and regional preferences, and standards of reporting etc. These issues led to the need to continuously update and maintain extensive manual rules. Lack of understanding of context for observations that spanned multiple sentences in a report led to either the need for a complex set of rules or manual intervention.

Machine learning techniques introduced a more data-driven approach by training models on annotated datasets of radiology reports (Pons et al., 2016). However, machine learning models had difficulty in generalizing across different institutions (Van Veen et al., 2023). The integration of medical ontologies and lexicons facilitated better identification and categorization of medical terms. These resources improved the semantic understanding of clinical text but still required substantial manual effort to maintain and update the lexicons (Mozayan et al., 2021).

Although LLM-based solutions solve most of the issues in structuring radiology reports, general-domain LLMs are limited in interpreting medical terminology. In addition, the significant cost of computational resources needed for LLM deployment make it difficult to deploy on standard hospital infrastructure or local systems.

5.3.2 Methodology

A recent study (Dhanaliwala et al., 2024) compared the performance of two LLM-based synoptic AI applications:

a. RadLing-System: A domain-adapted language model specifically fine-tuned for radiology tasks.
b. GPT-4-System: A general-purpose LLM using dynamic few-shot prompting for feature extraction and CDE mapping.

The study used 1,399 anonymized chest X-ray (CXR) reports, divided into 900 training reports and 499 test reports. Forty-four preselected features consisting of typical CXR findings were identified from the reports, such as the presence or absence of pleural effusion, cardiomegaly, and pulmonary nodules. Each feature was mapped to a CDE using RadElement. Both systems were tasked with extracting and standardizing the CXR features and mapping them to CDEs.

RadLing (Ghosh et al., 2023) is a domain-adapted version of the large ELECTRA pretrained language model (Clark et al., 2020), consisting of 335 million parameters and continuously pretrained on a million radiology reports. It incorporates domain-specific vocabulary and radiology ontologies and applies knowledge-aware masking during DAPT. This improves its ability to interpret radiology-specific complexities. RadLing is then fine-tuned on named entity recognition to find core concepts, locations, and modifiers in a radiology report. A sentence transformer is trained for the RadLing model using Sentence-BERT architecture (Reimers and Gurevych, 2019). This approach utilizes Siamese and triplet networks to train an encoder architecture that maps input sentences into a fixed-size vector space. By incorporating Sentence-BERT in RadLing, the model is trained to generate embeddings for sentences in radiology reports.

Both RadLing and GPT-4 systems perform the following two steps in synoptic AI:

1. Selection of feature class associated with text;
2. Mapping of the extracted features to relevant CDE attributes and their values.

5.3.3 Performance Evaluation

The performance of RadLing-system and GPT-4 system are evaluated on two aspects: Feature extraction and CDE standardization.

Feature Extraction: The domain-adapted RadLing-System outperformed GPT-4 in feature extraction, achieving an F1 score of 0.97 compared with 0.78 for GPT-4. This demonstrates the importance of domain-specific adaptations when dealing with highly specialized tasks like radiology report interpretation.

CDE Standardization: In the standardization task, RadLing-System outper formed GPT-4, achieving an F1 score of 0.98 compared with GPT-4's 0.94.

A major advantage of the RadLing-System is its capacity to accurately distinguish between absent and unspecified findings. This distinction refers to whether a feature class is explicitly stated as absent or if there is no mention of the findings in the report at all. RadLing-System achieved a 99% F1 score in differentiating absent findings, compared with only 64% for GPT-4.

5.3.4 Future Directions

Synoptic AI harnesses RadLing—a specialized, domain-adapted language model—to structure radiology data and outperform a more general GPT-4-based solution. As demonstrated in the previous section, RadLing's focus on radiology-specific terminology and ontologies gives it a distinct edge over GPT-4 when extracting and standardizing clinical features from free-text reports. When comparing an F1 score of 0.97 versus 0.78, the difference becomes striking in a real-world context. For instance, if a model processes 1,000 radiology reports, a system with an F1 of 0.78 might misclassify around 220 findings, whereas an F1 of 0.97 would reduce that figure to about 30—a roughly sevenfold decrease in errors. The impact is even more pronounced when distinguishing absent findings from unspecified ones: a model with an F1 of 64% could misclassify 36 out of every 100 such cases, whereas a 99% F1 system might get only one wrong. These discrepancies in performance can lead to unnecessary follow-up procedures, missed diagnoses, or confusion in the care process. Higher accuracy, therefore, translates into more reliable reports, improved patient outcomes, and greater workflow efficiency.

Going forward, synoptic AI should focus on overcoming challenges to achieve broader applicability and improved reliability in radiology. One priority is adapting the technology for imaging modalities other than CXR, which present distinct reporting patterns and terminology. This adaptation will be essential for achieving a robust, multimodal solution that can enhance the technology's utility across healthcare settings.

Consistency in representing uncertain or incomplete data is a critical need for integration into clinical automation workflow, especially as models are integrated into EHRs and shared across institutions. Future efforts should focus on further standardizing CDE frameworks like RadElement across healthcare systems to enhance interoperability and ensure a uniform approach to radiologic data structuring. With these refinements, synoptic AI has the potential to transform radiology reporting, enabling greater diagnostic precision, and streamlining clinical workflows.

Beyond synoptic AI, the process of generating radiology reports can also be influenced by LLMs—the next section takes a closer look at this topic.

5.4 RADIOLOGY REPORT GENERATION AND SUMMARIZATION

The cognitive task of interpreting medical images and writing the report remains the most critical and often time-consuming step in the radiology workflow. Thus, automating radiology report generation and summarization allows clinicians to interpret radiological images more easily, reduce their workload, and enhance patient care. Radiology reports include several sections, with the most important ones being the *Findings* and *Impressions*. The *Findings* describes abnormalities and diagnoses, while the *Impressions* summarizes the *Findings* and highlights major abnormalities and conclusions. In the report generation task, the objective is to create the *Findings* section based on the medical images. For report summarization, the goal is to generate the *Impressions* section based on the *Findings*.

Various efforts have been made on using AI models, such as neural sequence-to-sequence (Zhang et al., 2018), reinforcement learning, and pretrained language model for automatic report summarization (Jing, Xie and Xing, 2018). More recently, LLMs have been used for automatic radiology report *Findings* generation (Danu et al., 2023; Karn et al., 2022; Delbrouck et al., 2023). However, radiology being a highly specialized domain and radiology report generation a niche task, pretrained LLMs require efficient adaptation to achieve accurate performance in radiology tasks, which will be the focus of this section.

5.4.1 Challenges

Existing models for radiology report generation and summarization perform suboptimally due to the diversity in terms of imaging modalities and anatomies in available training data. Further challenges, such as the diversity of patients represented in the data and how it impacts the LLM's performance for underrepresented communities, exacerbates the situation. Moreover, different radiologists have distinct preferences and styles for writing reports, which must be considered when training models.

Previous studies in report generation and summarization have also highlighted the potential for missed abnormalities due to perceptual errors and hallucination of LLMs, which could lead to incorrect diagnoses. Careful consideration of LLM adaptation to these tasks is essential to mitigate these challenges and ensure effective integration into the clinical workflow.

Prior works such as Radiology-llama2 (Liu et al., 2023) and RadBloomz (Karn et al., 2023) have been proposed for the radiology domain and downstream tasks like report summarization for efficient and accurate performance. In the following subsections, we will focus on recent works in both report generation and summarization tasks.

5.4.2 Methodology

Among various methods for downstream task finetuning, multi-task IFT has been particularly successful. In recent works, domain-adaptation and finetuning of open-source LLMs, such as Bloomz model (Scao et al., 2022), to accomplish radiology report generation and summarization tasks have been proposed.

RadBloomz is a radiology-adapted LLM that is created by performing DAPT of Bloomz-7b1 using MIMIC-IV radiology reports dataset (Johnson et al., 2023). This dataset originally contains over 2.3 million radiology reports from 237,000 patients and amounts to approximately 616 million tokens using the Bloomz tokenizer (Muennighoff et al., 2022). The dataset was preprocessed and 1.4 million reports with 190 million tokens were utilized for DAPT of this model.

After the DAPT step, using standard prompt-based fine-tuning, *Findings* and 'TL;DR' were employed as prompts and the model was finetuned by comparing the auto-regressively generated summary tokens to ground-truth *Impressions* using cross-entropy loss. To prevent catastrophic forgetting, only the last layer of the model was finetuned for the downstream task. In all training stages, the same language modelling objective was used.

In addition, RadBloomz has been applied in report *Findings* generation (Danu et al., 2023). It has been integrated into a transfer learning framework and utilized for generation of radiology reports in CXR images. In this method, a domain-adapted vision model (Ghesu et al., 2022) is initially employed to detect abnormalities in the CXR images, followed by a fine-tuned RadBloomz to generate the *Findings* section of the report based on the detected abnormalities. RadBloomz is finetuned using the lung abnormalities in the MIMIC-CXR dataset (Johnson et al., 2019). The input prompts are created by concatenating a list of abnormalities detected in images with their corresponding probabilities into a single text and are used to train the model for *Findings* generation. Figure 5.2 shows the process of report findings generation and summarization.

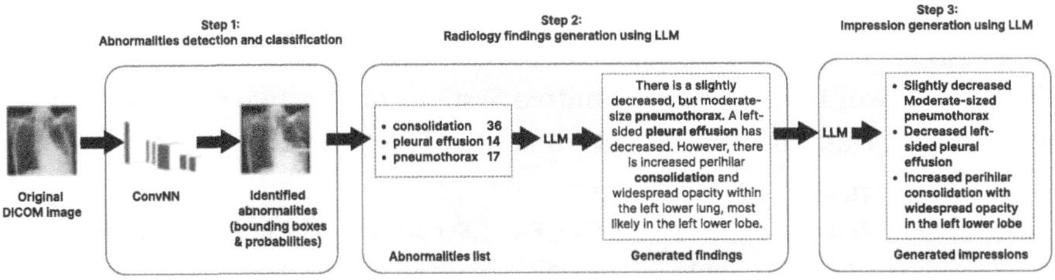

FIGURE 5.2 Radiology report Findings and Impressions generation using a domain-adapted LLM. Input image is encoded using an image encoder and abnormalities are identified. Then, a domain-adapted LLM generates Findings from the list of abnormalities and finally Impressions are generated from the findings section.

5.4.3 Performance Evaluation

A recent challenge on radiology report summarization, known as RadSum23, included MRI and CT datasets with various anatomies. The datasets were created by preprocessing MIMIC-III (Johnson et al., 2016) and multi-modal MIMIC-CXR datasets. Each dataset consists of a train, validation, hidden test, and open test split. The hidden test split is used during the RadSum23 challenge while the open test split was available for participating teams to evaluate their developed models. As an entry in RadSum23 with 10 other teams participating, RadBloomz ranked 1st and was the winner of the radiology report summarization task on MIMIC-III. To evaluate the model's performance, *Impressions* for given *Findings* in the test set are generated by the model and metrics are calculated.

Table 5.1 showcases RadBloomz's performance on MIMIC-CXR and MIMIC-III hidden test datasets compared with other submitted models. The MIMIC-III hidden dataset contains only reports, while MIMIC-CXR includes reports and images. In addition, to evaluate the effectiveness of the proposed training paradigm in RadBloomz, zero-shot evaluation on the open test split of the RadSum23 challenge is performed and compared with the standard DAPT followed by IFT approach on the same test splits. The average performance in zero-shot setting (without prompting any examples) was higher than the finetuned model, with average scores of 35.42 versus 35.03 in MIMIC-III and 51.95 versus 35.74 in MIMIC-CXR. This indicates that the DAPT approach is sufficient for achieving higher performance without requiring task-specific finetuning.

Error analysis on the open test datasets has shown that many generated *Impressions* receive low scores for both ROUGE and F1-RadGraph when the ground-truth radiology report *Impressions* does not mention any abnormalities. For instance, the generated *Impressions* "normal MRI of the cervical spine" and the ground truth *Impressions* "negative study" are semantically similar, but the n-gram overlap-based scores do not recognize their semantic relatedness. Also it was observed that similar *Findings* sometimes generate different *Impressions*, e.g., *Impressions* can be as detailed as: "near complete opacification of the ethmoid air cells and sphenoid sinuses, moderate air-fluid level with mucosal thickening of the right maxillary sinus, and moderate mucosal thickening of the left maxillary sinus." Meanwhile, the same *Findings* in another report may be followed with the *Impression* "pansinusitis, as described above."

TABLE 5.1
Performance of the Top Five Submitted Systems on the Two Categories of Hidden Test Data for Shared Task 1B at BioNLP 2023

Team	Hidden Test Set	BLEU4	ROUGE-L	BertScore	F1-CheXbert	F1-RadGraph
RadBloomz	MIMIC-III	**18.36**	**35.32**	**57.26**	N/A	**36.94**
Utsa-nlp		16.05	34.41	57.08	N/A	36.31
Aimi		16.61	33.43	55.54	N/A	35.12
Sinai		17.38	32.32	55.04	N/A	33.96
Knowlab		13.23	32.02	55.64	N/A	33.39
Dmis-msra	MIMIC-CXR	**18.62**	34.57	**55.90**	**72.36**	**43.20**
Utsa-nlp		16.33	**34.97**	55.54	69.41	42.86
Knowlab		14.41	33.63	54.72	67.20	39.98
RadBloomz		14.59	32.43	53.99	68.99	38.40
Aimi		5.15	31.84	47.83	64.18	32.05

Note: RadBloomz ranks first in MIMIC-III and fourth in MIMIC-CXR among all submitted systems, demonstrating the effectiveness of the introduced approach.

In the report *Findings* generation task, experiments conducted on the MIMIC-CXR dataset demonstrate that RadBloomz model achieves a score of 0.373, whereas the conventional models, CNN-RNN and CNN-Transformers, achieve 0.263 and 0.284, respectively. Qualitative analysis shows the ability of the model to generate accurate results, with little to no hallucinations.

5.4.4 Future Directions

Further refinement of the existing models is required to address several existing limitations. For instance, it has been observed that the verbosity and stylistic differences in *Findings* sometimes affect consistency in the generation of *Impressions*. Research should be conducted to quantify the consistency, uncertainty, and information richness of the generated *Impressions*, as well as to address issues in these areas.

Furthermore, to enhance the efficiency and accuracy of automated report generation in radiology, a future direction would be to equip the input prompts with more details about the detected *Findings*, such as the localization of the identified abnormalities. This will lead to a more comprehensive *Findings* section. Overall, hallucination and templated generation in generative LLMs are persistent issues that merit further investigation.

While this section focused on summarization of radiology reports and medical images, the next section explores use of LLMs for visual question answering based on these modalities in radiology.

5.5 MEDICAL VISUAL QUESTION-ANSWERING

Medical visual question-answering (MedVQA) in radiology involves answering questions related to a given radiology image, such as identifying abnormalities and determining their size and location. AI-based MedVQA systems have been developed as radiologist assistants to address repetitive tasks, such as image reading. Since MedVQA involves both images and text, multimodal LLMs, which get input data and generate output representing different modalities (image, text, video, audio), should be aptly used to perform this task. The selection of a multimodal model for an application depends on several factors, including the nature of the downstream task, domain of interest, required modalities, model accessibility, data privacy concerns, and their performance evaluation.

Among various multimodal LLMs, vision-language models (VLMs) are a prominent family of models that accept and generate image and text modalities and are very useful in MedVQA.

5.5.1 Challenges

Unlike LLMs, VLMs face unique challenges due to their multimodality focus. An important problem is the scarcity in multimodal datasets for adaptation to downstream tasks. In a radiology visual question-answering task, large amounts of radiology image-text pairs must be curated for training. Moreover, it is crucial to include a balanced representation of different modalities (X-ray, CT, MRI) and anatomies (abdomen, chest, brain etc.). Recent VLMs have been adapted by leveraging large radiology-specific multimodal datasets, like ROCO (Pelka et al., 2018) and ROCOv2 (Rückert et al., 2024).

While successful in general domain tasks, performance of general-domain VLMs, such as BLIP-2 (Li et al., 2023), may be limited when applied to highly specialized medical domain. These models struggle when performing medical tasks due to the complexity of medical questions and the need for a deep understanding of medical terminology and image context. Therefore, domain adaptation of VLMs must be considered to achieve accurate performance in radiology.

Moreover, semantic alignment between image and text is another key challenge. It requires an optimal implementation of a fusion module to match the semantic relationship between radiology text and image representation. Traditional MedVQA systems consider the task of question-answering as a classification task over a predefined set of answers. This approach restricts the ability of models to perform question-answering in a more natural way and generate free-form responses.

Key factors in enhancing the performance of VLMs for MedVQA include the training strategy, VLM architecture, and deployment of domain-specific LLMs customized for radiology. Considering these aspects, Ha et al. (2024) addressed the abovementioned challenges by introducing an efficient fusion module in VLMs and a parameter-efficient three-stage training strategy in radiology domain as described in the next section.

5.5.2 Methodology

Recent VLMs in Med-VQA typically comprise three key components: a vision encoder model, an LLM, and a fusion module for combining the information from these two models. The fusion module is used to map the encoded image features to the semantic space of the LLM and acts as a bridge between the vision encoder and LLM. A recent study (Ha et al., 2024) integrated a learnable query transformer architecture as the fusion module. It consists of a transformer architecture with a set of fixed queries as its input, which extracts a predetermined set of features from the output of the vision encoder. The query transformer output is projected to the embedding size of the LLM using a multi-layer perceptron network. These projected features are then combined with the embedded input question and fed into the LLM to generate the desired output. The proposed VLM architecture (Ha et al., 2024) is shown in Figure 5.3.

To explore the potential benefits of incorporating radiology domain-adapted LLM and vision models for radiology MedVQA, a decoder-only LLM, RadBloomz, is utilized. For more details about training and adaptation of RadBloomz, please refer to Section 5.4. Moreover, domain-adapted vision encoder models from PMC-CLIP (Lin et al., 2023) and BiomedCLIP (Zhang et al., 2023) are utilized. Both the vision encoder and LLM parameters remain frozen and the LoRA technique (Hu et al., 2021) is applied to the LLM to align it with the downstream MedVQA task while reducing the number of trainable parameters in the VLM.

The training approach comprises three stages. The first stage bridges the gap between the vision encoder and LLM, and pretraining the randomly initialized fusion module to align medical concepts with visual content. Therefore, this stage is formulated as a medical image caption prediction task, where the model predicts the next token in the caption given an input image. The PMC-OA dataset (Lin et al., 2023) containing 1.64 million image-caption pairs is used for the first stage of training.

The second stage is about adapting the model parameters for MedVQA. The PMC-VQA dataset (Zhang et al., 2023) with a wide range of questions that cover various aspects of medical images is used for this stage. It contains approximately 186,033 visual question-answer pairs.

The final stage is downstream task finetuning, where we fine-tune the model by utilizing the training split of two MedVQA benchmarks: VQA-RAD (Lau et al., 2018) and

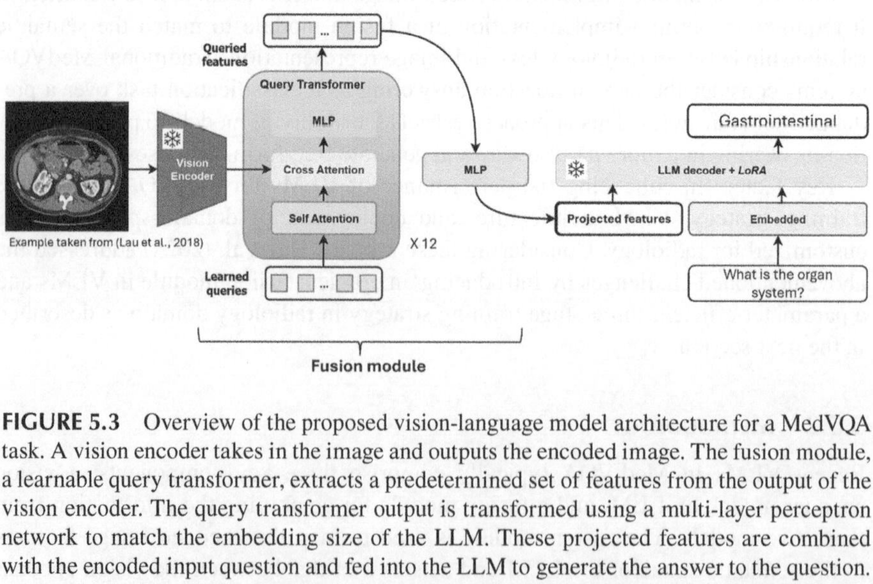

FIGURE 5.3 Overview of the proposed vision-language model architecture for a MedVQA task. A vision encoder takes in the image and outputs the encoded image. The fusion module, a learnable query transformer, extracts a predetermined set of features from the output of the vision encoder. The query transformer output is transformed using a multi-layer perceptron network to match the embedding size of the LLM. These projected features are combined with the encoded input question and fed into the LLM to generate the answer to the question.

SLAKE 1.0-English (Liu et al., 2021). This process helps further refine the model's performance. In both fine-tuning datasets, questions are categorized as either closed-ended or open-ended. Closed-ended questions are multiple-choice questions with a limited set of answers e.g., "yes/no" questions. Open-ended questions accept free-form answers. The loss function employed in all training stages is the sum of negative log-likelihoods of the correct next token in a given text sequence across all time stages.

5.5.3 Performance Evaluation

To evaluate the performance of the proposed method and the VLM architecture, experiments are conducted to compare with baseline VLMs, and accuracy metric is reported. To compute accuracy, the final answer is identified in the list of all possible answers from the training split of each dataset that is most similar to the answer generated by their model.

The V-ViT-RadBloomz model outperforms the previous approaches on SLAKE 1.0. The results illustrate the advantages of the training strategy and the utilization of a radiology domain-adapted RadBloomz language model in the MedVQA task (Table 5.2).

On the VQA-RAD dataset, the V-ViT-RadBloomz model outperforms PubMedCLIP and Biomed-CLIP models on the overall accuracy. However, it does not perform as well on open-ended questions, where it falls behind compared with the MedVInt-TD model. The lower performance on open-ended questions can be attributed to several factors like imbalanced dataset. This causes the VLM to become biased toward certain types of questions, reducing its ability to accurately answer the other types of questions that do not appear in the dataset often.

Another experiment to assess the influence of domain adaptation in the VLM using two LLMs was performed, using Bloomz-7b1 and RadBloomz-7b. V-ViT-RadBloomz

TABLE 5.2
Accuracy (%) Results of VLMs on SLAKE 1.0-English and VQA-RAD Datasets

	SLAKE 1.0			VQA-RAD		
Model	**Overall**	**Closed**	**Open**	**Overall**	**Closed**	**Open**
V-ViT-RadBloomz	**87.5**	**92.1**	**84.5**	73.2	83.5	57.5
V-ResNet-RadBloomz	82.5	88.5	78.6	67.6	79.4	49.7
MedVInT-TD (Zhang et al., 2023)	85.2	86.3	84.5	**81.6**	**86.8**	**73.7**
Biomed-CLIP (Zhang et al., 2023)	86.1	88.9	84.3	72.7	76.5	67.0
PubMedCLIP (Eslami, Meinel and de Melo 2023)	80.1	82.5	78.4	72.1	80.0	60.1
MUMC (Li et al., 2023)	84.9	–	–	79.2	84.2	71.5
PMC-CLIP (Lin et al., 2023)	84.3	88.0	81.9	77.6	84.0	67.0

Note: Performance on open-ended and closed-ended questions as well as overall performance are reported. VE represents a vision encoder. V-ViT-RadBloomz model achieves state-of-the-art performance on SLAKE 1.0, and outperforms PubMedCLIP and Biomed-CLIP models on VQA-RAD.

significantly outperforms its general domain counterpart, V-ViT-Bloomz, achieving an accuracy of 87.5% compared with 80.7% on Slake 1.0% and 73.2% compared with 68.3% on VQA-RAD.

Finally, the impact of training existing parameters in the fusion model is investigated. Using VLMs with trainable vision encoders, the results indicate that V-ViT-RadBloomz with a frozen vision encoder achieves higher accuracy (73.2%) while using fewer parameters than the model with a trainable vision encoder (71.4%) on VQA-RAD. These findings suggest that by incorporating an optimal fusion of a domain-adapted vision encoder and LLM, training the vision encoder becomes unnecessary. This approach results in a lightweight and efficient adaptation of the VLM.

5.5.4 Future Directions

There's a need to investigate approaches for an enhanced medical concept alignment, i.e., enhancing the model's ability to process complex interactions between visual and textual data in MedVQA. Additionally, the integration of these systems in the clinical workflow to improve the overall efficiency and expand their capabilities beyond question-answering to include personalized recommendation systems is a promising extension.

Until this section, we focused on LLMs that assist radiologists in clinical decision making. The next section will focus on the security posture of radiology equipment as medical devices and automation of vulnerability evaluation for these devices using LLMs.

5.6 LLMs FOR CYBERSECURITY OF MEDICAL DEVICES

Medical devices, including radiology devices and software, are susceptible to a range of security threats that can jeopardize patient safety, compromise privacy, undermine the integrity of medical data, and disrupt the device availability. The healthcare industry is currently facing an unprecedented increase in cybersecurity attacks, affecting millions of individuals (Identity Theft Resource Center, 2024). According to guidance from the U.S. Food and Drug Administration (U.S. Department of Health and Human Services, Food and Drug Administration, Center for Devices and Radiological Health, Office of the Center Director, Center for Biologics Evaluation and Research, 2016), medical device manufacturers (MDMs) are required to monitor third-party software components used in their devices for emerging vulnerabilities and assess their impact on device safety and security. The current manual monitoring methods demand trained manpower to address software vulnerabilities in medical devices promptly. The manual effort involves several key steps, including cross-referencing standard security databases, such as Common Vulnerabilities and Exposures (CVE[2]), which catalogues software vulnerabilities; Common Weakness Enumeration (CWE[3]), which describes software security weaknesses; and Common Attack Pattern Enumeration and Classification (CAPEC[4]), which details known attacks and mitigation patterns. Additionally, the vulnerability evaluation process entails examining the source code of the software and reviewing security whitepapers related to the devices.

Automated vulnerability evaluations serve two critical purposes: (1) it facilitates the quick and efficient management of the large volume of evaluations required,

TABLE 5.3
Details for Vulnerability Evaluation

Evaluation Types	Description	Value Type
VexCategory	Indicates whether the asset is affected by the notification	Binary categorical values
VexJustification	Provides further explanation if the asset is not affected	Multiclass categorical values
CVSS Vector	Evaluates the CVSS environmental metrics	Multiclass, multilabel categorical values
Internal Comment	Details the problem and advised internal solution	Text
Customer comment	Summarizes the notification's impact and solution directed toward the customer	Text

Note: Evaluation Types lists the different assessments done for vulnerability evaluation for a medical device with corresponding descriptions in the next column. The value type shows the expected output for the evaluation type and hence shows the type of machine learning problem the evaluation type pertains to.

ensuring consistent quality over time; and (2) it mitigates the risk associated with the potential unavailability of knowledgeable experts, who may not remain in the same role or organization throughout the entire post-market lifecycle of a device, which can extend up to 20 years. Given the current widespread application of LLMs, there exists a significant opportunity to use LLMs for automated vulnerability evaluation, thereby enabling faster mitigation of cybersecurity risks.

Vulnerability evaluation often involves the following assessments based on device details and the notification from third party component vendor: (1) VexCategory, (2) VexJustification, (3) CVSS Vector, (4) Internal comment, and (5) Customer comment. Table 5.3 provides a brief description of these evaluation types. From a machine learning perspective, the vulnerability evaluation tasks listed in Table 5.3 can be categorized into two primary tasks: text generation and text classification. Text generation involves the generation of Internal comment and Customer comment while text classification encompasses prediction of VexCategory, VexJustification, and Vector. Specifically, Vector classification is a multi-label, multi-class problem. VexCategory involves binary classification, while VexJustification is a multi-class classification task.

5.6.1 Challenges

Automated vulnerability evaluation in the medical device industry presents a range of challenges. Firstly, the diverse deployment environments of medical devices complicate the modelling of applicable vulnerabilities. Also, obtaining high-quality data that accurately reflect the impacts of various vulnerabilities can be difficult. In a rapidly evolving threat landscape, automated vulnerability evaluation demands substantial resources and continuous adaptation to newly identified vulnerabilities.

Additionally, the stringent compliance and regulatory requirements associated with vulnerability monitoring, combined with the sensitivity and inherent risks of such automation, hinder the adoption of automated vulnerability evaluation procedures in the medical device industry.

5.6.2 Methodology

Language models have been widely applied in vulnerability management for several purposes: (1) Quantification of severity of CVE using standard metrics like Common Vulnerability Scoring System (CVSS), (2) security database mapping, e.g., mapping vulnerabilities in the CVE database to corresponding attack tactics and techniques in the ATT&CK database, (3) vulnerability detection, and (4) vulnerability mitigation/repair.

Mapping CVE descriptions to standard CVSS metrics has traditionally employed machine learning methods, such as linear regression using Bag-of-Words models (Elbaz, Rilling and Morin, 2020) and neural networks like Doc2Vec (Vasireddy, Dale and Li, 2023). More recently, transformer-based classifiers, such as CVSS-BERT (Shahid and Debar, 2021), have been introduced. Automated mapping of security databases, such as CVE to ATT&CK, has utilized both encoder and decoder-based architectures. Encoder model techniques include transformer-based classification, e.g., RoBERTa (Liu et al., 2019)-based CVET model, and textual similarity-based clustering, as seen in the BERT-based SMET model (Abdeen et al., 2023). Decoder models for inter-database mapping have employed ChatGPT-based RAG techniques (Liu et al., 2023), as well as supervised finetuning of decoder models like Bloom, incorporating Chain-of-Thought instructions and domain knowledge (Zhang et al., 2024). Related research has also explored CVE vulnerability type classification using TF-IDF and traditional machine learning classifiers (Yosifova, Tasheva and Trifonov, 2021).

For vulnerability detection, encoder-based architectures have played a crucial role (Yin et al., 2020), particularly through pretrain-and-finetune approaches (Ameri et al., 2021). These methods have incorporated novel pretraining strategies and integrated deep learning paradigms like Graph Neural Networks (Sewak, Emani and Naresh, 2023) and Long Short-Term Memory networks (Hassanin et al., 2024). There has been a notable shift toward decoder-only LLMs (Zhou et al., 2024), where techniques like prompt tuning, finetuning, and RAG are prominently used for vulnerability detection.

Conversely, vulnerability remediation systems primarily leverage decoder-based LLMs (Pearce et al., 2023). These systems often utilize finetuning, transfer learning, RAG, or reinforcement learning (Islam et al., 2024). Recent advancements also explore combining these approaches with ensemble learning methods, meta-learning strategies, and the integration of domain-specific knowledge to enhance vulnerability management in complex software systems.

We developed a vulnerability evaluation system called CVE-LLM (Ghosh et al., 2025), trained on proprietary historical vulnerability evaluations using *pretrain-then-finetune* paradigm. We selected an open-source LLM, MPT7B (MosaicML NLP Team, 2023) and adapted it to the cybersecurity domain with vocabulary augmentation using a DAPT dataset. The dataset is a combination of a publicly available CVE database and vulnerability notifications from MDM third party vendors.

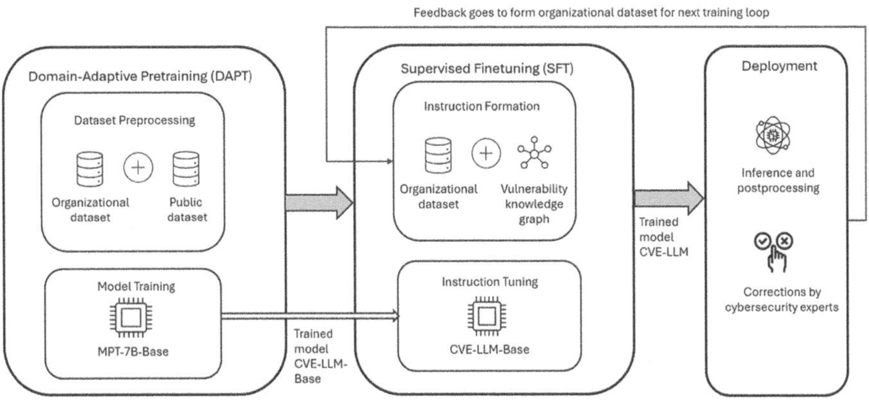

FIGURE 5.4 Schematic diagram showing the training and deployment process of CVE-LLM. Public vulnerability datasets are combined with organizational datasets to train a domain-adapted vulnerability model CVE-LLM-Base from MosaicML's MPT-7B-Base model. The organizational dataset is combined with the vulnerability knowledge graph to form instructions for finetuning CVE-LLM-Base to form CVE-LLM. This model is deployed in an offline environment and feedback is collected from cybersecurity experts, which becomes a part of the organizational database.

CVE-LLM is further finetuned using a dataset which contains proprietary vulnerability evaluations. Figure 5.4 shows the overview of the training and deployment process.

The DAPT dataset consists of 350 K vulnerability description documents, whereas the IFT dataset consists of 750 K unique instructions made from 208 K proprietary vulnerability evaluations for 1.7 K devices from a single MDM and 145 K notifications from other vendors. The vulnerability evaluations are converted into an instruction format, where the instruction pertains to the generation of one of the evaluation types in Table 5.3, given MDM product and a relevant third-party notification. The instructions are enriched using the Vulnerability Knowledge Graph called SEPSES CSKG (Kiesling et al., 2019) which contains linked CVE and CAPEC databases that provide further insights into the vulnerabilities. CVE-LLM is deployed into an offline production environment where the model is served using vLLM (Kwon et al., 2023) and the generated model output is processed using predefined rules to avoid anticipated erroneous generations (Ghosh et al., 2024). The model output is integrated into the Cybersecurity Monitoring System of an MDM and verified by cybersecurity professionals and product experts. Their inputs are logged into the system for future retraining of the model.

5.6.3 Performance Evaluation

The metrics used for performance evaluation in vulnerability assessment systems are determined by the modality and criteria of the assessment. Table 5.4 illustrates the various metrics employed by vulnerability evaluation solutions that incorporate NLP tools. Systems that map CVE descriptions to CVE scores typically report errors in severity prediction using metrics associated with regression tasks, such as MAE (Vasireddy, Dale and Li, 2023) or error intervals (Elbaz, Rilling and Morin, 2020).

TABLE 5.4
Description of Performance Metrics across Different Studies in Studies that Report on Text-based Vulnerability Assessment

Metric	Metric Description	Study	Reported Metric Value	Additional Information
Error interval	Range of errors in CVSS V3 score prediction	Elbaz, Rilling and Morin (2020)	−2.6 to 2.1 for 80% of CVSS V3 score prediction	Bag-of-words-based approach
MAE (mean absolute error)	Measure of the average absolute difference between predicted and actual values of CVSS V3 score	Vasireddy, Dale and Li (2023)	0.42	Doc2Vec-based approach
		CVE-LLM	<0.1	LLM-based approach
F1 score (CVSS vector categories)	Harmonic mean of precision and recall for CVSS vector categories	CVSS-BERT Shahid and Debar (2021)	0.83–0.95 (across base and temporal vector categories)	BERT-based approach
		CVE-LLM	0.97 (environmental vector)	LLM-based approach
ROUGE-L score (explanation comments)	Measure of overlap between predicted and actual text, focusing on the longest common subsequence	CVE-LLM	0.7–0.88 (across internal and customer comments)	LLM-based approach

In contrast, studies focusing on the determination of CVSS vectors typically report F1 scores for each CVSS vector category. BERT-based studies (Shahid and Debar, 2021) report a range of F1 scores from 0.83 to 0.95 across CVSS base and temporal vector categories, while CVE-LLM reports 0.88–0.97 F1 across environmental vector categories. Lastly, vulnerability assessments that generate comments explaining a vulnerability commonly utilize textual comparison metrics, such as the ROUGE-L score.

5.6.4 Future Directions

Current research on LLMs in the context of cybersecurity for medical devices demonstrates significant potential for automating vulnerability evaluation. Future enhancements could involve the incorporation of more accurate cybersecurity data, more information from cybersecurity databases, and insights from reputable blogs, thereby strengthening the knowledge accessible to the LLMs. Additionally, integrating product knowledge and source code, along with collaboration with threat intelligence platforms, could further enhance the capabilities of these automated systems. Moreover, developing human-assisted systems and implementing reinforcement learning algorithms will help address the generalization and scale-up of LLM-drive automated vulnerability evaluation utilized across devices from various MDMs.

5.7 OTHER APPLICATIONS OF LLM IN RADIOLOGY

Beyond structured reporting, report generation, summarization, and MedVQA, LLMs are finding broader use in radiology. Many of these applications aim to streamline workflow, reduce errors, and enhance clinical decision-making, ultimately improving both patient care and the radiologist's day-to-day experience.

5.7.1 Clinical Decision Support

LLM-assisted clinical decision support systems (CDSS) have shown considerable promise in assisting healthcare professionals by providing evidence-based recommendations and aiding in decision-making (Lammert et al., 2024). CDSS use LLMs to extract useful information from medical literature, patient records, and clinical notes (Huang, 2019). This knowledge is further used to predict clinical outcomes such as disease progression (Lu and Naseem, 2024), re-admission risks (Chung et al., 2024), risk stratification (Grandinetti et al., 2024), or likelihood of complications (Hsueh et al., 2024), and generate an optimal course of treatment (Goh et al., 2024), differential diagnoses (McDuff et al., 2023), and personalized treatment plans (Kirk et al., 2023). LLMs can be integrated in CDSS in co-pilot mode with supervising medical and pharmaceutical practitioners to provide real-time guidance, responding to clinician queries, offering evidence-based recommendations, continuous updates based on the latest medical research (Ong et al., 2024).

However, there are several limitations in adapting LLMs for CDSS solutions in clinical settings. Firstly, hallucinations and lack of sufficient reasoning capabilities of LLMs have resulted in diagnostic inaccuracy, particularly in complex cases (Hager et al., 2024). Furthermore, LLMs exhibited overconfidence in their incorrect diagnoses, posing significant risks to patient safety. Inflexibility in LLM reasoning capabilities gets highlighted in cases where prior experience hinders problem-solving, leading to fixation on familiar patterns and overlooking novel solutions, also known as "Einstellung effect" (Kim et al., 2025). This leading to low performance on clinical reasoning-based datasets like Medical Abstraction and Reasoning Corpus (M-ARC). Secondly, sensitivity of LLMs to changes in input phrasing, information quantity, and order also undermines the reliability and consistency needed in clinical settings. Inability to interpret laboratory results, inconsistency in adhering to clinical guidelines and lower accuracy in real-world, unseen cases impairs their ability to make accurate clinical assessments and recommendations (Kim et al., 2024). Thirdly, LLMs trained on biased datasets may reinforce existing healthcare disparities, leading to inaccurate recommendations for underrepresented populations (Liu et al., 2023). Finally, there are other practical challenges, e.g., high computational resource utilization, medical compliance issues, ethical issues and so on, that require concentrated effort by the medical community in adapting LLMs for clinical decision-making.

5.7.2 Error Detection and Quality Assurance

Integrating LLMs into radiology workflows has shown promise in enhancing the accuracy and consistency of radiology reports. LLMs have demonstrated proficiency in identifying errors stemming from speech recognition inaccuracies in radiology reports (Schmidt et al., 2024). For example, if a radiologist dictates "right-sided pneumothorax," but the transcription records "write-sided pneumothorax," an LLM can flag this as a likely error. In addition, LLMs have been effective tools in ensuring consistency and coherence by cross-verification across sections of radiology reports (Gertz et al., 2024), contradictory information identification (Salam et al., 2025), and numerical inconsistency identification (Zech et al., 2019). Although these tools do not replace a final human review, studies indicate that an AI "proofreading" step can reduce oversights in clinical notes.

Although LLM-based error detection is promising, there are challenges in understanding complex medical terminology. While supervised training methods and knowledge infusion will solve such issues, this is time- and training-intensive (Siwicki et al., 2024). Final accountability for confirming or dismissing flagged issues remains with the radiologist, underscoring the need for clear oversight to avoid complacency (Zanardo et al., 2024).

5.7.3 Triage and Prioritization

LLMs can enhance triage and prioritization in radiology workflows by automating report analysis, flagging urgent cases, and optimizing radiologist workload (Haim et al., 2024). LLMs can analyze radiology reports to do automated flagging of critical or abnormal reports (Mansoor et al., 2024) and prioritize reports to reduce delays in diagnosing life-threatening conditions (D'Antonoli et al., 2024). The integration of AI-based imaging tools can validate or enhance prioritization decisions in emergency cases (Pham et al., 2024). However, the dependence on multiple models introduces more errors, overdependence on textual data, lack of contextual understanding, and reliability concerns (Bhayana, 2024). Many solutions remain at a proof-of-concept stage, lacking robust integration with PACS or radiology information systems (RIS).

5.7.4 Patient Communication and Explanation

LLMs are increasingly being integrated into the radiology department to improve patient communication and explanation. They offer a range of capabilities that can enhance the interaction between healthcare providers and patients, ensuring that patients are better informed about their diagnoses and the imaging procedures they undergo. Simplification of complex medical jargon (Goldsack et al., 2024), automation of routine communication tasks such as explaining the purpose of imaging procedures to patients and informing them about preprocedural instructions (Arnold et al., 2025), integration into patient portals for answering patient questions, and follow-up care assistance are some of the most common patient-centric LLM-powered applications

(Laymouna et al., 2024). In addition, there is ongoing research on emotional patient support (Baek et al., 2025), translation services (Raido, 2024), and interactive decision support that may have far-reaching impact on patient care. While LLMs hold promise in improving patient communication in radiology, there are significant issues regarding accuracy and misinterpretation of medical information, lack of human empathy, over-reliance, data security, cultural sensitivity, and personalization.

5.8 CONCLUSION

So far, this chapter has provided an overview of applications of LLMs in the field of radiology, encompassing clinical and operational processes, providing a deeper understanding of the potential of LLMs in the field. The evaluation results offered in the sections further support the integration of LLMs into existing workflows toward enhancing productivity without introducing significant risk and ensuring that radiologists maintain autonomy and oversight. Implementing LLMs in radiology workflow software (e.g. PACS, RIS) could enable efficient human-in-the-loop validation, allowing identification and subsequent correction for hallucinations. Unlike general domain AI systems that often thrive on creativity, clinical AI applications require precise and consistent reasoning, with a strong emphasis on continuous testing, validation, and upgrades to meet healthcare standards. Based on the initial research questions posed in Section 5.1, we can concisely enumerate the answers as follows.

Analysis: Sections 5.3 and 5.4 described prevailing challenges in radiology that can be addressed by leveraging LLMs in automating clinical decision-making and improving workflow efficiency while reducing clinicians' workload. The use of LLMs as radiologist assistants in repetitive tasks to help clinicians is shown as a promising area of research. Specifically, Section 5.5 focused on the development of such technology. Similarly, Section 5.6 contrasts leveraging proprietary LLMs typically hosted in the cloud and associated with some risk of breaches in patient data privacy and safety.

Limitations: Section 5.3 focused on challenges posed by the lack of standardized data representation across different healthcare systems. Moreover, it highlights the significant manual effort required for data analysis and extraction of critical information, which is prone to error. Similarly, Section 5.6 investigated using LLMs to enhance the resilience of medical devices against cybersecurity attacks, thereby avoiding disruption to patient care and ensuring patient safety. Automated vulnerability evaluation demonstrates the capability of LLMs to learn from expert-curated data to efficiently assess if a medical device has an exploitable vulnerability toward unravelling information on patching and remediations to address the vulnerability.

Data: The development of LLM-based clinical applications is often the scarcity of relevant data for domain and task adaptation. To overcome this challenge, close collaboration with clinical experts to create high-quality data is essential. Furthermore, clinical applications demand that LLMs be as accurate and validated based on requirements from regulatory authorities. Furthermore, to address data privacy concerns in clinical settings, the highlighted applications tend toward local or on-premises implementation of open-source, LLM-based solutions, as opposed to cloud-based deployments.

Performance Improvements: Section 5.4 described LLM-based applications involving domain adaptation of the models for specific tasks in radiology. It involves the DAPT of instruction-tuned LLMs for the task of *Impressions* generation. Domain-adaptation improves parametric knowledge of the model in accomplishing radiology-specific tasks and consequently reduces hallucinations. Based on our evaluation results, we are convinced that this approach can be generalized and applied to various tasks in radiology and across other healthcare professions. Section 5.5 detailed a domain-adapted multimodal LLM for question answering over medical images. For this purpose, a novel training strategy with PEFT of a multimodal LLM was applied that achieved outstanding performance on public benchmarks for MedVQA. As multimodal LLMs evolve, generalizing them toward interactive multi-turn conversation between AI systems and physicians may become mainstream in clinical settings.

In addition to the application of LLMs in radiology described in this chapter, other use cases beyond the radiology domain include clinical decision support for non-radiology clinical specialties (Poulain, Fayyaz and Beheshti, 2024) assistants for personalised treatment plans (Benary et al., 2023), interactive patient-focused summarization of imaging findings (Amin et al., 2023), and clinical training and education (Trivedi and Wawira Gichoya, 2024).

Finally, it is crucial to consider ethical aspects of deploying LLMs in healthcare, including designing AI-based solutions with positive impact on patient care, sustainability, and humanity in general. Moreover, we must continue to aim for trustable LLM-based applications with careful consideration of short-term and long-term benefits to support broader adoption in the healthcare industry.

DISCLAIMER

The concepts and information presented in this paper/presentation are based on research results that are not commercially available. Future commercial availability cannot be guaranteed.

NOTES

1 https://www.radelement.org/.
2 www.mitre.org.
3 https://cwe.mitre.org/.
4 https://capec.mitre.org/.

REFERENCES

Abdeen, Basel, Ehab Al-Shaer, Anoop Singhal, Latifur Khan, and Kevin Hamlen. 2023. "Smet: Semantic Mapping of CVE to Attack and its Application to Cybersecurity." In: *IFIP Annual Conference on Data and Applications Security and Privacy,* Sophia Antipolis, France, pp. 243–260.

Agarwal, Vibhor, Yiqiao Jin, Mohit Chandra, Munmun De Choudhury, Srijan Kumar, and Nishanth Sastry. 2024. "MedHalu: Hallucinations in Responses to Healthcare Queries by Large Language Models." https://arxiv.org/abs/2409.19492.

Ameri, Kimia, Michael Hempel, Hamid Sharif, Juan Lopez Jr, and Kalyan Perumalla. 2021. "Cybert: Cybersecurity Claim Classification by Fine-Tuning the Bert Language Model." *Journal of Cybersecurity and Privacy*, 1(4), 615–637.

Amin, Kanhai, Pavan Khosla, Rushabh Doshi, Sophie Chheang, and Howard Paul Forman. 2023. "Focus: Big Data: Artificial Intelligence to Improve Patient Understanding of Radiology Reports." *The Yale Journal of Biology and Medicine*, 96(3), 407–417.

Arnold, Philipp, Maurice Henkel, Fabian Bamberg, and Elmar Kotter. 2025. *Integration of Large Language Models into the Clinic: Revolution in Analysing and Processing Patient Data to Increase Efficiency and Quality in Radiology*. Radiologie, Heidelberg, Germany.

Baek, Gumhee, Chiyoung Cha, and Jin-Hui Han. 2025. "AI Chatbots for Psychological Health for Health Professionals: Scoping Review." *JMIR Human Factors,* 12, e67682.

Benary, Manuela, Xing David Wang, Max Schmidt, Dominik Soll, Georg Hilfenhaus, Mani Nassir, Christian Sigler, Maren Knodler, Ulrich Keller, and Dieter Dieter. 2023. "Leveraging Large Language Models for Decision Support in Personalized Oncology." *JAMA Network Open,* 6(11), e2343689.

Bhayana, Rajesh. 2024. "Chatbots and Large Language Models in Radiology: A Practical Primer for Clinical and Research Applications." *Radiology,* 310, e232756.

Chronopoulou, Alexandra, Christos Baziotis, and Alexandros Potamianos. 2019. "An Embarrassingly Simple Approach for Transfer Learning from Pretrained Language Models." arXiv preprint arXiv:1902.10547.

Chung, Hyung Won, Le Hou, Shayne Longpre, Barret Zoph, Yi Tay, William Fedus, Yunxuan Li, Xuezhi Wang, Mostafa Dehghani, Siddhartha Brahma, Albert Webson, Shixiang Shane Gu, Zhuyun Dai, Mirac Suzgun, Xinyun Chen, Aakanksha Chowdhery, Alex Castro-Ros, Marie Pellat, Kevin Robinson, Dasha Valter, Sharan Narang, Gaurav Mishra, Adams Yu, Vincent Zhao, Yanping Huang, Andrew Dai, Hongkun Yu, Slav Petrov, Ed Chi, Jeff Dean, Jacob Devlin, Adam Roberts, Denny Zhou, Quoc Le, and Jason Wei 2024. "Scaling Instruction-Finetuned Language Models." *Journal of Machine Learning Research,* 25, 1–53.

Chung, Philip, Christine T Fong, Andrew McMillan Walters, Nima Aghaeepour, Meliha Yetisgen, and Vikas Navin O'Reilly-Shah. 2024. "Large Language Model Capabilities in Perioperative Risk Prediction and Prognostication." *JAMA Surgery*, 159, 928–937.

Clark, Kevin, Minh-Thang Luong, Quoc Viet Le, and Christopher David Manning. 2020. "ELECTRA: Pre-training Text Encoders as Discriminators Rather than Generators." In: *8th International Conference on Learning Representations, ICLR 2020,* Addis Ababa, Ethiopia.

D'Antonoli, Tugba Akinci, Arnaldo Stanzione, Christian Blüthgen, Federica Vernuccio, Lorenzo Ugga, Michail E. Klontzas, Renato Cuocolo, Roberto Cannella, and Burak Koçak. 2024. "Large Language Models in Radiology: Fundamentals, Applications, Ethical Considerations, Risks, and Future Directions." *Diagnostic and Interventional Radiology,* 30(2), 80–90.

Danu, Manuela Daniela, George Marica, Sanjeev Kumar Karn, Bogdan Georgescu, Awais Mansoor, Florin Ghesu, Lucian Mihai Itu, Constantin Suciu, Sasa Grbic, Oladimeji Farri, and Dorin Comaniciu. 2023. "Generation of Radiology Findings in Chest X-Ray by Leveraging Collaborative Knowledge." In: *Tenth International Conference on Information Technology and Quantitative Management (ITQM 2023), Procedia Computer Science*, Oxford, UK, pp. 1102–1109.

Delbrouck, Jean Benoit, Maya Varma, Pierre Chambon, and Curtis Langlotz. 2023. "Overview of the RadSum23 Shared Task on Multi-modal and Multi-anatomical Radiology Report Summarization." In: *The 22nd Workshop on Biomedical Natural Language Processing and BioNLP Shared Tasks,* Association for Computational Linguistics, Toronto, Canada, pp. 478–482.

Delbrouck, Jean-Benoit, Pierre Chambon, Christian Bluethgen, Emily Tsai, Omar Almusa, and Curtis Langlotz. 2022. "Improving the Factual Correctness of Radiology Report Generation with Semantic Rewards." In: Yoav Goldberg, Zornitsa Kozareva, and Yuan Zhang (Eds.), *Findings of the Association for Computational Linguistics: EMNLP 2022*. Association for Computational Linguistics, Dubrovnik, Croatia, pp. 4348–4360.

Dhanaliwala, Ali Haider, Rikhiya Ghosh, Sanjeev Kumar Karn, Poikavila Ullaskrishnan, Oladimeji Farri, Dorin Comaniciu, and Elliott Charles Kahn. 2024. "General-Purpose vs. Domain-Adapted Large Language Models for Extraction of Structured Data from Chest Radiology Reports." arXiv:2311.17213.

Elbaz, Clement, Louis Rilling, and Christine Morin. 2020. "Fighting N-day Vulnerabilities with Automated CVSS Vector Prediction at Disclosure." In: *Proceedings of the 15th International Conference on Availability, Reliability and Security*, University College Dublin, Dublin, Ireland. Association for Computing Machinery, New York, pp. 1–10. doi:10.1145/3407023.3407038.

Eslami, Sedigheh, Christoph Meinel, and Gerard de Melo. 2023. "PubMedCLIP: How Much Does CLIP Benefit Visual Question Answering in the Medical Domain?" In: Timothy Baldwin, Marco Lapata, and Ivan Titov (Eds.), *Findings of the Association for Computational Linguistics: EACL 2023.* Association for Computational Linguistics, Dubrovnik, Croatia, pp. 1181–1193.

Feng, Chao, Xinyu Zhang, and Zichu Fei. 2023. "Knowledge Solver: Teaching Llms to Search for Domain Knowledge from Knowledge Graphs." arXiv preprint arXiv:2309.03118.

Gertz, Roman Johannes, Thomas Dratsch, Alexander Christian Bunck, and Jonathan Kottlors. 2024. "Potential of GPT-4 for Detecting Errors in Radiology Reports: Implications for Reporting Accuracy." *Radiology,* 311(1), e232714.

Ghesu, Florin Constantin, Bogdan Georgescu, Awais Mansoor, Youngjin Yoo, Dominik Neumann, Pragneshkumar Patel, Rudregowda Sarojamma Vishwanath, James Michael Balter, Yue Cao, Sasa Grbic, and Dorin Comaniciu. 2022. "Self-supervised Learning from 100 Million Medical Images." arXiv preprint arXiv:2201.01283.

Ghosh, Rikhiya, Oladimeji Farri, Hans-Martin von Stockhausen, Martin Schmitt, and George Marica Vasile. 2024. "CVE-LLM: Automatic Vulnerability Evaluation in Medical Device Industry Using Large Language Models." arXiv preprint arXiv:2407.14640.

Ghosh, Rikhiya, Hans-Martin von Stockhausen, Martin Schmitt, George Marica Vasile, Sanjeev Kumar Karn, and Oladimeji Farri. 2025. "CVE-LLM: Ontology-Assisted Automatic Vulnerability Evaluation Using Large Language Models." In: *Innovative Applications of Artificial Intelligence Conference (IAAI),* Philadelphia, PA. doi:10.1609/aaai.v39i28.35139.

Ghosh, Rikhiya, Oladimeji Farri, Sanjeev Kumar Karn, Manuela Danu, Ramya Vunikili, and Larisa Micu. 2023. "RadLing: Towards Efficient Radiology Report Understanding." In: *The 61st Annual Meeting of the Association for Computational Linguistics*, Toronto, Canada.

Goh, Ethan, Robert Gallo, Jason Hom, Eric Strong, Yingjie Weng, Hannah Kerman, Josephine Andree Cool, Zahir Kanjee, Andrew Stephen Parsons, Neera Ahuja, Eric Horvitz, Andrew Rodman, Jonathan H. Chen. 2024. "Large Language Model Influence on Diagnostic Reasoning: A Randomized Clinical Trial." *JAMA Network Open,* 7(10), e2440969.

Goldsack, Tomas, Carolina Scarton, Matthew Shardlow, and Chenghua Lin 2024. "Overview of the Biolaysumm 2024 Shared Task on the Lay Summarization of Biomedical Research Articles." arXiv preprint arXiv:2408.08566.

Grandinetti, Jace, Cole Friedes, Russell J.L. Maxwell, Zayne Belal, John Nicholas Lukens, Alexander Lin, William Ross Green, and Rafe McBeth. 2024. "Pathology AI for Risk Stratification (PAIRS): Predicting Adverse Pathological Features in Post-Operative Head and Neck Cancer with Large Language Models." *International Journal of Radiation Oncology, Biology, Physics,* 120(2), e626–e627.

Ha, Cuong, Shima Asaadi, Sanjeev Kumar Karn, Oladimeji Farri, Tobias Heimann, and Thomas Runkler. 2024. "Fusion of Domain-Adapted Vision and Language Models for Medical Visual Question Answering." In: *Proceedings of the 6th Clinical Natural Language Processing Workshop,* Mexico City, Mexico. Association for Computational Linguistics, pp. 246–257.

Hager, Paul, Friederike Jungmann, Robbie Holland, Kunal Bhagat, Inga Hubrecht, Manuel Knauer, Jakob Vielhauer, Marcus Makowski, Rickmer Braren, Georgios Kaissis, and Daniel Rueckert. 2024. "Evaluation and Mitigation of the Limitations of Large Language Models in Clinical Decision-Making." *Nature Medicine,* 30(7), 2613–2622.

Haim, Gal Ben, Mor Saban, Yiftach Barash, David Cirulnik, Amit Shaham, Ben Zion Eisenman, Livnat Burshtein, Orly Mymon, and Eyal Klang. 2024. "Evaluating Large Language Model-Assisted Emergency Triage: A Comparison of Acuity Assessments by GPT-4 and Medical Experts." *Journal of Clinical Nursing,* 33(1), 1–10.

Hassanin, Mohammed, Marwa Keshk, Sara Salim, Majid Alsubaie, and Dharmendra Sharma. 2024. "PLLM-CS: Pre-trained Large Language Model (LLM) for Cyber Threat Detection in Satellite Networks." arXiv preprint arXiv:2405.05469.

Hsueh, Jessica Y, Daniel Nethala, Shiva Singh, William Marston Linehan, and Mark Wayne Ball. 2024. "Investigating the Clinical Reasoning Abilities of Large Language Model GPT-4: An Analysis of Postoperative Complications from Renal Surgeries." *Urologic Oncology: Seminars and Original Investigations*, 42(9), 292.e1–292.e7.

Hu, Edward J., Yelong Shen, Phillip Wallis, Zeyuan Allen-Zhu, Yuanzhi Li, Shean Wang, Lu Wang, and Weizhu Chen. 2021. "Lora: Low-Rank Adaptation of Large Language Models." arXiv preprint arXiv:2106.09685.

Huang, Kexin, Jaan Altosaar, and Rajesh Ranganath. 2019. "Clinicalbert: Modeling Clinical Notes and Predicting Hospital Readmission." arXiv preprint arXiv:1904.05342.

Identity Theft Resource Center. 2024. Annual Data Breach Report 2023. https://www.idtheftcenter.org/wp-content/uploads/2024/01/ITRC_2023-Annual-Data-Breach-Report.pdf (accessed October 10, 2024).

Islam, Nafis Tanveer, Joseph Khoury, Andrew Seong, Gonzalo De La Torre Parra, Elias Bou-Harb, and Peyman Najafirad. 2024. "LLM-Powered Code Vulnerability Repair with Reinforcement Learning and Semantic Reward." arXiv preprint arXiv:2401.03374.

Jing, Baoyu, Pengtao Xie, and Eric Poe Xing. 2018. *On the Automatic Generation of Medical Imaging Reports.* Association for Computational Linguistics, Melbourne, Australia, pp. 2577–2586.

Johnson, Alistair Edward William, Tom Joseph Pollard, Seth Joshua Berkowitz, Nathaniel R. Greenbaum, Matthew P. Lungren, Chih-ying Deng, Roger Greenwood Mark, and Steven Horng. 2019. "MIMIC-CXR, a De-Identified Publicly Available Database of Chest Radiographs with Free-Text Reports." *Scientific Data,* 6(1), 317.

Johnson, Alistair Edward William, Tom Joseph Pollard, Lu Shen, Li-wei H. Lehman, Mengling Feng, Mohammad Ghassemi, Benjamin Moody, Peter Szolovits, Leo Anthony Celi, and Roger Greenwood Mark. 2016. "MIMIC-III, a Freely Accessible Critical Care Database." *Scientific Data*, 3, 160035.

Johnson, Alistair, E. W., Tom Pollard, Steven Horng, Leo Anthony Celi, and Roger Mark. 2023. "MIMIC-IV-Note: Deidentified Free-Text Clinical Notes (Version 2.2)." *PhysioNet.* https://doi.org/10.13026/1n74-ne17.

Karn, Sanjeev Kumar, Rikhiya Ghosh, Padmanabha Kusuma, and Oladimeji Farri. 2023. "Shs-Nlp at RadSum23: Domain-Adaptive Pre-training of Instruction-tuned LLMs for Radiology Report Impression Generation." In: *The 22nd Workshop on Biomedical Natural Language Processing and BioNLP Shared Tasks,* Toronto, Canada. Association for Computational Linguistics, pp. 550–556.

Karn, Sanjeev Kumar, Ning Liu, Hinrich Schuetze, and Oladimeji Farri. 2022. "Differentiable Multi-Agent Actor-Critic for Multi-Step Radiology Report Summarization." In: *Proceedings of the 60th Annual Meeting of the Association for Computational Linguistics,* Dublin, Ireland (Vol. 1, Long Papers). Association for Computational Linguistics, pp. 1542–1553.

Kiesling, Elmar, Andreas Ekelhart, Kabul Kurniawan, and Fajar Ekaputra. 2019. "The SEPSES Knowledge Graph: an Integrated Resource for Cybersecurity." In: *International Semantic Web Conference,* Springer, New York, pp. 198–214. doi:10.1007/978-3-030-30796-7_13.

Kim, Jonathan, Anna Podlasek, Kie Shidara, Feng Liu, Ahmed Alaa, and Danilo Bernardo. 2025. "Limitations of Large Language Models in Clinical Problem-Solving Arising from Inflexible Reasoning." https://arxiv.org/abs/2502.04381.

Kim, Su Hwan, Severin Schramm, Lisa Christine Adams, Rickmer Braren, Keno Kyrill Bressem, Matthias Keicher, Claus Zimmer, Dennis Martin Hedderich, and Benedikt Wiestler. 2024. "Performance of Open-Source LLMs in Challenging Radiological Cases: A Benchmark Study on 4,049 Eurorad Case Reports." *medRxiv*. https://doi.org/10.1101/2024.09.04.24313026.

Kirk, Hannah Rose, Bertie Vidgen, Paul Rottger, and Scott Arthur Hale. 2023. "Personalisation within Bounds: A Risk Taxonomy and Policy Framework for the Alignment of Large Language Models with Personalised Feedback." https://arxiv.org/abs/2303.05453.

Koenecke, Allison, Anna Seo Gyeong Choi, Katelyn Xiaoying Mei, Hilke Schellmann, and Mona Sloane. 2024. "Careless Whisper: Speech-to-Text Hallucination Harms." In: *Proceedings of the 2024 ACM Conference on Fairness, Accountability, Transparency,* Rio de Janeiro, Brazil, pp. 1672–1681.

Kwon, Woosuk, Zhuohan Li, Siyuan Zhuang, Ying Sheng, Lianmin Zheng, Cody Hao Yu, Joseph Gonzalez, Hao Zhang, and Ion Stoica. 2023. "Efficient Memory Management for Large Language Model Serving with Pagedattention." In: *Proceedings of the 29th Symposium on Operating Systems Principles*, Charlottesville, Virginia, USA, pp. 611–626.

Lammert, Jacqueline, Tobias Dreyer, Sonja Mathes, Leonid Kuligin, Kai J Borm, Ulrich Andreas Schatz, Marion Kiechle, Alisa Martina Lorsch, Johannes Jung, and Sebastian Lange. 2024. "Expert-Guided Large Language Models for Clinical Decision Support in Precision Oncology." *JCO Precision Oncology,* 8, e2400478.

Lau, Jason J., Soumya Gayen, Asma Ben Abacha, and Dina Demner-Fushman. 2018. "A Dataset of Clinically Generated Visual Questions and Answers about Radiology Images." *Scientific Data,* 5, 180251.

Laymouna, Moustafa, Yuanchao Ma, David Lessard, Tibor Schuster, Kim Engler, Bertrand Lebouche. 2024. "Roles, Users, Benefits and Limitations of Chatbots in Health Care: Rapid Review." *Journal of Medical Internet Research,* 26, e56930.

Lee, Harrison, Samrat Phatale, Hassan Mansoor, Kellie Lu, Thomas Mesnard, Colton Bishop, and Victor Carbune, Abhinav Rastogi. 2023. "Rlaif: Scaling Reinforcement Learning from Human Feedback with Ai Feedback." arXiv preprint arXiv:2309.00267.

Lewis, Patrick, Ethan Perez, Aleksandra Piktus, Fabio Petroni, Vladimir Karpukhin, Naman Goyal, Heinrich Kuttler, Mike Lewis, and Wen-Tau Yih, Tim Rocktäschel, Sebastian Riedel, Douwe Kiela. 2020. "Retrieval-Augmented Generation for Knowledge-Intensive Nlp Tasks." *Advances in Neural Information Processing Systems,* 33, 9459–9474.

Li, Junnan, Dongxu Li, Silvio Savarese, and Steven Hoi. 2023. "Blip-2: Bootstrapping Language-Image Pre-Training with Frozen Image Encoders and Large Language Models." *International Conference on Machine Learning,* Honolulu, Hawaii, PMLR, pp. 19730–19742.

Li, Pengfei, Gang Liu, Jinlong He, Zixu Zhao, and Shenjun Zhong. 2023. "Masked Vision and Language Pre-training with Unimodal and Multimodal Contrastive Losses for Medical Visual Question Answering." In: Hayit Greenspan, Anant Madabhushi, Parvin Mousavi, Septimiu Salcudean, James Duncan, Tanveer Syeda-Mahmood, Russell Taylor (Eds.), *Medical Image Computing and Computer Assisted Intervention—MICCAI 2023.* Springer Nature Switzerland, Cham.

Lin, Chin-Yew. 2004. "ROUGE: A Package for Automatic Evaluation of Summaries." In: *Presented at the Workshop on Text Summarization Branches Out.* Association for Computational Linguistics, Barcelona, Spain, pp. 74–81.

Lin, Weixiong, Ziheng Zhao, Xiaoman Zhang, Chaoyi Wu, Ya Zhang, Yanfeng Wang, and Weidi Xie. 2023. "PMC-CLIP: Contrastive Language-Image Pre-Training Using Biomedical Documents." In: *Medical Image Computing and Computer Assisted Intervention—MICCAI 2023*. Springer Nature Switzerland, Cham, pp. 525–536. doi:10.1007/978-3-031-43993-3_51.

Liu, Bo, Li-Ming Zhan, Li Xu, Lin Ma, Yan Yang, and Xiao-Ming Wu. 2021. "Slake: A Semantically-Labeled Knowledge-Enhanced Dataset for Medical Visual Question Answering." In: *2021 IEEE 18th International Symposium on Biomedical Imaging (ISBI),* Virtual Conference, IEEE, pp. 1650–1654.

Liu, Mingxuan, Yilin Ning, Salinelat Teixayavong, Mayli Mertens, Jie Xu, Daniel Shu Wei Ting, Lionel Tim-Ee Cheng, Jasmine Chiat Ling Ong, Zhen Ling Teo, Ting Fang Tan, Narrendar RaviChandran, Fei Wang, Leo Anthony Celi, Marcus Eng Hock Ong, and Nan Liu. 2023. "A Translational Perspective towards Clinical AI Fairness." *NPJ Digital Medicine,* 6(1), 172.

Liu, Pengfei, Weizhe Yuan, Jinlan Fu, Zhengbao Jiang, Hiroaki Hayashi, and Graham Neubig. 2023. "Pre-Train, Prompt, and Predict: A Systematic Survey of Prompting Methods in Natural Language Processing." *Association for Computing Machinery,* 55(9), 1–35.

Liu, Xin, Yuan Tan, Zhenghang Xiao, Jianwei Zhuge, and Rui Zhou. 2023. "Not the End of Story: An Evaluation of Chatgpt-Driven Vulnerability Description Mappings." In: Dan Jurafsky, James H. Martin, Sam Bowman, & Luke Zettlemoyer (Eds.), *Findings of the Association for Computational Linguistics: ACL* 2023. Association for Computational Linguistics, Barcelona, Spain, pp. 3724–3731.

Liu, Yinhan, Myle Ott, Naman Goyal, Jingfei Du, Mandar Joshi, Danqi Chen, Omer Levy, Mike Lewis, Luke Zettlemoyer, and Veselin Stoyanov. 2019. "Roberta: A Robustly Optimized BERT Pretraining Approach." arXiv preprint arXiv:1907.11692.

Liu, Zhengliang, Yiwei Li, Peng Shu, Aoxiao Zhong, Longtao Yang, Chao Ju, Zihao Wu, Chong Ma, Jie Luo, Cheng Chen, Sekeun Kim, Jiang Hu, Haixing Dai, Lin Zhao, Dajiang Zhu, Jun Liu, Wei Liu, Dinggang Shen, Tianming Liu, Quanzheng Li, and Xiang Li. 2023. "Radiology-Llama2: Best-in-Class Large Language Model for Radiology." https://arxiv.org/abs/2311.08692.

Lu, Haohui and Usman Naseem. 2024. "Can Large Language Models Enhance Predictions of Disease Progression? Investigating Through Disease Network Link Prediction." In: *Proceedings of the 2024 Conference on Empirical Methods in Natural Language Processing,* Miami, Florida, USA, pp. 17703–17715.

Lu, Keming, Hongyi Yuan, Runji Lin, Junyang Lin, Zheng Yuan, Chang Zhou, and Jingren Zhou. 2023. "Routing to the Expert: Efficient Reward-Guided Ensemble of Large Language Models." https://arxiv.org/abs/2311.08692.

Mansoor, Awais, Ingo Schmuecking, Florin Constantin Ghesu, Bogdan Georgescu, Sasa Grbic, Rudregowda SarojammaVishwanath, Oladimeji Farri, Rikhiya Ghosh, Ramya Vunikili, Mathis Zimmermann, James Sutcliffe, Steven L. Mendelsohn, Dorin Comaniciu, and Warren B. Gefter. 2024. "Large-Scale Study on AI's Impact on Identifying Chest Radiographs with No Actionable Disease in Outpatient Imaging." *Academic Radiology,* 31, 5300–5313.

McDuff, Daniel, Mike Schaekermann, Tao Tu, Anil Palepu, Amy Wang, Jake Garrison, Karan Singhal, Yash Sharma, Shekoofeh Azizi, Kavita Kulkarni, Le Hou, Yong Cheng, Yun Liu, S. Sara Mahdavi, Sushant Prakash, Anupam Pathak, Christopher Semturs, Shwetak Patel, Dale R. Webster, Ewa Dominowska, Juraj Gottweis, Joelle Barral, Katherine Chou, Greg S. Corrado, Yossi Matias, Jake Sunshine, Alan Karthikesalingam, and Vivek Natarajan. 2023. "Towards Accurate Differential Diagnosis with Large Language Models." https://arxiv.org/abs/2312.00164.

MITRE Corporation. 2024. Common Attack Pattern Enumeration and Classification. https://capec.mitre.org/.

MITRE Corporation. 2024. Common Vulnerabilities and Exposures. www.mitre.org.

MITRE Corporation. 2024. Common Weakness Enumeration. https://cwe.mitre.org/.

MosaicML NLP Team. 2023. Introducing MPT-7B: A New Standard for OpenSource, Commercially Usable LLMs. https://www.databricks.com/blog/mpt-7b.

Mozayan, Ali, Alexander Richard Fabbri, Michelle Maneevese, Irena Tocino, and Sophie Chheang. 2021. "Practical Guide to Natural Language Processing for Radiology." *RadioGraphics,* 41(5), 1453–1465.

Muennighoff, Niklas, Thomas Wang, Lintang Sutawika, Adam Roberts, Stella Biderman, Teven Le Scao, Maruf Saiful Bari, Sheng Shen, Zheng-Xin Yong, Hailey Schoelkopf, Xiangru Tang, Dragomir Radev, Alham Fikri Aji, Khalid Almubarak, Samuel Albanie, Zaid Alyafeai, Albert Webson, Edward Raff, and Colin Raffe. 2022. "Crosslingual Generalization through Multitask Finetuning." arXiv preprint arXiv:2211.01786.

Omar, Mahmud, Vera Sorin Sr, Jeremy Collins, David Reich, Robert Freeman, Nicholas Gavin, and Alexander Charney 2025. "Large Language Models Are Highly Vulnerable to Adversarial Hallucination Attacks in Clinical Decision Support: A Multi-Model Assurance Analysis." *medrXiv.* https://www.medrxiv.org/content/10.1101/2025.03.18.25324184v1.full.pdf.

Ong, Jasmine Chiat Ling, Liyuan Jin, Kabilan Elangovan, Gilbert Yong San Lim, Daniel Yan Zheng Lim, Gerald Gui Ren Sng, Yuhe Ke, Joshua Yi Min Tung, Ryan Jian Zhong, Christopher Koh, Keane Lee, Xiang Chen, Jack Chng, Aung Than, Ken Goh, and Daniel Tin. 2024. "Development and Testing of a Novel Large Language Model-Based Clinical Decision Support Systems for Medication Safety in 12 Clinical Specialties." arXiv preprint arXiv:2402.01741.

Papineni, Kishore, Salim Roukos, Todd Ward, and Wei-Jing Zhu. 2002. "Bleu: a Method for Automatic Evaluation of Machine Translation." In: *Proceedings of the 40th Annual Meeting of the Association for Computational Linguistics,* Association for Computational Linguistics, Philadelphia, Pennsylvania, USA, pp. 311–318.

Pearce, Hammond, Benjamin Tan, Baleegh Ahmad, Ramesh Karri, and Brendan Dolan-Gavitt. 2023. "Examining Zero-Shot Vulnerability Repair with Large Language Models." In: *2023 IEEE Symposium on Security and Privacy (SP)*, San Francisco, CA, USA, IEEE, pp. 2339–2356.

Pelka, Obioma, Sven Koitka, Johannes Rückert, Felix Nensa, and Christoph M. Friedrich. 2018. "Radiology Objects in COntext (ROCO): A Multimodal Image Dataset." In: Danail Stoyanov, Zeike Taylor, Simone Balocco, Raphael Sznitman, Anne Martel, Lena Maier-Hein, Luc Duong, Guillaume Zahnd, Stefanie Demirci, Shadi Albarqouni, Su-Lin Lee, Stefano Moriconi, Veronika Cheplygina, Diana Mateus, Emanuele Trucco, Eric Granger, and Pierre Jannin (Eds.), *Intravascular Imaging and Computer Assisted Stenting and Large-Scale Annotation of Biomedical Data and Expert Label Synthesis*. Springer International Publishing, Cham, pp. 180–189. doi:10.1007/978-3-030-01364-6_19.

Pham, Justin H., Charat Thongprayoon, Jing Miao, Supawadee Suppadungsuk, Priscilla Koirala, Iasmina M Craici, and Wisit Cheungpasitporn. 2024. "Large Language Model Triaging of Simulated Nephrology Patient Inbox Messages." *Frontiers in Artificial Intelligence,* 1452–1469. doi:10.3389/frai.2024.1452469

Pons, Ewoud, Loes MM Braun, Maria Gabriëlla Myriam Hunink, and Jan A Kors. 2016. "Natural Language Processing in Radiology: A Systematic Review." *Radiological Society of North America,* 279(2), 329–343.

Poulain, Raphael, Hamed Fayyaz, and Rahmatollah Beheshti. 2024. "Bias Patterns in the Application of LLMs for Clinical Decision Support: A Comprehensive Study." arXiv preprint arXiv:2404.15149.

Raido, Vanessa Enriquez. 2024. "Artificial Intelligence in Healthcare Translation: A Contemporary Systematic Review." In E. Monzó-Nebot & V. Tasa-Fuster (Eds.), *The social impact of automating translation: An ethics of care perspective on machine translation* (pp. 123–146). Routledge. https://doi.org/10.4324/9781003465522-7.

Reimers, Nils, Iryna Gurevych. 2019. "Sentence-BERT: Sentence Embeddings using Siamese BERT-Networks." In: *Proceedings of the 2019 Conference on Empirical Methods in Natural Language Processing and the 9th International Joint Conference on Natural Language Processing (EMNLP-IJCNLP,* Hong Kong, China. Association for Computational Linguistics, pp. 3982–3992.

Robertson, Stephen, Hugo Zaragoza, and Michael Taylor. 2004. "Simple BM25 Extension to Multiple Weighted Fields." In: *Proceedings of the Thirteenth ACM International Conference on Information and Knowledge Management,* Washington, DC, USA. Association for Computing Machinery, pp. 42–49.

Rubin, Daniel Lee Jr and Charles Edward Kahn. 2017. "Common Data Elements in Radiology." *Radiology.* 283(3), 837–844. doi:10.1148/radiol.2016161553.

Rückert, Johannes, Louise Bloch, Raphael Brüngel, Ahmad Idrissi-Yaghir, Henning Schäfer, Cynthia S. Schmidt, Sven Koitka, Obioma Pelka, Asma Ben Abacha, Alba G. Seco de Herrera, Henning Müller, Peter A. Horn, Felix Nensa, and Christoph M. Friedrich.abrina 2024. "ROCOv2: Radiology Objects in COntext Version 2, an Updated Multimodal Image Dataset." *Scientific Data,* 11, 688.

Salam, Babak, Claire Stuwe, Sebastian Nowak, Alois Martin Sprinkart, Maike Theis, Dmitrij Kravchenko, Narine Mesropyan, Tatjana Dell, Christoph Endler, Claus Christian Pieper, Daniel L. Kuetting, Julian Alexander Luetkens, and Alexander Isaak. 2025. "Large Language Models for Error Detection in Radiology Reports: A Comparative Analysis Between Closed-Source and Privacy-Compliant Open-Source Models." *European Radiology,* 35(8), 4549–4557.

Scao, Teven Le et al. 2022. "Bloom: A 176bparameter Open-Access Multilingual Language Model." arXiv preprint arXiv:2211.05100.

Schmidt, Reuben A, Seah, Chen Yi Jarrel, Ke Cao, Lincoln Lim, Wei Lim, and Justin Yeung. 2024. "Generative Large Language Models for Detection of Speech Recognition Errors in Radiology Reports." *Radiology: Artificial Intelligence,* 6(2), e230205.

Sewak, Mohit, Vamsi Emani, and Annam Naresh. 2023. "CRUSH: Cybersecurity Research using Universal LLMs and Semantic Hypernetworks." In: R. Gupta & S. Srinivasa (Eds.), *Proceedings of the Workshop on Enterprise Knowledge Graphs using Large Language Models (EKG-LLM 2023),* co-located with the *32nd ACM International Conference on Information, Knowledge Management (CIKM 2023),* Birmingham, UK, October 22, 2023. CEUR Workshop Proceedings, Volume 3532. CEUR-WS.org. https://ceur-ws.org/Vol-3532/paper5.pd

Shahid, Mustafizur Rahman, and Herve Debar. 2021. "Cvss-Bert: Explainable Natural Language Processing to Determine the Severity of a Computer Security Vulnerability from its Description." In: *2021 20th IEEE International Conference on Machine Learning and Applications (ICMLA*), Virtual, IEEE, pp. 1600–1607.

Siwicki, Bill. 2024. What LLMs Can Do for Radiologists and for the Radiologist Shortage. 23 August. Accessed March 31, 2025. https://www.healthcareitnews.com/news/what-llms-can-do-radiologists-and-radiologist-shortage.

Sun, Liwen, James Zhao, Megan Han, and Chenyan Xiong. 2024. "Fact-Aware Multimodal Retrieval Augmentation for Accurate Medical Radiology Report Generation." arXiv preprint arXiv:2407.15268.

Topol, Eric. 2023. Ground Truths—A substack from Eric Topol. https://erictopol.substack.com/p/geoffrey-hinton-large-language-models.

Trivedi, Hari and Judy Wawira Gichoya. 2024. "The LLM Will See You Now: Performance of ChatGPT on the Brazilian Radiology and Diagnostic Imaging and Mammography Board Examinations." *Radiology: Artificial Intelligence*, 6(1), e230568.

U.S. Department of Health and Human Services, Food and Drug Administration, Center for Devicesand Radiological Health, Office of the Center Director, Center for Biologics Evaluation and Research. 2016. "Postmarket Management of Cybersecurity in Medical Devices: Guidance for Industry and Food and Drug Administration Staff." https://www.fda.gov/media/95862/download.

Van Veen, Dave, Cara Van Uden, Maayane Attias, Anuj Pareek, Christian Bluethgen, Malgorzata Polacin, Wah Chiu, Jean-Benoit Delbrouck, Juan Manuel Zambrano Chaves, Curtis P, Langlotz, Akshay S. Chaudhari; John Pauly. 2023. "RadAdapt: Radiology Report Summarization Via Lightweight Domain Adaptation of Large Language Models." arXiv preprint arXiv:2305.01146.

Vasireddy, Dinesh Thyagarajan, Dakota S Dale, and Qinghua Li. 2023. "CVSS Base Score Prediction Using an Optimized Machine Learning Scheme." In: *2023 Resilience Week (RWS)*, pp. 1–6. IEEE. https://doi.org/10.1109/RWS58133.2023.10284627.

Vishwanath, Prathiksha Rumale, Simran Tiwari, Tejas Ganesh Naik, Sahil Gupta, Dung Ngoc Thai, Wenlong Zhao, Sunjae Kwon, Victor Ardulov, Karim Tarabishy, Andrew McCallum and Wael Salloum. 2024. "Faithfulness Hallucination Detection in Healthcare AI." In: *Artificial Intelligence, Data Science for Healthcare: Bridging Data-Centric AI, People-Centric Healthcare*, Barcelona, Spain.

Whitehouse, Chenxi, Monojit Choudhury, and Alham Fikri Aji. 2023. "Llm-Powered Data Augmentation for Enhanced Crosslingual Performance." arXiv preprint arXiv:2305.14288.

Woznicki, Piotr, Caroline Laqua, Ina Fiku, Amar Hekalo, Daniel Truhn, Sandy Engelhardt, Jakob Kather, Sebastian Foersch, Tugba Akinci D'Antonoli, Daniel Pinto dos Santos, Bettina Baeßler, and Fabian Christopher Laqua. 2024. "Automatic Structuring of Radiology Reports with on-Premise Open-Source Large Language Models." *European Radiology,* 35(4), 2018–2029.

Xie, Qianqian, Jiayu Zhou, Yifan Peng, and Fei Wang. 2023. "FactReRanker: Fact-Guided Reranker for Faithful Radiology Report Summarization." arXiv preprint arXiv:2303.08335.

Xu, Lingling, Haoran Xie, Si-Zhao Joe Qin, Xiaohui Tao, and Fu Lee Wang. 2023. "Parameter-Efficient Fine-Tuning Methods for Pretrained Language Models: A Critical Review and Assessment." arXiv preprint arXiv:2312.12148.

Yin, Jiao, MingJian Tang, Jinli Cao, and Hua Wang. 2020. "Apply Transfer Learning to Cybersecurity: Predicting Exploitability of Vulnerabilities by Description." *Knowledge-Based Systems,* 210, 106529.

Yosifova, Veneta, Antoniya Tasheva, and Roumen Trifonov. 2021. "Predicting Vulnerability Type in Common Vulnerabilities and Exposures (cve) Database with Machine Learning Classifiers." In: *12th National Conference with International Participation (ELECTRONICA),* Sofia, Bulgaria. IEEE.

Zanardo, Moreno, Jacob Johannes Visser, Anna Colarieti, Renato Cuocolo, Michail Emmanouil Klontzas, Daniel Pinto dos Santos, Francesco Sardanelli, and European Society of Radiology (ESR). 2024. "Impact of AI on Radiology: A EuroAIM/EuSoMII 2024 Survey among Members of the European Society of Radiology." *Insights into Imaging,* 15, 240.

Zech, John, Jessica Forde, Joseph James Titano, Deepak Kaji, Anthony Costa, and Eric Karl Oermann. 2019. "Detecting Insertion, Substitution and Deletion Errors in Radiology Reports Using Neural Sequence-To-Sequence Models." *Annals of Translational Medicine,* 7, 233.

Zhang, Chenhui, Le Wang, Dunqiu Fan, Junyi Zhu, Tang Zhou, Liyi Zeng, and Zhaohua Li. 2024. "VTT-LLM: Advancing Vulnerability-to-Tactic-and-Technique Mapping through Fine-Tuning of Large Language Model." *Mathematics,* 12, 240.

Zhang, Sheng, Yanbo Xu, Naoto Usuyama, Hanwen Xu, Jaspreet Bagga, Robert Tinn, Sam Preston, Rajesh Rao, Mu Wei, Naveen Valluri, Cliff Wong, Andrea Tupini, Yu Wang, Matt Mazzola, Swadheen Shukla, Lars Liden, Jianfeng Gao, Matthew Lungren, Tristan Naumann, Sheng Wang, and Hoifung Poon. 2023. "BiomedCLIP: A Multimodal Biomedical Foundation Model Pretrained from Fifteen Million Scientific Image-Text Pairs." arXiv preprint arXiv:2303.00915.

Zhang, Tianyi, Varsha Kishore, Felix Wu, Kilian Quirin Weinberger, and Yoav Artzi. 2020. "BERTScore: Evaluating Text Generation with BERT." In: *International Conference on Learning Representation,* Addis Ababa, Ethiopia.

Zhang, Xiaoman, Chaoyi Wu, Ziheng Zhao, Weixiong Lin, Ya Zhang, Yanfeng Wang, and Weidi Xie. 2023. "PMC-VQA: Visual Instruction Tuning for Medical Visual Question Answering." ArXiv abs/2305.10415v1.

Zhang, Yuhao, Daisy Yi Ding, Tianpei Qian, Christopher David Manning, and Curtis P. Langlotz. 2018. "Learning to Summarize Radiology Findings." In: *Proceedings of the Ninth International Workshop on Health Text Mining and Information Analysis,* Brussels, Belgium. Association for Computational Linguistics, pp. 204–213.

Zhou, Xin, Sicong Cao, Xiaobing Sun, and David Lo. 2024. "Large Language Model for Vulnerability Detection and Repair: Literature Review, Roadmap." arXiv preprint arXiv:2404.02525.

6 Transforming Visual Communication

The Role of Large Language Models in Media Content Creation

Tina Korani

6.1 INTRODUCTION: WHERE HUMAN CREATIVITY MEETS AI

> A designer types a sentence and a visual world appears. But who, or what, is the author?

Artificial intelligence (AI) tools are advancing so fast that we can barely keep up. What may have started as some techies' late night experiments has pushed the entire world into uncharted territories. AI has integrated into mainstream creative tools at a mind-blowing pace. This is prompting ongoing questions about authorship, creative agency, and who has created what in the design and teaching practices. But this isn't new, technology has disrupted creative work before. As Epstein et al. (2023) observed, AI may seem to be replacing artistic production, but we saw the same thing happen when photography seemed to threaten the concept and livelihood of painters. It didn't end painting by any means, but it changed the landscape of how we looked at art. Instead it "liberated it from realism, giving rise to Impressionism and the Modern Art movement," spinning the idea that AI is not the death of the arts, but an evolving medium.

These tools are already reshaping how we create and teach design. Here we'll cover how this transformation has caught the attention of the creative world. The *if* of AI is no longer a question. *How* this change is already unfolding across every educational arena and professional pipeline is simply by removing human limits and augmenting the creative process efficiently and effectively; indeed, AI offers an amazing promise of how technology can empower humanity, while it also challenges clear boundaries, shifting continuously as technologies advance. The results, however, are empowering creative practices, and the partnership between AI and humanity can be extremely beneficial, as will be demonstrated in the examples provided in this chapter.

These changes are evident not only in the educational field but also in overall industry practices. As Jay Rodge, Senior Developer Advocate at NVIDIA, notes

DOI: 10.1201/9781003492252-6

when interviewed, "LLMs usually help me a lot by just brainstorming what I should cover and then refining it" (Jay Rodge, Personal Communication, February 5, 2025). His professional experience is an example of how creative work systems are being reshaped through the careful use of AI tools with the human touch.

Through trial-and-error research, faculty surveys, classroom experiments, and industry insights, this chapter will explore four different approaches to using AI in design education. These case studies range from a poster design to comparative methods, timed experiments, and thorough production pipelines. You'll see consistent patterns showing how successful human-AI collaboration works. This isn't a story of technology replacing humans. Done properly, the readjustment of creative labor between machines and humans will complement rather than hinder the project.

Our findings challenge popular narratives about AI's impact on creative fields. AI is mind bogglingly fast, but it still can't replace human emotion or make the words flow like they should. There are technical limitations to be aware of as well as amazing opportunities at the fingertips of humanity. Being aware of the best and the worst AI has to offer will likely complement nearly any project. The pros and cons suggest that the best use of AI applications is when these tools are used with well-conceptualized and creative workflows that complement human strengths and take advantage of machine intelligence.

Moving away from the theory, this chapter offers steps for educators and designers alike, to utilize across visual communication studies. Students can be critical collaborators instead of passive consumers of technology. They need to maintain a foundation of design skills while embracing new tech opportunities, while they prioritize the process over the product. When a creative learns to redistribute a portion of the creative labor while maintaining their ethics, design educators will likely have succeeded in preparing media creators to flourish in an increasingly AI-saturated landscape. A creative's work must possess an authentic human experience to thrive in a tech-driven world.

6.2 THE COLLABORATIVE CREATIVE PROCESS: A FRAMEWORK FOR AI-HUMAN DESIGN

Our research shows that effective integration of AI in design practice is key. Replacement of human creativity by using solely AI has been shown to be blatantly obvious and lacks authenticity. As students worked with AI tools on a variety of educational projects, a consistent pattern emerged in how this collaboration developed.

Figure 6.1 illustrates this AI-human design pipeline that was observed across multiple case studies and developed by the author. This workflow model identifies four key phases in the design process, showing the distinct roles and communications between human designers and AI tools at each stage. Rather than a human-to-machine handoff, a continuous cycle of direction, generation, evaluation, and refinement was clear with different responsibilities during each phase.

The AI-human design pipeline shows how creative people and machines are working together in real-world design practices. In the human direction phase, the creative carefully chooses the references, molding the vision, and crafting the prompts with

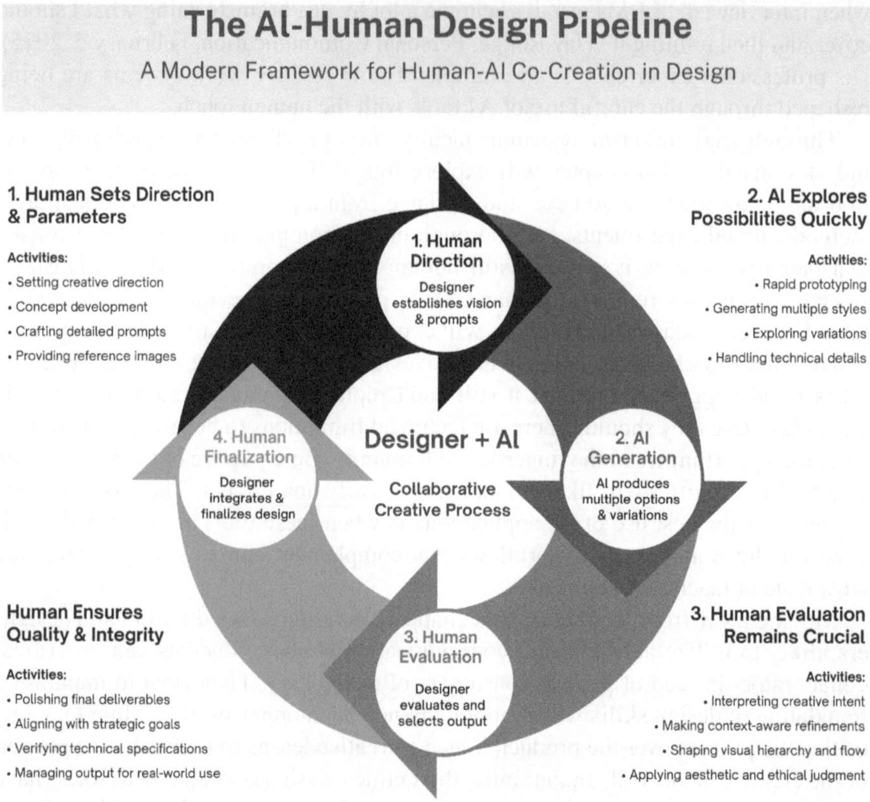

FIGURE 6.1 The artificial intelligence (AI)-human design pipeline. A circular flowchart diagram showing the AI-human design pipeline with four interconnected phases. The diagram shows "1. Human Direction" at the top (designer establishes vision and prompts), "2. AI Generation" on the right (AI produces multiple options and variations), "3. Human Evaluation" at the bottom (designer selects and refines output), and "4. Human Finalization" on the left (designer integrates and finalizes design). Each phase includes detailed activities listed around the circle, illustrating the collaborative nature of AI-Human design work.

care. It's a hands-on process that shows what Hwang and Wu (2025) highlighted as a necessary new skillset: learning to communicate with AI by careful and well-chosen keywords that come from thorough research.

This may sound like a quick fix, but it's more complicated than simply typing in a few words. One participant stated the challenge of finding "the appropriate keywords to describe my thoughts" and trying to create "my own AI vocabulary" by "designing keywords that correspond to my design style through the keywords of others" to enhance "the efficiency of finding keywords" (Hwang & Wu, 2025). It's far more complicated than most people initially think. More than one student said that "the difficulty lies in how to transform the thoughts in my mind into prompts for text" and then recognizing that "I see expressing thoughts and keywords from my mind as a new challenge for designers" (Hwang & Wu, 2025). This groundwork clearly shows that "preliminary design processes such as trend analysis and research have become more crucial" because "comprehensive understanding of text, content, and context, influences the outcome of the generated images," changing educational priorities at the ground level from original, traditional output-focused teaching to developing process-oriented skills (Hwang & Wu, 2025).

The **AI generation** phase is where things get fast and AI can do things traditional design software just can't. This is what Epstein et al. (2023) identify as the prime shift for the creative. AI tools can create a staggering amount of ideas per minute allowing the designer to focus on the processes that require human touch and engagement.

Carefully crafted prompts are only possible with what Epstein et al. (2023) describe as "meaningful human control." Using AI will require that "human creators can creatively express themselves through the generative system, leading to an outcome that aligns with their intentions and carries their personal, expressive signature." The goal is not for technology to replace people but instead it will enable designers to maintain their creative agency while using AI's astounding potential.

This is where we'll see the designer's role transform from a hands-on approach to a carefully directed automated process. Students will need to understand that "using personalized prompts based on pre learned content produced distinct graphic design results compared to when generic prompts were used."

This clearly shows that the quality of the person's direction influences the output of what AI generates (Hwang & Wu, 2025).

The **human evaluation** phase appears to require the most thought and input from the designer while assessing the combined ideas that AI has produced while maintaining their personal standards. Hwang and Wu's (2025) research shows, "creators first select a collection of images to supply (pre-curation) and provide these inputs to generative AI algorithms seeking to mimic them. Then, in the final stage, they curate and select the final image from numerous output images within seconds (post-curation)." This style of decision-making is required throughout this pipeline example.

This process matters more as nearly all companies are using some AI integration. Research shows that 89% of companies are now either using AI in their operations or considering AI projects. The need for effective AI–human collaboration has never been more necessary (NVIDIA, 2025).

In the evaluation phase, students also described that they had to learn from "the 'mistakes' of AI" which "can trigger a designer's creativity and inspiration." This suggests that proper evaluation recognizes unexpected outcomes that could help with the creative direction (Hwang & Wu, 2025). The people that participated in the research stated that this evaluation process requires you to understand "the stylistic characteristics I desire for the poster during the process," making creatives specifically state what they want for an outcome and also apply their own aesthetic judgment in the live judgement of multiple AI created options (Hwang & Wu, 2025). The balance needed between AI and a person's judgment reflects what industry analysts term "intelligent creativity." This is a combination of automation and human intuition that recognizes there's a place for both in marketing experiences (Pattisall, 2024).

The **human finalization** phase demonstrates that the combination of machine and human can't rely solely on a machine. It doesn't work. Students have discovered that successfully working with AI requires "personal research" and creative vision or all the outputs be similar (Hwang & Wu, 2025). As one student said, "AI exists as just a tool" that can "greatly improve efficiency compared to traditional image production methods" but people keep the ultimate creative authority, "with just a simple explanation, we can create various draft images" but the creative vision, aesthetics, and end-result creative decisions are still human responsibilities (Hwang & Wu, 2025). Also, the finalization phase combines what Hwang and Wu (2025) call a "graphic storytelling" approach, where students "tend to make efforts to derive creative keywords to imbue a scene with a unique narrative and story." That way, AI-generated elements serve the human ideas instead of functioning as solely automated outputs.

This framework helped create educational approaches examined in the following case studies. Whether focusing on a graphic design, an experiment with a tight deadline, comparative methods, or a production pipeline, each different approach is geared toward preparing students to navigate the workflow effectively.

6.3 RESEARCH METHODOLOGY: FACULTY SURVEY AND CLASSROOM EXPERIMENTS

We reached out to over 30 universities and these 36 faculty members were the ones willing to dive into the messy realities of teaching with AI. The findings presented here are being published for the first time. Their responses revealed the complexity of this moment, as one faculty member put it, large language models (LLMs) can "break the blank canvas syndrome and allow students to see their ideas take form," while another worried they could "inspire laziness of thought" or "siphon jobs from working creatives and homogenize output." These weren't abstract opinions but real concerns from educators actively wrestling with how to prepare students for an uncertain future.

As shown in Figure 6.2, the faculty reported a variety of ways they incorporated AI into their teaching practices. Most of them were using LLMs, such as ChatGPT and Claude, for brainstorming and ideas (79.3%), followed by demonstrations of AI tools in design workflows (65.5%), active assignments involving LLMs (62.1%), and classroom discussions about AI's impact on design and creativity (58.6%). While LLMs were the most commonly used, some faculty also integrated diffusion or vision models such as DALL.E, FireFly, or Midjourney for generating visual content. Faculty are taking a comprehensive approach, balancing practical use with critical examination of AI's creative implications.

Current Use Cases: Faculty are primarily using LLMs for brainstorming and ideation (79.3%), providing examples and demonstrations (58.6%), and teaching prompt engineering (51.7%)

Ethical Guidelines: 61.1% of faculty provide guidelines or support for students using LLMs ethically and legally.

AI–Human Partnership: 97.2% of faculty (35 out of 36) believe that LLMs will not eventually replace human visual communication designers.

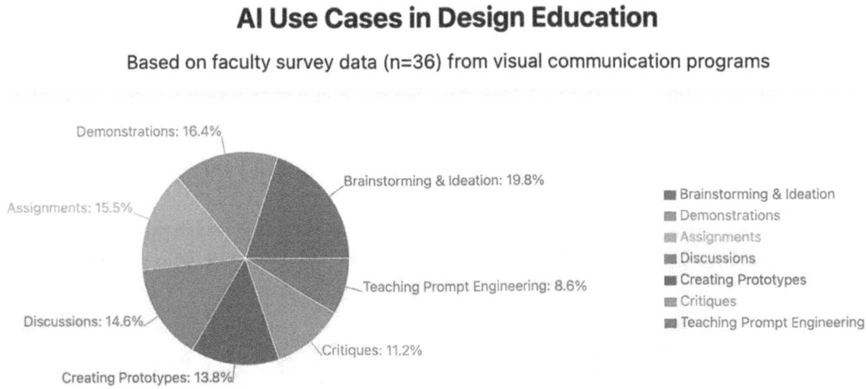

FIGURE 6.2 Faculty AI use cases in design education by percentage. A pie chart displaying AI use cases in design education based on faculty survey data ($n=36$). The chart shows "Brainstorming & Ideation" as the largest segment at 19.8%, followed by "Demonstrations" at 16.4%, "Assignments" at 15.5%, "Discussions" at 14.6%, "Creating Prototypes" at 13.8%, "Critiques" at 11.2%, and "Teaching Prompt Engineering" at 8.6%.

Institutional Policies: Most institutions (44.4%) are still developing formal policies, while 22.2% leave AI use to instructor discretion.

The survey also reflects faculty attitudes toward AI that ranged from enthusiastic adoption to cautious skepticism. As one person responded, LLMs can "break the 'blank canvas' syndrome and allow students to see their ideas take form," highlighting the technology's benefit of potentially overcoming creative blocks. Others were concerned that LLMs could "inspire laziness of thought" or "siphon jobs from working creatives and homogenize output," reflecting broader anxieties about the negative impact of AI on creative industries.

From the broader faculty survey, two distinct approaches to classroom integration of AI emerged, represented by the work of Joshua Duttweiler who developed the "*Something Old, Something New*" methodology, and Jason Porter who created the "*Prompt to Production*" approach. Both Joshua and Jason are faculty members who have been pioneering innovative pedagogical methods for integrating AI into design education. Additionally, we conducted two classroom experiments at San José State University, focusing on festival poster design and time-constrained design tasks. These four case studies together provide a comprehensive view of current approaches to AI integration in design education represented through these methodological approaches.

Case Study 6.1: Evaluating AI Image Generators in Festival Poster Design

CONTEXT AND METHODOLOGY

In February 2025, advertising students explored five AI image generators (DALL-E, Adobe Firefly, Midjourney or Stable Diffusion, Canva Magic Design (a design platform that incorporates AI-powered features), and Microsoft Designer) by creating music festival posters as part of a comparative analysis. This systematic evaluation went far beyond simple tool comparison. It explored the evolving relationship between the creative designer and the machine during this creative process.

Students developed comprehensive creative briefs for existing or fictional music festivals. They established clear parameters around the festival names, taglines, dates, locations, and stylistic direction. This prep work mirrored typical design methods,

FIGURE 6.3 Example of student submission for AI poster assignment. Four AI-generated space-themed music festival posters for "The Milky Wave" festival. The posters feature cosmic imagery with galaxies, planets, and celestial elements in black and white/grayscale tones. All show "August 3-5" and "San Francisco" with various interpretations of the space theme, demonstrating different AI-generated approaches to the same prompt.

emphasizing conceptual development before executing the visual product. Students crafted the consistent prompts across all five platforms and generated roughly 20 poster ideas per tool, for a total of 100 designs per student.

This approach positioned students to think as critical evaluators instead of passive users. They compared outputs while acknowledging each unique "language" and the ability of each AI program (Figure 6.3).

Prompt used: "Design a poster for a space themed music festival called Milky Wave in San Francisco with the dates being August 3-5th add date to bottom of the poster. make sure to include the location."

6.4 WHAT WE LEARNED

Here's the thing these were mostly junior designers in their first typography class, so they immediately zeroed in on the text problems. And boy, were there problems. Typography was a mess across almost every AI platform we tested. Students consistently reported misspellings, illegible text, and garbled formatting. Student feedback consistently highlighted typography problems, with participants reporting that "Words were misspelled or distorted which made it difficult to look at or even read."

Since the completion of this study, newer updates to several tools have significantly improved text and typography rendering, particularly for shorter phrases.

This finding highlights an important limitation. AI systems are great at generating compelling visuals and styles, but they struggle with semantic and typographic details that professional designers know are essential. As Epstein et al. (2023) note in their broad study, "generative models are trained by reducing error on reconstruction tasks on training data. As such, they are fundamentally bound by reproducing what they have already seen." That may explain why AI tools generate variations using familiar patterns but struggle with the proper execution of specific elements like typography because AI doesn't have exact semantic understanding.

The case study also revealed remarkable differences in how different AI platforms interpret and execute aesthetic concepts. Some tools produced visually coherent compositions that resembled authentic festival posters, while others generated what students described as "bland" and "uninspiring" results despite identical prompt guidance. This inconsistency across platforms underscores that AI tools are not neutral instruments but complex systems with distinct "personalities," biases, and interpretive tendencies.

Perhaps most importantly, the case study demonstrates that prompt engineering is an essential skill for design. Students found that the quality of their AI-generated output directly correlated with the depth of their prompts. The more detailed the direction regarding mood, composition, style, and color, the better the results compared to using minimal prompts. Students quickly learned that better prompts meant better results. As one participant put it, "to obtain better outcomes, I widened my prompts by providing more detailed details about the posters' mood, style, and layout."

The students concluded that AI tools offered great value for ideas and exploration instead of replacements for human decision-making. Participants emphasized "AI is not a substitute for a designer's vision. To achieve the best results, it still requires guidance, vision, and adjustments."

The case study also reveals critical ethical issues regarding AI-generated imagery. One student noted disturbing stereotypes in response to prompts for a "Mexican Punk Rock festival." AI relied heavily on "Mexican calavera with guns around it" imagery and even labeled one of the elements as "slaves." This demonstrates how AI can produce and amplify harmful cultural stereotypes embedded in their training data. In fact, this has a disturbing potential to cause more damage than good based on characteristics like race and gender (Epstein et al., 2023) because AI is trained using data that mirror existing social prejudices (Breazu & Katsos, 2024; Epstein et al., 2023).

AI-driven image search results for occupations continually come up short due to human input (Epstein et al., 2023). Because algorithms are developed by humans, they show the biases that people express and replicate them during training. Students and teachers need to be critical when examining the output from AI tools.

6.5 TEACHING IMPLICATIONS

This case study reveals important shifts happening in design education as AI tools become part of the creative process. What I found most striking was how students naturally moved away from focusing on software mastery toward developing judgment skills: deciding what to ask for, how to refine it, and when the output was actually useful.

The students who reflected most thoughtfully on their work were those who could articulate how AI changed their thinking, not just their output. This suggests that we need to build more intentional reflection into our pedagogy, asking students to examine their own creative decision-making as it evolves with these tools.

Perhaps most telling were the persistent typography problems across different AI platforms. Despite rapid technological advances, students still needed solid design fundamentals to recognize and correct these issues. This reinforces my belief that we're not replacing traditional design education but rather adding new layers of complexity to it.

The ethical dimensions were equally revealing. Students varied widely in their ability to spot problematic representations in AI-generated content, particularly around cultural imagery. This wasn't just about technical training. It required the kind of critical thinking that develops through sustained engagement with visual culture and social responsibility.

What we're seeing is a pedagogical challenge that goes beyond simply adding new software to the curriculum. We're asking students to develop what I'm calling critical algorithmic literacy: the ability to work productively with AI while maintaining both creative agency and ethical awareness. This requires rethinking not just what we teach, but how we help students understand their own evolving role as designers in an AI-augmented world.

Case Study 6.2: How Time Pressure Shapes AI Adoption in Design: A Mixed-Methods Experiment Context and Methodology

We conducted a 45-minute design experiment conducted in ADV 95: Working with Type at San José State University. The focus was on how students from different academic backgrounds, including Advertising, Business Administration, Creative

Advertising, and Visual Communication, worked on the logotype design challenges under time pressure.

The exercise required students to create a logotype for UPS in 15 minutes, making sure it aligned with the brand's existing core values, which are trustworthiness, professionalism, and simplicity. After the completion, they spent 30 minutes completing a pointed survey about their design process, tools they used, challenges, and thoughts on AI in graphic design.

Before diving into what we found, I should be upfront about the scope of this study. We're looking at a relatively small group (20 students total), and most of them were junior designers with very little professional experience. It was their first assignment in the course, so they were still getting comfortable with the classroom dynamic and expectations. Also, the vast majority had never used AI for image generation before this. These factors definitely shaped these findings. But even with that said, the patterns shown were enough to warrant a deeper dive.

Time pressure did create patterns in how students approached tools, but the unexpected findings proved more interesting. There wasn't a uniform resistance or acceptance of AI, but three different groups of students who made well-considered, value-based decisions about this technology while they were under pressure.

The three distinct groups were students who embraced AI tools, those who tried AI but abandoned it, and others who avoided AI entirely. This showed how individuality, experience, comfort with risk, technology attitudes, and ethics played out under the pressure of a tight deadline. Students self-reported their tool usage and approach in the post-exercise survey, and their responses were categorized based on their described interaction with AI tools during the 15-minute design task.

About four students actually used AI tools during their entire design process. They leveraged platforms like Adobe, Midjourney, and various AI features for specific design elements, to refine their existing designs or generate new ideas. One student said that AI helped them "generate four to five different logo ideas in just one minute," dramatically speeding up the concept phase. These students saw AI as a brainstorming partner that helped them explore more options quickly.

But here's where it gets interesting. Several students tried AI at the beginning but abandoned it when they didn't like their results. One second-year advertising student who taught herself design said this perfectly: "I wanted to use a design or lettering that would show motion. I used AI in the beginning but it was really bad at showing my ideas so I switched to Canva. I used a simple design and I think it could have been much better." Her experience shows that using AI goes beyond simple access or time pressure. It's about how quickly the tool delivers on expectations. When AI didn't produce what she envisioned in a snap, she moved to a familiar platform instead of tackling the learning curve during the timed exercise.

The final group of students avoided AI entirely. They stuck with traditional tools like Adobe Creative Suite, Canva, and manual sketching. Their concerns weren't just about unfamiliarity but about professional identity, creative authenticity, and ethical considerations.

6.6 VALUES-DRIVEN DECISION MAKING

Students split into three clear camps, and it wasn't what we expected. About 60% saw AI as helpful for brainstorming, but 30% had serious ethical concerns that went way beyond just "will this take my job?" The remaining 10% of students avoided AI for reasons other than ethics, such as unfamiliarity or time constraints. It's important to note that some overlap existed between categories, as some of the 60% who saw AI as helpful for brainstorming were among those who initially tried but then abandoned AI tools during the exercise.

A participant noted: "I have seen brands begin to use AI in advertisements and designs, which places real people and designers in a bad position. While I believe that human creativity is something that cannot be perfectly replicated by computers, with the way that generative AI works, it can get pretty close, which may threaten the field of graphic design. I think that AI can be a helpful tool for generating ideas, but it gets to a point when AI if the only way for people to start their creative process, taking away what makes design 'human.'"

Another student was even more direct about the ethical implications, arguing that "AI in graphic design, I think should be limited to use as 'inspiration' and never to be claimed as your own work. AI scrapes other artists' work and merges just to answer the prompt. I think instead of creative jobs, it should be applied in other places that require more technical and monotonous work. If AI is continuously used, then you would only see bland, everything is the same gunk that someone who never has ever appreciated the time and skill it takes to make designs." Because of this student's commitment to these principles, she chose to work entirely by hand. She ended up submitting only a sketch because she ran out of time to create a finished product.

As shown in Figure 6.4, on the right is a student's hand-drawn sketch concept, created under time constraints without AI tools. On the left is a refined version, using the original sketch as input, generated by ChatGPT in under 10 seconds. While the AI version captures the basic concept, it still needs to be refined and shows the potential and limitations of AI as a design development tool.

6.7 WHAT WE LEARNED

These weren't just classroom discussions. Students were making real decisions about their majors and careers based on AI anxiety. One student stated candidly: "I switched my major from graphic design to creative advertising in my first year because I wanted something more broad to open the door to new opportunities." She continued, "I think the field of graphic design will definitely worsen, so I think it's important to hone in more skills besides graphic design so that you aren't left with nothing valuable if AI does progress." Students making significant choices based on their thoughts about AI's impact was eye-opening. And, often they were doing it without having enough hands-on experience with the technology itself.

The career anxiety wasn't only negative, though. Students with more design experience tended to be more optimistic and used AI for efficiency, seeing AI as a tool for repetitive tasks and to offer inspiration. They felt like it could free them up for higher-level creative decisions. The students with less experience had more anxiety,

FIGURE 6.4 Hand-drawn sketch versus AI-refined logo comparison. A side-by-side comparison showing a hand-drawn UPS logo sketch on lined paper (top right) and an AI-refined version below it. The original sketch shows a simple oval with "UPS" text and an arrow, while the AI version presents a more polished black and white logo with the same basic concept, illustrating AI's ability to refine rough concepts.

which makes sense since they're still developing their professional identity. One participant said, "Human designers are still needed to curate, refine, and ensure the quality and ethical integrity of the final product." Another student had a different take: "I think that eventually AI will replace designers, as AI will do for the majority of jobs. However, I think humans will still be needed to tell the AI what to do and iterate their drafts. So I think that if graphic designers get familiar with AI tools to help them with designs that would be ideal."

After analyzing the student responses, it became clear that the student who adopted AI did so because of ease, rather than basic resistance to technology. Students who found it difficult to communicate effectively with AI through prompts often abandoned AI, not because of philosophical objections, but because they didn't have enough time to figure it out. This showed that the very constraint we thought might drive efficiency actually discouraged experimenting with a potentially efficient tool for some students, while others took it and ran with it.

One student who used AI during the entire process wrote: "When I'm trying to design (redesign), I will get stuck at the beginning. Although I have little thought about that, I don't actually have any graphics and images. After I use AI, I can use some words or letters to describe and then can show something to let my ideas be more clear." This student understood AI's potential to help turn vague ideas into visuals, and recognized the creativity AI can provide during ideation phases. The gap between students with no prior experience in AI image generation and ones with experience showed that some were willing to invest the time while others weren't.

The students who successfully integrated AI into their workflow during the exercise showed us what's possible when the technology works as intended. They used AI to accelerate ideation while maintaining creative control, leveraged automation for efficiency while preserving human oversight, and viewed AI as a creative partner rather than a replacement. One noted, "I feel like AI is not perfect which is why a lot of people often have to edit their prompts to make it be the way they want it to." Rather than seeing this iterative process as a limitation or inefficiency, these students viewed it as part of the creative workflow, similar to how they might iterate on sketches or refine designs in traditional software.

6.8 TEACHING IMPLICATIONS

Several important patterns emerged from this small, yet revealing study. Multiple things became clear about how we need to think about AI in design education, particularly with junior designers who are still developing their toolkit. First of all, the time constraints affected students differently depending on their current familiarity with AI tools and their risk tolerance for experimentation. The students that had some familiarity with AI or were open to experimenting found it invaluable for rapid ideation and exploring concepts. Those with no AI experience or who were concerned about ethics felt the time pressure made experimenting too risky or inappropriate. These findings show that AI literacy isn't just about knowing the tools are available but about having enough knowledge to use them effectively under pressure.

Second, the ethical concerns students raised about authorship, originality, and professional authenticity aren't just barriers to overcome through better tool design or marketing. They're real questions about creative practice that need serious consideration. Students are thinking about what's authentic design work, how AI-generated content affects human creativity and its value, and whether using AI tools aligns with their personal and professional values. These are important conversations that need to happen in design education instead of being dismissed or ignored.

AI adoption in the design world is far more complicated than simple acceptance or resistance. Junior designers are encountering these tools for the first time in an academic setting. Students are making decisions based on their experience, comfort with risk, career concerns, ethics, and their current practical needs. The time constraint element was key because it created real-world pressure. That revealed how students actually behave when efficiency matters, rather than how they might theoretically approach integrating AI use while in a low-stakes situation.

Some found AI invaluable. Rapid exploration and overcoming creative blocks were benefits, while some found the learning curve with tight deadlines too much, or the ethical considerations were unclear when considering their values. Students must think seriously about AI's role in their futures. They need to actively try to understand how to best position themselves in an ever-changing field instead of passively accepting or rejecting AI. The career anxiety is real and valid. It represents strategic thinking around their professional development in uncertain times. Some conclusions led to defensive rather than proactive skill development approaches and each student will have to decide where AI fits into their lives.

However, there are significant limitations to this study. The small sample size was of 20 students, mostly junior designers with limited professional experience using AI tools for the first time. This was also their first assignment in the course. The findings might look quite a bit different with seasoned designers, larger sample sizes, or students with prior exposure to AI. Also, the constraints of creating a logo design within a 15-min timeframe have design challenges and different time pressures.

The question for educators isn't whether or not to include AI in design curriculum. Students are already using AI and making decisions based on their personal values and future concerns. The question is how to provide the right learning environment to help students navigate these complex technologies. They're already building on the sophisticated thinking they've already demonstrated during this study. They are addressing legitimate concerns and the knowledge gaps they see as this study revealed. What we need is education that prepares students. Not just to use AI tools but to carefully think about when, how, and why to use them in ways that position them with their creative values and professional goals.

Case Study 6.3: Something Old, Something New: Comparing Human and AI Instruction Context and Methodology

The *Something Old, Something New* project came about as a direct response to growing anxiety among design students about AI. After numerous informal discussions with his advanced typography class at Texas State University, Professor Joshua Duttweiler found that many students had deep concerns about their professional futures. The increasing presence of AI in the industry was frightening.

Rather than dismissing these anxieties or reinforcing them, Duttweiler had an assignment that would place students in conversation with AI tools. He positioned them to be critical collaborators, not as passive bystanders of technological change. They could leverage their existing knowledge while exploring new possibilities.

What made this approach particularly effective was its affirmation of students' current expertise. Instead of presenting AI as an intimidating new skill set to master, the project framed it as a comparative exercise: How does your existing knowledge of a subject compare to AI generated content?

The project rolled out in several carefully planned phases, each one building on the last to help students engage more thoughtfully with both the AI tools themselves and the bigger questions about what it all means.

Phase 1: **Getting the Ground Rules Clear**: The project started with a careful look at the Communication Design department's AI policy, which they'd adapted from Rhode Island School of Design's guidelines (Rhode Island School of Design, 2025). This initial discussion was important for establishing the ethical boundaries, questions about attribution, copyright, and academic integrity that students needed to understand upfront. After laying this groundwork, Duttweiler opened a conversation about where students stood on the topic and the use of AI at that moment.

Phase 2: **Getting Their Hands Dirty**: Next came structured experimentation through a hands-on AI workshop. Using resources from Harvard's metaLAB AI Pedagogy Project, students worked with multiple platforms at the same time, comparing their interfaces, what they could do, and where they fell short in real time (metaLAB (at) Harvard, 2025). A large part of this workshop was learning to refine prompts collaboratively, figuring out together how to get better results through comparing results and constructive feedback.

Phase 3: **Side-by-Side Documentation**: With both the technical skills and conceptual framework in place, students moved into the heart of the project:

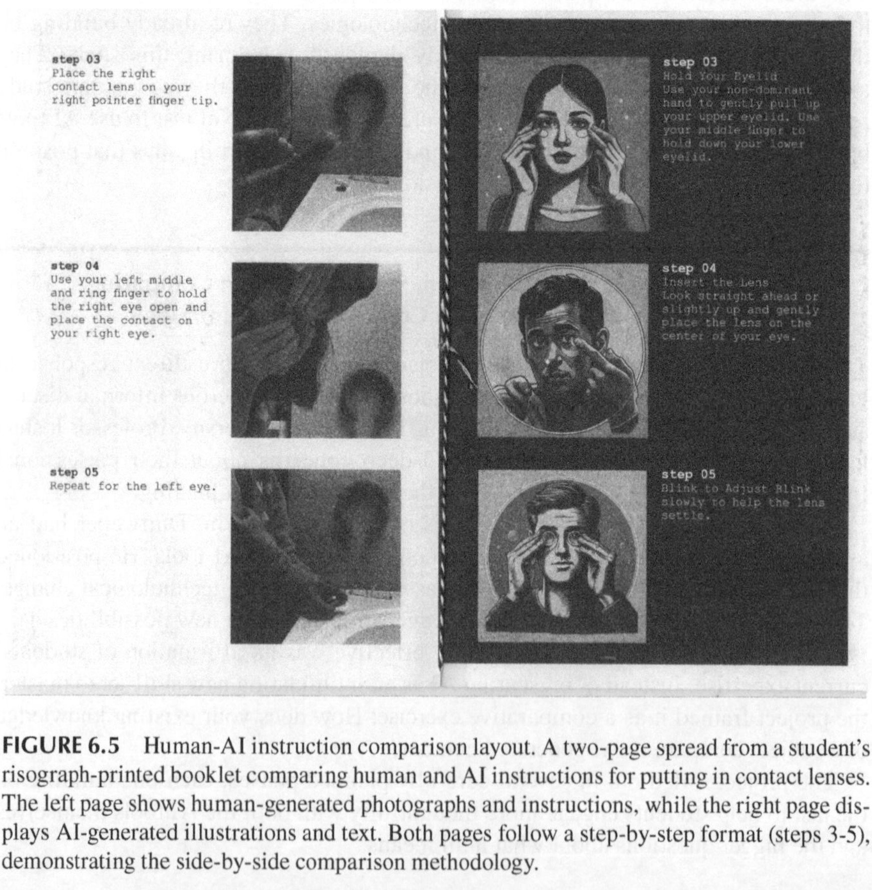

FIGURE 6.5 Human-AI instruction comparison layout. A two-page spread from a student's risograph-printed booklet comparing human and AI instructions for putting in contact lenses. The left page shows human-generated photographs and instructions, while the right page displays AI-generated illustrations and text. Both pages follow a step-by-step format (steps 3-5), demonstrating the side-by-side comparison methodology.

1. Each student picked something they already knew how to do well. Everything from everyday tasks like putting in contact lenses to more specialized processes like making a Spotify playlist.
2. They documented their own process step by step, capturing both the explicit instructions and photographs showing how it worked.
3. Using whichever AI tool they preferred, they then asked the system to generate instructions and visuals for the same task.
4. In structured class presentations, they shared both sets of instructions, pointing out the differences, limitations, and surprising discoveries.

Phase 4: **From Analysis to Design**: The final phase turned this comparative work into a design challenge. Students created risograph-printed booklets that put the two sets of information side by side, their own human-derived instructions alongside the AI-generated alternatives. The constraint of two-color risograph printing added both a practical and conceptual element to the project. Each student picked two colors, one to represent human knowledge and one for AI-generated content, which gave them a visual framework for thinking about the differences (Figure 6.5).

Figure 6.5 shows a publication example where the student used a more traditional side-by-side layout to compare results for "how to put in contact lenses." Like the eye chart you see during a visit to the optometrist, easy readability was a key part of the student's design concept when highlighting how similar the AI and human instructions were to each other. In this figure, images and text on the right page were produced by AI and laid out by the student.

6.9 WHAT WE LEARNED

The presentation phase revealed some fascinating patterns that showed up across almost all the student projects. Nearly every student reported that AI-generated instructions missed the nuances that come from doing something yourself. The AI also failed to include crucial contextual information, those "tricks of the trade" that experienced people develop through trial and error. It's what Michael Polanyi called "tacit knowledge," the stuff you know but can't easily put into words.

One student captured this perfectly during our discussion: "The AI instructions were technically correct but would be hard to follow without already knowing how to do it. It's like it knows the what but not the why."

What made some of these comparisons particularly revealing was how they highlighted language differences. One student discovered that the AI couldn't handle colloquial phrases, like describing a folding technique as creating a "hot dog" shape, a common way teachers explain things to kids (Figure 6.6). The AI got completely confused by this kind of everyday language that humans use instinctively to make instructions clearer and more relatable.

Figure 6.6 demonstrates how, after comparing instructions for folding an origami crane, the student developed two characters to represent the AI and a human to help visualize their differences and eventual annoyance with one another when trying to complete the same task. Using the characters to act as narrators for the publication,

FIGURE 6.6 Student publication comparing AI and human communication styles. A creative two-page spread featuring illustrated cartoon characters representing AI and human perspectives in an origami crane folding tutorial. The pages show step-by-step instructions with speech bubbles, including the human character saying, "Grab a piece of paper and fold in half hamburger style" and references to "hot dog" folding techniques that confused the AI system.

the student was able to communicate their own frustration with the AI technology which could not create images using colloquial language and direction such as "hot dog" when describing how to fold the paper in half. In this figure, images and text on both pages were produced by AI and laid out by the student alongside original character illustrations. The project revealed several insights that go well beyond this specific assignment to bigger questions about how we teach design when AI is part of the mix:

Challenging Visual Stereotypes of "Machine Aesthetics": Duttweiler noticed that many students automatically reached for stereotypical "computer" looks when they were trying to show AI generated content, pixel fonts, rigid grid systems, mechanical imagery. This visual shorthand revealed something deeper about how they were thinking about AI as basically alien or futuristic, separate from human creativity instead of woven together with it.

Assessing Process Over Product: Instead of just grading the final designs, Duttweiler focused on how students were thinking about their work and their changing relationship with the technology. This was a significant shift from how design projects usually get evaluated. Students were assessed on their design decisions, sure, but also on how they were processing and reflecting on what it meant to work with AI.

Materiality as Conceptual Framework: The choice to use risograph printing turned out to be more meaningful than just an aesthetic decision. The two-color limitation forced students to figure out visual systems for distinguishing between human and AI content. The physical, slightly imperfect qualities of risograph printing created an interesting contrast with the often seamless, perfectly rendered look of digital AI outputs. What started as a practical constraint became a way of thinking about the differences between human and machine creation.

6.10 PEDAGOGICAL IMPLICATIONS

The "Something Old, Something New" project shows us what happens when we stop treating AI like either a miracle cure or an existential threat and start thinking about it more practically. Instead of setting up some grand battle between humans and machines, the assignment created space for students to figure out how they might work together. It's what you might call collaborative rather than competitive.

This fits with how we've always learned and created, really. Thinking has never happened in isolation; we've always used tools, systems, and cultural knowledge to extend what our minds can do. When you frame AI as just another one of these mediating technologies, powerful but not magical, it becomes less mysterious without losing its importance.

Duttweiler puts it well in his reflection: "I am reminded that creativity is not just about output. As we navigate a world where AI generated content is designed to produce more and more and faster, I remind my students that their true strength lies in their creative process. Their intuition, combined with an understanding of social, historical, and emotional contexts, is their secret power, something they can integrate with any technology."

This perspective resonates with nearly all the faculty we surveyed. Out of 36 respondents, 35 don't believe generative AI will eventually replace human visual communication designers. One faculty member captured it perfectly: "discernment is a human trait." That's the kind of thinking that can't be automated, no matter how sophisticated the technology gets.

Case Study 6.4: Prompt to Production: Teaching the AI Creative Pipeline

CONTEXT AND METHODOLOGY

The *Prompt to Production* workshop series was developed in response to the growing integration of AI tools into professional creative pipelines. Rather than approaching AI as a standalone novelty, Porter designed the series to position generative tools into a cohesive, multi-tool workflow, mirroring real-world production cycles in advertising, branding, and content creation. The pedagogical goal was deliberately structured to help students develop what Porter calls "AI art direction skills": the ability to guide GenAI with intention, precision, cohesion, and strategic oversight.

What set this approach apart was its emphasis on process over product. Students were not simply learning to work with AI tools; they were being trained to think across a production pipeline. The scaffolded workshops move from understanding how AI "sees" and interprets visuals to directing aesthetic consistency across multiple outputs, to integrating GenAI materials into video-based outputs.

The workshop series was structured as an escalating sequence of creative challenges, with each session building upon the skills developed in the previous one:

Workshop 1: Reverse Engineering AI to Create Better Images: This opening session introduced students to the concept of "AI visual literacy" through reverse-engineering exercises. Instead of beginning with prompts, students first analyzed how an LLM describes and interprets visual content.

In one particularly effective exercise, students were presented with an image (fruit and a smoothie on a table) and asked to describe it in their own words. These descriptions were then compared with an AI-generated description of the same image. The comparison revealed key differences between how humans and LLMs see and describe visual content, sparking class discussion about design language, detail, and mood.

Informed by this observation, Porter guided students in developing what he called an "Advanced Prompt Formula" that incorporated industry-standard design language:

- Adjective-Descriptive words
- Main Subject. What is the focus of the image?
- Action. If applicable, what is happening?
- Setting. Where is this taking place?
- Camera/Lens. What camera or film? What lens?
- Lighting/Mood. What lighting or feeling does this image convey?
- Depth of Field. How much blur?
- Composition/Perspective. Describe the layout (USE DESIGN LANGUAGE)
- Color Grading. What tones or palettes?
- Cinematic Style. General artistic direction
- Branding Aesthetic. What use case? Aesthetic style.

The formula breaks prompts into key components, giving students a structured yet flexible system for prompt clarity and creative control.

Workshop 2: Directing AI to Create Consistent Visual Campaigns: While the first workshop focused on prompt engineering, the second expanded students' understanding of how to achieve consistency and control in AI-generated imagery, a critical skill for professional applications.

Porter introduced students to what he termed the "GenAI Triangle of Influence" (Figure 6.7), a conceptual model illustrating the three primary factors that shape AI image generation:

- Composition (through uploaded sketches or reference images)
- Style (through uploaded reference images or preset options)
- Text Prompts (through detailed written descriptions)

Through a series of experiments, including a playful exercise bringing monster design sketches to life, students explored how varying the strength of each input influenced the outcome. As their sketches became designed monsters, a critical discovery emerged: only two of the three inputs could effectively dominate the

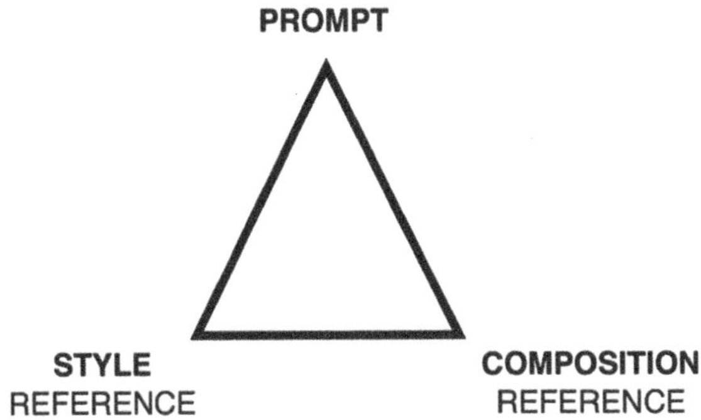

FIGURE 6.7 GenAI triangle of influence conceptual model. A simple triangular diagram illustrating the GenAI Triangle of Influence concept. The triangle has three vertices labeled "PROMPT" at the top, "STYLE REFERENCE" at the bottom left, and "COMPOSITION REFERENCE" at the bottom right, demonstrating the three key factors that influence AI image generation.

generation process at a time. The relationship between composition, style, and text prompts wasn't simply additive but required strategic balancing for effect control and consistency in GenAI image creation.

Figure 6.7 shows the GenAI Triangle of Influence. Of the three primary influences in GenAI image creation, only two can be used to effectively influence at a time. So, for example, if *Prompt* and *Style* are the main influences, *Composition* will be mostly ignored.

Workshop 3: Producing Social-Ready Video Content with AI: The final session integrated still image creation with video and motion design, reflecting industry trends toward short-form, multi-platform video content.

Students were guided through a comprehensive post-production workflow:

- **Asset Preparation**: Refining and upscaling images in Adobe Firefly and Photoshop using generative expand and correction tools.
- **Image-to-Video Generation**: Using these refined image assets as "first frames" in AI video generation tools, applying their prompt engineering skills to direct motion and camera behavior.
- **Post-Production Integration**: The GenAI video clips were then imported into traditional editing software, where final compositions were edited and refined combining traditional video skills with AI-assisted assets for final output (Figure 6.8).

This integrated approach reinforced a crucial lesson: AI tools are most effective when they're part of a thoughtfully designed creative process that includes both generative and traditional techniques. This approach mirrors professional applications like Coca-Cola's use of NVIDIA Omniverse and NVIDIA NIM microservices to scale global campaigns through their "Production Studio" system, which generates multilingual text, images, and videos adapted to over 100 markets (Mills, 2024).

FIGURE 6.8 Students' GenAI-assisted marketing materials examples. Two examples of student GenAI-assisted marketing campaigns. The top example shows a step-by-step breakdown of a "poppi" soda advertisement, progressing from original studio photograph to Photoshop composite, AI-generated splash effects, and final polished design with the tagline "soda, but better." The bottom example demonstrates the GenAI Triangle of Influence in practice, showing how composition reference (50% influence) and style reference (100% influence) affect the final output of a woman drinking water in workout clothing, with progression from reference images to the final GenAI output and video product.

6.11 KEY FINDINGS AND PEDAGOGICAL IMPLICATIONS

The project yielded several notable outcomes that extended beyond technical skill development:

Changing Relationship to Creative Tools: Students began to conceptualize AI not as a replacement for design thinking but as a collaborative material. One student noted: "It's like having a fast junior designer who needs very specific direction." This framing helped students maintain creative control while embracing the generative capacity of the tools.

Development of Strategic Thinking: Moving away from the overwhelming number of possibilities GenAI tools present, students began to distinguish between using GenAI for open-ended exploration versus precise execution; an important step toward becoming art directors rather than tool operators.

Recognition of AI's Current Limitations: Through the trial and error of experimentation, students identified common weak points in different AI systems, such as handling of complex hand positions, certain aesthetic styles, or cinematic camera logic. These insights fostered a healthy skepticism and improved evaluation skills.

Pipeline Integration Skills: What really matters is that students developed the ability to think in terms of creative pipelines rather than isolated tools. They learned to purposefully sequence AI and non-AI tools to achieve cohesive outcomes, recognizing that effective design in an AI augmented landscape requires a combination of old and new techniques.

In his reflection on the workshop series, Porter emphasizes that the goal was never to replace traditional design skills but to expand students' creative toolbox: "We're not training students to be prompt engineers, we're developing designers who can thoughtfully incorporate AI into their practice while maintaining a clear sense of creative control and ownership."

This approach reflects what industry experts are calling the emergence of a new job profile: the "AI designer" individuals who possess a unique skill set combining technical expertise in AI with creative design sensibilities, who are responsible for building, tuning, and managing AI-driven creative projects (Design Wizard, n.d.).

6.12 INDUSTRY PERSPECTIVES: LLMs TRANSFORMING MEDIA PRODUCTION

To understand how LLMs are transforming professional media workflows beyond the classroom, we interviewed Jay Rodge, a Senior Developer Advocate at NVIDIA who specializes in creating AI powered applications that utilize Generative AI models. His insights provide a valuable industry perspective on how LLMs are being integrated into commercial visual communication pipelines.

6.13 LLMs IN PROFESSIONAL CONTENT CREATION WORKFLOWS

Rodge describes himself as "an advanced user of large language models in my productivity setup, at my job, as well as in my personal life" (personal communication, February 5, 2025). In his role teaching developers how to optimize and accelerate LLM workflows, he regularly creates educational content including blog posts and tutorial videos.

His content creation process exemplifies how media professionals use LLMs to streamline workflows: "Often I have to create video briefs about what I'm going to talk about in my next video, which would be where I basically lay out the plan for what would be the video about, who's the target audience, what would be the length of the video, what are the topics that I would cover" (personal communication, February 5, 2025).

This practical application directly mirrors the "Creative Direction" phase of our AI-human design pipeline (Figure 6.1), where LLMs assist in establishing parameters for content development while maintaining human oversight of strategic objectives.

For enterprises producing marketing content, retrieval augmented generation (RAG) provides particularly powerful capabilities. As Rodge explains: "It is important because LLMs are trained on internet data and it might not have your specific enterprise or your company's data" (personal communication, February 5, 2025). He provides a concrete example of how RAG enables context-aware content creation for product launches in the cosmetics industry, where LLMs can access company-specific information to generate appropriate content.

6.14 TECHNICAL FOUNDATIONS: GPU ACCELERATION AND MODEL OPTIMIZATION

The performance improvements made possible by GPU acceleration are transforming what's possible in media workflows: "GPUs in general accelerate the whole pipeline. Because I've seen... people creating a storyboard where they just ask to generate a video brief and also create some images which represent what the video is going to be about" (personal communication, February 5, 2025).

This acceleration enables rapid iteration, a critical advantage in time-sensitive creative industries that aligns with the "Generation" phase of our AI-human design pipeline. With GPU acceleration, the speed of ideation increases exponentially compared with running models on CPUs.

One of the biggest technical challenges in optimizing LLMs for visual communication is speed, particularly with vision language models. To address these challenges, NVIDIA has developed specific software solutions like NVIDIA NIM™ and TensorRT-LLM that help optimize open source models with remarkable performance improvements, "If you run, a model a year ago versus the same model now on the same hardware, like you will see a performance boost without doing anything" (personal communication, February 5, 2025).

6.15 DEVELOPER TOOLS AND FRAMEWORKS FOR MEDIA PROFESSIONALS

Rodge highlights several developer tools and frameworks that are particularly effective for media professionals working with AI, including LM Studio for desktop applications, NVIDIA's ChatRTX for less technical users, and a testing environment at build.nvidia.com where developers can experiment with over 100 models in a playground setting.

6.16 THE FUTURE OF VISUAL COMMUNICATION

Looking ahead, Rodge sees LLMs changing the ideation and iteration processes for media professionals: "This would improve the workflows of the people who are working behind the scenes in creating these kinds of media assets. And what this would mean is they would have faster iteration times on what they are trying to communicate. So, they could have 5-7 ideas, running or working on them simultaneously and have different workflows" (personal communication, February 5, 2025).

The emergence of video language models promises even more change in the future, where clear prompts could generate 10-15 second clips as starting points for video production.

These insights from the industry align with what we observed in our classroom case studies. The collaborative design pipeline we documented is not merely an educational framework but reflects emerging professional practices in AI-augmented visual communication.

6.17 CONCLUSION: WHERE WE GO FROM HERE

Five years ago, freelance designers were competing for $500 logo jobs; now they're declining six-figure campaigns because she's booked solid, armed with nothing but her laptop and AI assistants that cost her less than her monthly coffee budget.

This transformation spans every creative industry. In rural Kenya, a filmmaker produces Hollywood quality visuals without studio backing. Teenagers in Lagos create animated series that rival major studio productions. What initially appeared to be job elimination has revealed itself as job evolution.

Integrating AI into creative workflows will require rethinking the division of tasks and the redistribution of creative labor, emphasizing collaboration instead of eliminating human involvement (Beckett & Yaseen, 2023; Xue & Huang, 2025). When used with intention, AI tools will support and enhance the creative process.

Yet even years after these tools become widely available, many creators remain caught between fear and curiosity, worried about job security while unsure how to bridge the knowledge gap. Since today's design students will become tomorrow's professionals, we surveyed both faculty and students to understand not only how this uncertainty is being addressed in the classroom but also whether educators themselves are using these tools and how they're integrating them into their teaching practices.

What we discovered reveals that the most effective approaches to bringing AI into design education go beyond simply adding some new software to the curriculum. Across the four case studies/classroom experiments, we've seen the same themes emerge, providing practical direction for design educators trying to professionally and carefully navigate this challenging technological shift.

Central to this approach is positioning students as critical collaborators instead of passive consumers of AI tools. When students are encouraged to evaluate what AI can do, compare machine outputs with their own knowledge, and make decisions about the use of these tools, something important happens. AI is no longer an intimidating, potentially threatening tool but a medium they can actually engage with creatively. The most effective approaches we found were focused on the process over the product. Emphasizing students' changing relationship with the technology, learning to refine prompts, and their critical reflections rather than focusing solely on the final product. This will prepare students for ongoing tech changes instead of trying to perfect their use of current systems.

Equally important is that effective AI teaching must directly address ethics. Cultural biases in generated imagery, questions of authorship and originality, and ethical considerations can't be separated from technical instruction. We must instill ethical awareness in every aspect of the curriculum to help students develop what might be called algorithmic critical consciousness.

This ethical foundation becomes even more crucial as we witness rapid technological advancement. Just four months ago, typography was a glaring weakness in AI-generated content, now LLMs like ChatGPT handle typography with near-perfect accuracy. Similarly, while AI videos were easily identifiable as artificial, tools like Google Veo 3 now produce footage that's very hard to distinguish from reality. This rapid advancement suggests these technical limitations will continue to diminish. We may see the development of entirely new tools for visual expression as AI-human collaboration evolves. AI tools have the potential to make design more accessible to people with diverse backgrounds and abilities, leading to a more inclusive and varied visual culture. The rise of virtual influencers and AI-augmented visual communication hints at entirely new categories of digital experiences becoming possible through AI-Human Collaboration. As our case studies show, effective integration requires careful ongoing consideration at every step regarding data privacy, copyright, ownership, and potential bias.

As these tools continue to evolve, one thing becomes clear: the creative professionals who will thrive are those who understand that their value lies not in competing with machines, but in directing them.

AI integration is redefining the creative process itself. By helping students embrace these emerging tools while maintaining the fundamentals of design, prioritizing process over product, and addressing ethical issues, design instructors can prepare the next generation of visual communicators. They will be able to adapt and actively shape their future while using AI in design. As we prepare students, we need to remember that their unique strength isn't their ability to write perfect prompts or navigate the latest tools. Their intuition, empathy, contextual understanding, and ethical reasoning make them irreplaceable. By helping students recognize and develop these capabilities, we can empower them to create visual communication that is innovative, meaningful, inclusive, and aligned with human values and needs.

ACKNOWLEDGMENTS

The author would like to acknowledge several individuals whose contributions made this chapter possible. Special thanks to Joshua Duttweiler and Jason Porter, the two faculty members whose innovative case studies are featured in this chapter, for their willingness to share their pedagogical approaches and insights. Gratitude to Dr. Phylis Johnson for her thoughtful review and editorial guidance throughout the development of this work. Thanks also to Julio Solar, who helped brainstorm the conceptual framework for Figure 6.1, the AI-human design pipeline. Their collective expertise and collaboration significantly enriched this research.

REFERENCES

Beckett, C., & Yaseen, M. (2023). *Generating Change: A Global Survey of What News Organisations Are Doing with Artificial Intelligence.* JournalismAI, Department of Media and Communications, The London School of Economics and Political Science. https://www.journalismai.info/s/Generating-Change-_-The Journalism-AI-report-_-English.pdf.

Breazu, P., & Katsos, N. (2024). ChatGPT-4 as a journalist: Whose perspectives is it reproducing? *Discourse & Society*, *35*(6), 687–707. https://doi.org/10.1177/09579265241251479.

Chen, L., Cai, Z., Jiang, Z., Luo, J., Sun, L., Childs, P., & Zuo, H. (2024). AskNatureNet: A divergent thinking tool based on bio-inspired design knowledge. *Advanced Engineering Informatics*, *62*, 102593. https://doi.org/10.1016/j.aei.2024.102593.

Design Wizard. (n.d.). How generative AI is affecting graphic design. *Design Wizard Blog*. https://designwizard.com/blog/how-generative-ai-is-affecting-graphic-design/.

Epstein, Z., Hertzmann, A., Herman, L., Mahari, R., Frank, M. R., Groh, M., Schroeder, H., Smith, A., Akten, M., Fjeld, J., Farid, H., Leach, N., Pentland, A., & Russakovsky, O. (2023). Art and the science of generative AI: A deeper dive. arXiv preprint arXiv:2306.04141.

Epstein, Z., Hertzmann, A., & Investigators of Human Creativity. (2023). Art and the science of generative AI: Understanding shifts in creative work will help guide AI's impact on the media ecosystem. *Science*, *380*(6650), 1110–1111. https://doi.org/10.1126/science.adh4451.

Epstein, Z., Smith, A., Cook, M., Schroeder, H., Gold, T., & Lippman, A. (2023). Art and the science of generative AI. *Science Perspectives*. https://doi.org/10.1126/science.adh4451.

Hwang, Y., & Wu, Y. (2025). Graphic design education in the era of text-to-image generation: Transitioning to contents creator. *International Journal of Art & Design Education*, *44*(1), 239–253. https://doi.org/10.1111/jade.12558.

metaLAB (at) Harvard. (2025). *AI Pedagogy Project*. Berkman Klein Center for Internet & Society. https://aipedagogy.org/.

Mills, J. (2024, July 29). Recipe for MAGIC: WPP and NVIDIA omniverse help the Coca-Cola Company scale generative AI content that pops with brand authenticity. *NVIDIA Blog*. https://blogs.nvidia.com/blog/coca-cola-wpp-omniverse-generative-ai/.

NVIDIA. (2025, January 10). AI gets real for retailers: 9 out of 10 retailers now adopting or piloting AI, latest NVIDIA survey finds. *NVIDIA Developer Blog*. https://developer.nvidia.com/blog/ai-gets-real-for retailers-9-out-of-10-retailers-now-adopting-or-piloting-ai-latest-nvidia-survey-finds/.

Park, H. (2025). Researching character design with generative AI: Focusing on creating five probiotic characters. In C. Stephanidis, M. Antona, S. Ntoa, & G. Salvendy (Eds.), *HCI International 2024—Late Breaking Posters* (p. 295). Springer, Switzerland.

Pattisall, J. (2024, November 19). Coca-Cola's AI generated holiday ads approach a creative tipping point. *Forrester*. https://www.forrester.com/blogs/coca-colas-ai-generated-holiday-ads-approach-acreative-tipping-point/.

Rhode Island School of Design. (2025). *Artificial Intelligence (AI)*. Teaching and Learning Lab. https://teachingandlearninglab.risd.edu/teaching-support/tech/ai.

Xue, Y., & Huang, S. (2025). *Household Dehumidifier Based on AHP and QFD*. In: C. Stephanidis et al. (Eds.): HCII 2024, CCIS 2320, pp. 161–173. Springer Nature Switzerland AG (2025).

7 Transforming Software Engineering and Software Acquisition with Large Language Models

John E. Robert, Ipek Ozkaya, and Douglas C. Schmidt

7.1 INTRODUCTION

We depend on software for our daily lives. Software forms key parts of mobile, transportation, healthcare, entertainment, and national defense systems. Software engineering—the discipline of creating, deploying, and maintaining the software that we depend upon—has similarly become critical to building trust and assurance in software through engineering rigor.

Creating and maintaining mission- or safety-critical software is hard. Engineers must identify, architect, and build systems that meet quality attribute requirements, such as safety, predictability, or reliability. Tests must be performed to ensure systems meet their functional requirements and security standards. All these activities are performed as part of the software development lifecycle (SDLC) that guides software engineers in orchestrating iterative processes of software development, including bug fixes, new capability updates, and verification before software deployment. Many of these steps are repetitive, error-prone, and time-consuming, which motivates software engineers to devise ways to automate key SDLC activities intelligently and reliably.

Artificial intelligence (AI) is both a construct of software engineering and a tool that can enable better software engineering. AI systems are built using software and data to create complex responses in many types of application domains, from product recommendation engines to self-driving cars. Applying AI concepts to the SDLC has been discussed for years (Barstow, 1987), with early examples applying machine learning (ML) to estimate software development around 1995 (Srinivasan and Fisher, 1995). The advent of powerful and accessible large language models (LLMs) has the potential of reshaping SDLC, with new opportunities and risks for software engineers (Ozkaya, 2023).

This chapter examines when—and how—LLMs can be applied effectively throughout the SDLC. It also explores the potential benefits and challenges of LLMs

DOI: 10.1201/9781003492252-7

in specific software engineering and software acquisition activities, such as development and testing, and evaluates their overall impact on the SDLC.

The remainder of this chapter is organized as follows: Section 7.2 discusses the expanding role of AI in the SDLC and shows how it is ushering in a new era of software engineering; Section 7.3 identifies key SDLC activities, opportunities, and risks for leveraging LLMs and summarizes recent research that examines the suitability of using LLMs in different SDLC activities; and Section 7.4 presents concluding remarks and outlines future research and advancements.

7.2 EXPLORING THE EXPANDING ROLE OF AI IN THE SDLC

The impact of AI on the SDLC has multiple dimensions. The discipline of software engineering has experienced several transformative changes in recent decades, ranging from the adoption of higher-order functional programming languages to advances in automation using DevSecOps tools and processes.[1] A consistent theme during these times of transformation is that it's often hard to foresee how far changes will go as the hope (and hype) meets the realities of the technical and programmatic limitations.

AI is influencing software engineering both in terms of replacing or augmenting capabilities with AI-augmented approaches and increasing intelligent automation to support software engineering activities. Figure 7.1 expands upon a vision presented in the book *Architecting the Future of Software Engineering: A National Agenda for Software Engineering Research & Development (Carleton et al., 2021).* This figure depicts key dimensions of delivering AI capabilities and intelligently automating the SDLC by applying AI augmentation in both system operations and software engineering activities, ranging from conventional to fully AI-augmented methods (Robert et al., 2024).

Each portion of Figure 7.1 is summarized below, starting with the lower-left quadrant.

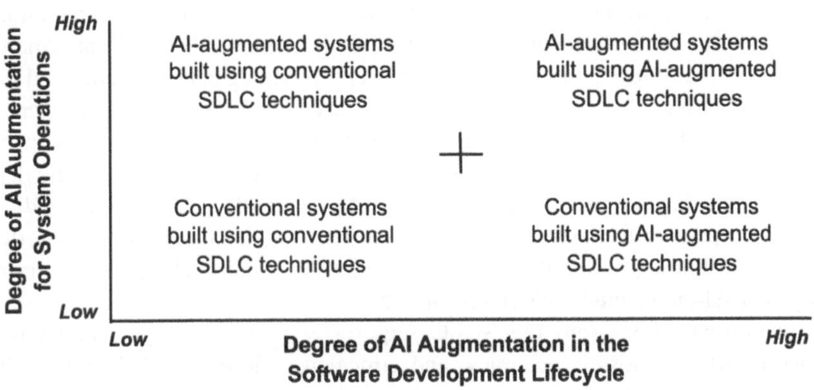

FIGURE 7.1 Taxonomy of artificial intelligence (AI) augmentation for system operations and software development lifecycle (SDLC) activities.

- **Conventional Systems Built Using Conventional SDLC Techniques**: This quadrant represents a low degree of AI augmentation for both system operations and the SDLC, which is the baseline for most software-reliant projects heretofore. An example is an avionics mission computing system that operates via distributed object computing middleware and rate monotonic scheduling (Harrison et al., 1997) and is developed using conventional SDLC processes without any AI-augmented tools or methods.
- **Conventional Systems Built Using AI-Augmented Techniques**: This quadrant represents an emerging area of research, development, and practice in the software engineering community, where system operations have a low degree of AI augmentation, but AI-augmented tools and methods are used in the SDLC. An example is a website hosting service where the content is not augmented by AI, but the development process employs AI-augmented code generators (such as GitHub Copilot), code review tools (such as Codiga), and/or testing tools (such as DiffBlue Cover).
- **AI-Augmented Systems Built Using Conventional SDLC Techniques**: This quadrant represents a high degree of AI augmentation in systems, especially in their runtime operations, but uses conventional methods in the SDLC. An example is a recommendation engine in an e-commerce platform that employs ML algorithms to personalize recommendations, but the software itself is developed, tested, and deployed using conventional Agile methods and the React.js and Node.js frameworks.
- **AI-Augmented Systems Built Using AI-Augmented Techniques**: This quadrant represents the pinnacle of AI augmentation, with a high degree of AI augmentation for both systems operations and the SDLC. An example is a self-driving automotive system that operates via ML algorithms for navigation and decision making while also using AI-driven code generators, code review and repair tools, unit test generation, and DevOps tools for software development, testing, and deployment.

Although the majority of SDLC activities today are performed in the lower-left quadrant (*i.e.*, conventional systems built using conventional SDLC techniques), the trend and innovation trajectory is toward the upper-right quadrant, where an AI-augmented SDLC is used to build and deploy AI-augmented operational capability. This trend has accelerated over the last decade, starting with increased reliance on AI-augmented capabilities as part of systems and recently with more AI-augmented tools coming to market and being applied to develop, test, and deploy software. In that context, however, a range of new challenges have emerged, such as the fast pace of technology changes, managing modifications to software development workflows, and understanding potential bias and mistakes in the training corpus of AI-augmented tools (Panyam, n.d.).

Integrating LLMs into the SDLC requires a measured approach, balancing concerns like disclosure, accuracy, and ethical use (Robert and Schmidt, 2023). Success hinges on developing organizational policies for these concerns and adapting to evolving governance and regulations. An empirical understanding of workflow alterations and data collection helps inform decisions about the success of new approaches. Moreover, traditional practices, such as code reviews with customized checklists, may even regain prominence, providing humans in the loop with tools and methods needed to accelerate the reliability and testability of code and systems developed by leveraging the assistance of LLMs.

7.3 APPLYING LLMS THROUGHOUT COMMON SDLC PHASES

A recent literature survey (Hou et al., 2024) on applying LLMs to software engineering reviewed 395 research papers from January 2017 to January 2024, categorizing where the applications of research are in applying LLMs in software engineering. Their findings indicate that 62% of the investigations focus on software implementation-related activities. Given our summary of opportunities for applications of LLMs to different activities across the SDLC, studying the limits of how to leverage LLMs is an ongoing research area with implications in tool and method development and application.

We need to assess the following when considering whether LLMs can assist with software engineering tasks to enhance SDLC processes that improve system capabilities:

1. *How LLMs can improve a single task*, focusing largely on doing something existing tools and engineers can do better, *e.g.*, code review or code completion,
2. *How applying LLMs on single or integrated task workflows can help consistency* with respect to other artifacts and activities related to that task and its inputs/outputs, *e.g.*, propagating a refactoring change consistently to the rest of a system or ensuring consistency of artifacts across multiple artifacts, and
3. *How applying LLMs may result in workflow shifts resulting in further improvements* by understanding the dependencies of a single task on related artifacts and task workflows, e.g., enabling different orchestrations of activities and a new SDLC augmented by LLMs.

As research and data on the outcomes of these applications continue to grow, we expect they will influence the development of new SDLC processes (Ozkaya, 2023).

This chapter focuses largely on the impact of LLMs on single SDLC activities, discusses implications on the dependencies of these tasks to other activities, and describes how LLMs can be applied to enable orchestration across multiple SDLC activities. Although applying LLMs in the SDLC involves AI performing some work traditionally done by people, many basic principles and practices remain relevant. In particular, many tools, techniques, and procedures used to ensure confidence in the validity and verification of software still apply, although LLMs and associated generative AI tools will now perform more of the workload.

The remainder of this section summarizes recent research that examines the application of LLMs in different software engineering tasks mapped to the SDLC phases shown in Figure 7.2 below. Understanding the limits of LLMs applied to software engineering tasks is a fast moving area of model development, tool development, and research. This section thus provides examples of ongoing work and is not intended as a comprehensive survey of the state of research on LLMs to date.

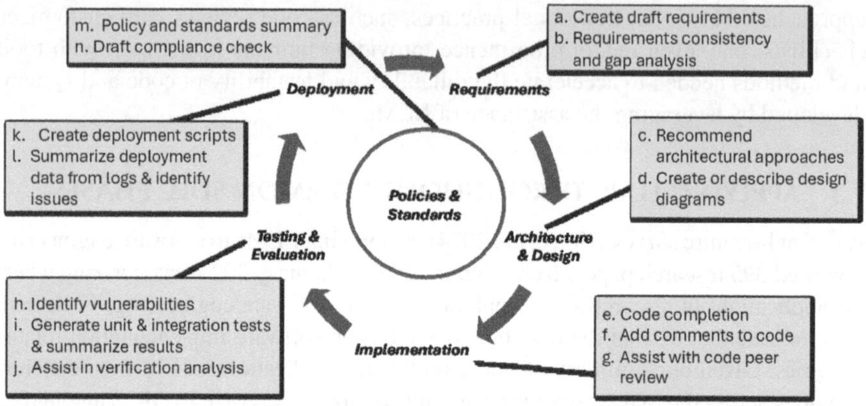

FIGURE 7.2 Opportunities to apply large language models (LLMs) to SDLC activities.

7.3.1 Assessing the Opportunities for Applying LLMs in SDLC Phases

Contemporary SDLC models are iterative, reflecting a continuous cycle of software professionals utilizing tools that enable greater automation. Figure 7.2 visualizes a high-level generalization of some common activities in SDLC phases, focusing on where LLMs can be—and are being—applied at each phase. While the orchestration of these activities may vary depending on what SDLC model a team uses, all software projects execute these activities in some order to deliver software.

We next describe each SDLC phase and summarize research on applying LLMs to improve specific activities by accelerating the efficiency of generation while not compromising on quality. An implicit assumption in applying LLMs to any activity, including software engineering activities, is that the LLM has been trained with the relevant data, enabling it to produce relevant outcomes. The relevance of the training corpus of an LLM will vary from domain to domain and activity to activity.

For example, systems that have been the subject of public discussion and debate (*e.g.*, patient enrollment websites associated with the Affordable Care Act in the USA) may have more data to draw from for LLM requirements. Conversely, activities that are not consistently represented by artifacts in the public domain may incur lower rates of LLM accuracy. As a case in point, while there are many good open-source software repositories, there are few examples of good open-source software architecture documents to train LLMs upon.

7.3.1.1 Requirements

Activities in the *Requirements* phase involve collecting, specifying, and understanding the needs, use cases, and constraints of a software-reliant system. During this phase, specifications are created that define what the software should do, including functional requirements (*e.g.*, features and functionality) and quality attribute requirements (*e.g.*, performance, security, and usability). These requirements can be documented in many formats, such as documents listing specifications, use cases describing expectations of the system, compliance and policy documents providing constraints, and/or diagrams describing expected interactions with users. This phase includes a significant amount of discovery and interaction with the key stakeholders and the analysis needed to ensure that requirements are clear, complete, and supported by stakeholders. These requirements are then prioritized with stakeholder feedback to ensure the most valued requirements are implemented and improved in subsequent development cycles.

Ongoing research investigates using LLMs to (1) elicit requirements, including examining requirements documents, feature requests, and user feedback, and (2) identify themes or areas of high importance. For example, an area of ongoing research elicits requirements by generating initial drafts of requirements based on project descriptions or stakeholder inputs (Abbas, 2023). Recent research (Tikayat Ray et al., 2023) has found that LLMs can accelerate requirements by converting requirements to machine-readable formats. Prompt engineering patterns have been identified to accelerate the elicitation of requirements (White et al., 2023) and assist in prioritization of requirements (Sami et al., 2024). LLMs can also be used to help identify contradictions in software specifications (Gärtner and Göhlich, 2024).

A detailed document called the *Software Requirement Specification* (SRS) is typically output from the *Requirements* phase, especially in regulated environments. Other outputs include iteratively identified use cases, formal specifications, and architecturally significant requirement documents. There are multiple elicitation and documentation activities that are expressed using natural language, which make LLMs well-suited to assist software engineers during the *Requirements* phase. As shown in Figure 7.2, some examples include the following:

a. **Create Draft Requirements or Use Cases**: LLMs can rapidly convert short text summaries of ideas or requests into more readable requirement descriptions. Prompt engineering can provide additional structure and guidance to create draft requirements statements that are relevant for the software domain and also structured in a desired format. LLMs can also generate alternative use cases that describe how software will interact with users or other systems. It is important to highlight that these will only be draft requirements because human software engineers who understand the software domain, standards, and risks should review, edit, and approve the final requirements. However, using LLMs to create draft requirements quickly—and in some cases with alternatives—can accelerate the requirements elicitation process, provide examples of edge scenarios, and accelerate the pace of iteration, resulting in improved draft documents during the *Requirements* phase. Moreover, LLMs can automatically generate use cases, scenarios, and requirement specifications from natural language inputs.
b. **Requirements Consistency and Gap Analysis**: LLMs can assist in interpreting and analyzing stakeholder inputs, analyzing an SRS for completeness or consistency, and/or converting these stakeholder inputs into formatted requirements documents through summarization. LLMs can help users identify common concepts across requirements or categorize requirements. They can also suggest requirements that may be missing or are incomplete by processing vast amounts of data, including previous project requirements, industry standards, policy documents, and best practices.

Leveraging LLMs throughout the *Requirements* phase can enable increasing the pace of idea generation and requirements elicitation. This approach, however, requires teams to understand their baseline and goals so that they can properly assess the quality of LLM outputs.

The *Requirements* phase informs the *Architecture and Design* phase since a system should be architected and designed to meet the requirements it must achieve.

7.3.1.2 Architecture and Design

The *Architecture and Design* phase defines and describes the structure, components, and interfaces of a system's software. This phase includes high-level architectural planning, architectural analysis, prototyping to compare trade-offs or assess response measures, and detailed design specifications. During this phase, engineers make architectural decisions to satisfy quality attribute requirements (such as performance and security) and perform experimentation or analysis to verify that the architecture and design perform as intended. These architectural decisions are captured in architecture design records, architectural diagrams, design diagrams, prototypes, and accompanying text that describes the architectural elements and interfaces.

The output of the *Architecture and Design* phase is a set of architecture decisions, diagrams detailing these decisions, and the descriptions of key constraints, requirements, and trade-offs that provide the context for the decisions. These artifacts are typically captured in a *Software Architecture Document* (SAD) that provides an overall description of the software architecture and each architecture element that needs further refinement, including its responsibilities and interfaces. Some teams apply lightweight documentation approaches and capture these details in diagrams or *architecture design records* (ADRs). Large organizations and highly-regulated projects may need to develop a *Detailed Design Document* that describes each component, as well as an *Interface Control Document* (ICD) that provides data exchange details for all interfaces.

There have not been many investigations to date in applying LLMs to support software architecture and design tasks broadly. The area seeing most attention is architecture knowledge generation (Ozkaya, 2023). An exploratory empirical study shows that while LLMs can document and formulate design decisions in the structure of ADRs for a given context, their quality is not commensurate to those generated by architects, although fine-tuning can improve their results (Dhar et al., 2024). Likewise, experiments with high-level architecture-relevant prompts show that software architects value the quick alternative generation ability of LLMs, but also consider trickier issues, such as copyright concerns and hallucinations, as critical barriers to adoption (Jahic and Sami, 2024).

There are also ongoing investigations to generate code from given images in the domain of control software. For example, (Koziolek and Koziolek, 2024) show how to generate IEC 61131-3 Structured Text control logic source code from piping-and-instrumentation diagrams with LLM-trained image recognition. As these capabilities mature, generating code from image recognition use cases can be generalized to other situations where code is generated from architecture diagrams, provided these diagrams have sufficient detail and formal semantics that an LLM can map to patterns of implementation details.

As shown in Figure 7.2, examples to leverage LLMs in the *Architecture and Design* phase include the following:

a. **Recommend Architectural Approaches**: A strength of LLMs is their access to general knowledge. A significant amount of information across domains and systems exists about common architecturally-significant requirements, such as performance, security, availability, and scalability, and experiences in implementing these requirements also exist. Using LLMs to brainstorm architectural design alternatives to given problems can thus improve a team's ability to analyze their designs and supplement their knowledge with existing patterns and tactical information. This analysis can take the form of providing sufficient context to an LLM and asking it to generate output relevant for similar systems, patterns, tactics, and technologies. Architects can use this LLM output to explore design spaces during the *Architecture and Design* phase.
b. **Create or Describe Design Diagrams**: LLMs can assist in generating diagrams of architecture documentation during the *Architecture and Design* phase. For example, LLMs can take high-level descriptions of system components, their relationships, and interactions as input and automatically generate visual representations, such as UML diagrams, flowcharts, and data flow diagrams. By interpreting the textual descriptions of the architecture and translating them into structured visual outputs, LLMs can facilitate refining existing diagrams by suggesting improvements based on patterns they learned from vast amounts of architecturally-relevant training data. This capability can enable teams to quickly prototype and iterate on designs. When diagrams are involved as part of the tasks, the underlying assumption is that an LLM is multi-modal, *i.e.*, it can work with images, video, and/or audio, in addition to text.

LLMs can also provide detailed descriptions and explanations of design diagrams created during the *Architecture and Design* phase. By analyzing the structural elements and relationships depicted in diagrams, LLMs can generate initial narratives that elucidate the purpose, functionality, and interactions of various components within a system. These descriptions can explain a system's modules, data flows, and interface connections, as well as how the design addresses specific requirements, such as scalability, security, and performance. Moreover, LLMs can identify potential patterns reflected in the diagrams, offering insight that can aid in understanding and refining the architecture. This capability makes LLMs valuable tools for documenting, summarizing, and communicating design concepts, assisting stakeholders in establishing an understanding of system software architectures.

The designs generated during the *Architecture and Design* phase should be reified during the *Implementation* phase to build the actual software, thereby ensuring it meets functional and quality attribute requirements, such as safety, security, reliability, and reusability.

7.3.1.3 Implementation

The *Implementation* phase involves coding and developing the software components according to requirements and within the constraints identified during the *Architecture and Design* phase. Programming is performed in this phase to create new software capability, fix previously identified bugs, and/or perform other modifications. This phase includes not just writing code, but also commenting the code, integrating the new code into revision control systems and DevSecOps pipelines, and conducting software quality control activities, such as software peer reviews and static/dynamic tool analysis.

Software developers are increasingly using LLMs to generate code, usually in response to prompts in a browser window. Moreover, coding assistants, such as GitHub Copilot and Amazon CodeWhisperer, are being merged with popular integrated development environments, such as IntelliJ, Android Studio, Visual Studio, and Eclipse. In many cases, creating code from prompts can increase developer productivity. Moreover, these AI code assistants include other capabilities, such as code refactoring to modify existing code. There is also ongoing work focused on investigating code transformations, translating code into different programming languages, programming language versions, and/or computing platforms (Pan et al., 2024).

As recent systematic literature reviews have demonstrated, code generation is where LLMs often excel, especially for popular programming languages like Java and Python. For example, LLMs have demonstrated effectiveness at auto code completion (Bird et al., 2023). LLMs can help developers learn new programming languages or frameworks by providing instant code examples and explanations (Lones, 2024). Likewise, LLMs enhance code documentation by generating summaries and explanations, making it easier for developers to understand and maintain codebases. A recent example (Geng et al., 2024) highlights the benefits of code comment generation, which is a common use case for software engineers. LLMs also enable easy and rapid creation of sample code or prototypes that contribute to learning and applying software concepts rapidly (Jiang et al., 2022). Other ongoing research includes creating reusable libraries for specific domains (Grand et al., 2024), supporting code reviews (Li et al., 2022), and refactoring code when integrated with static analysis tools (Pomian et al., 2024).

Common outputs from the *Implementation* phase include new source code files, code comments, and code commit history into version control systems, as well as code peer review or other quality control summaries. As shown in Figure 7.2, there are multiple opportunities to leverage LLMs in the *Implementation* phase, including the following:

a. **Code Completion**: Code completion is an activity that software engineers are utilizing LLMs for effectively to date. This activity leverages LLM strengths in pattern matching to suggest recommended ways for completing implementations as developers use IDEs. In such activities, the context scope is narrow and relevant data are available, which minimizes the likelihood of LLM hallucinations and other implementation mistakes.

b. **Adding Comments to Code**: LLMs can be used to add comments to software written by programmers, enhancing code readability and maintainability. By analyzing the structure and functionality of the code, LLMs can automatically generate meaningful comments that describe the purpose of methods and classes, the role of variables, and the logic behind specific code blocks. These comments can be tailored to various levels of detail, from high-level summaries of entire modules to line-by-line explanations of complex algorithms. Moreover, LLMs can be integrated into IDEs and development workflows to ensure comments remain up-to-date as code evolves by suggesting updates to comments after modifications. This automation reduces the burden on developers to document their code manually, enabling them to focus on writing efficient, high-quality software while ensuring the codebase remains well-documented for future developers and maintainers.

c. **Assist with Code Review and Bug Fixing**: LLMs can be applied to review code as a peer review assistant (Li et al., 2022). Recent LLMs have been trained extensively on valid code and can assist with bug fixing (Xia et al., 2023). LLMs can also summarize the results of multiple peer reviews (including those by humans) to identify trends or common issues from reviews. Likewise, LLMs can compare code commits to planned code additions or modifications to identify code changes.

The *Implementation* phase is followed by *Testing and Evaluation* activities, where the developed code is compared and assessed against relevant requirements and design specifications.

7.3.1.4 Testing and Evaluation

The *Testing and Evaluation* phase focuses on conducting activities to ensure (1) software meets its specified requirements and (2) implementations adhere to commonly agreed-upon standards to avoid mistakes resulting in unintended bugs and security issues. This phase includes multiple testing levels, such as unit, integration, system, and acceptance testing. Thorough testing is crucial for evolution and sustainment since it validates that the software is reliable and ready for deployment and ongoing maintenance. Testing commonly includes functional and non-functional aspects of the system, as well as user testing.

The ability of LLMs to convert natural language prompts into software testing activities reduces the manual effort required for these important tasks (J. Wang et al., 2023). For example, LLMs can also accelerate test automation and development of testing sequences (Liu et al., 2024a).

Outputs of the *Testing and Evaluation* activities typically include test plans, test cases or scenarios, and test reports that summarize all testing activities. Opportunities to leverage LLMs in the *Testing and Evaluation* phase have a similar structure to some activities discussed in the *Requirements* and *Implementation* phases, ranging from document generation and analysis to code generation in the form of unit tests. As shown in Figure 7.2, some examples include the following:

a. **Identify Vulnerabilities**: LLMs can review code and help humans identify possible vulnerabilities or insecure programming practices. Secure coding practices are documented (*SEI CERT Coding Standards*) and are always evolving, making it hard for humans to keep abreast of the latest practices. When given examples, LLMs can examine code to identify possible vulnerabilities and in multiple programming languages (Purba et al., 2023). Early LLMs, such as ChatGPT-3.5, demonstrated mixed results in detecting vulnerabilities (*Using ChatGPT to Analyze Your Code?*), but results continue to improve with good prompt engineering (Liu et al., 2024b) and with the latest improvements in LLMs (Sherman, 2024).
b. **Generating Unit and Integration Tests and Summarizing Results**: Applying LLMs to create test cases is a rapidly evolving area of research. These tools can enable interactions with software engineers and analysts to explore code interactively in ways relevant to testing activities, including asking for code summaries, checking compliance with coding standard(s), and/or exploring how code relates to specific considerations, such as safety, security, or performance (Freeman et al., 2025a, 2025b). Generation of unit tests is an activity with a known structure and limited context scope, which is a good fit for LLMs.
c. **Assist with Verification Analysis**: LLMs can help identify inconsistencies in verification analysis during the *Testing and Evaluation* phase, leveraging their strengths in document processing and summarization. Software engineers can accelerate their analysis of test cases, test results, and the corresponding requirements and design specifications by using summarization prompts. For example, LLMs can compare expected and actual test outcomes, potentially alerting to these discrepancies. Likewise, LLMs can analyze text in documentation and test reports to identify patterns that do not match, potentially detecting conflicting interpretations or ambiguities that might yield inconsistencies. After these gaps are identified, developers can also utilize LLMs to suggest possible corrective actions or refinements to both the code and test cases. Section 7.4 explores these capabilities in more detail and provides a case study example.

The *Testing and Evaluation* phase is followed by the *Deployment* phase, where tested code is released into operational environments and systems. DevSecOps tools and techniques connect SDLC testing and deployment activities into incremental and automated iterations that enable rapid, reliable, and repeatable releases of new software.

7.3.1.5 Deployment

The *Deployment* phase involves releasing software into operational systems. This phase also collects data related to operational effectiveness to inform maintenance and enhancements post-deployment, thus ensuring systems remain aligned with user needs and adapt to changing requirements. Moreover, the *Deployment* phase identifies bug fixes, updates, and new features to provide feedback from this phase to other phases for future software releases.

LLMs can play an important role in supporting software engineers during the *Deployment* phase, which is an area being increasingly automated. One example is program repair (Jin et al., 2023), which fixes issues found as deployment scripts are run to perform software updates. DevSecOps pipelines are used extensively across software development, operation, and management, and LLMs provide new opportunities to automate and scale software pipeline activities (*How to Put Generative AI to Work in Your DevSecOps Environment*, 2024). Early examples (Gurunathan, 2024) integrate LLMs into software repositories and include creating YAML scripts.

Artifacts created during the *Deployment* phase typically include deployment plans (such as rollback procedures, updated deployment scripts, and configuration files) and release notes documenting changes to software, as well as minimal requirements for hardware and software environments. As shown in Figure 7.2, multiple activities can leverage LLMs in the *Deployment* phase, including the following:

a. **Create Deployment Scripts**: This activity targets generating code relevant to deployment automation, which is another example of code generation where LLMs excel. Given specific requirements of a deployment environment as input prompts, LLMs can generate relevant automation scripts that support setting up servers, configuring environments, and deploying software components. These generated scripts handle a wide range of tasks, including database migrations, environment variable configurations, load balancing, and rollback procedures when deployments fail, all of which have existing patterns and exemplar reusable code that enable LLMs to generate robust code. When guided with structured prompts, LLMs can also adapt these scripts to different deployment platforms, such as cloud services, on-premises servers, or containerized environments, ensuring the deployment process is smooth and consistent across heterogeneous infrastructure. Similar to other code generation tasks, software engineers can apply LLMs to guide them through processes that optimize deployment scripts by identifying potential inefficiencies and suggesting improvements. For example, LLMs can reduce the time and effort to write scripts manually and minimize the risk of errors during deployment.

b. **Summarize Deployment Data from Logs and Identify Issues**: By processing large volumes of log data generated during deployment, LLMs can extract key information, such as successful operations, warnings, and errors. They can then generate concise summaries that highlight important events and patterns, making it easier for developers and operations teams to understand the overall status of a deployment. Likewise, LLMs can analyze log data to help detect anomalies, potential bottlenecks, or recurring issues that might indicate underlying problems. Employing LLMs in these summarization tasks can help humans examine large volumes of data and possibly identify problems more rapidly than conventional manual reviews, thereby enabling teams to take corrective actions swiftly and ensuring a smoother deployment process.

7.3.1.6 Policies and Standards

Policies and Standards have implications across all SDLC phases, as shown in Figure 7.2 (which places these artifacts at the center of the SDLC). These artifacts include software standards, such as coding or architectural standards, as well as standards that set expectations for quality attributes, such as safety, security, and reliability. Software must also adhere to policies established by organizations that develop software and may also include policies set by government regulators or acquisition programs. Compliance with software standards and policies is important for each phase of the SDLC, as well as for each iteration of software releases.

The powerful features provided by LLMs motivate a focus on generative AI capability models that characterize levels of functionality provided by LLMs to assist organizations in adapting to using generative AI services intentionally and avoiding their pitfalls. Much like software capability maturity models were used to assess the ability of organizations to create robust software solutions, LLM capability models are being proposed to help organizations measure their ability to create or deploy LLMs in their business processes effectively. Several capability models have been suggested, including models that describe how to leverage LLMs for applications (*Generative AI Capability Model*, n.d.), as well as models that explore the use of generative AI more broadly across organizations (*The MITRE AI Maturity Model and Organizational Assessment Tool Guide*, 2023).

Figure 7.2 shows the following ways that LLMs can be leveraged to create *Policies and Standards*:

a. **Policy and Standards Summary**: LLMs can be used to analyze policy and standards documents and provide summaries for humans (Bright et al., 2024; Robert and Schmidt, 2023). Many software engineering activities require conforming to governance and regulatory documentation across the SDLC. For example, software engineers perform many other tasks beyond coding, such as participating in regulatory compliance meetings, examining regulatory documents, or interacting with different industry or government standards stakeholders. These activities historically require humans to inspect and summarize reams of documentation manually. LLMs can help humans perform those activities more efficiently and accurately, as well as help improve the quality and efficiency of humans involved with government software acquisition activities and policies. LLMs can also be used for document summarization and decision support, though workflows must ensure that human judgement is central to—and designed into—these processes due to limitations with conventional LLMs.

b. **Draft Compliance Check**: The pattern-matching capabilities of LLMs enable them to process and understand large volumes of textual data. These capabilities can be leveraged to analyze software documentation, code, and design artifacts automatically and then ensure they adhere to established organizational policies and industry standards. These LLMs can compare the content of such artifacts against predefined requirements, such as coding standards, security protocols, or regulatory guidelines, to identify discrepancies or non-compliant elements. Using LLMs to draft compliance checks from predefined templates can help users focus on higher-value tasks, reduce the risk of missed compliance issues, and ensure that quality and regulatory requirements are consistently met throughout the development lifecycle. The example LLM activities shown in Figure 7.2 and described above are relevant to current and emerging software engineering use cases. However, leveraging the knowledge of experienced software engineers remains vital to avoid overreliance on generative AI tools that have not reached high levels of maturity and trust. What is new is the interactivity provided by LLMs that enables software engineers to explore answers to questions and iteratively develop solutions to common SDLC problems.

7.3.2 Assessing the Risks for Applying LLMs in SDLC Phases

Determining whether to apply LLM capabilities throughout the SDLC requires a framework for assessing the risks. One framework for assessing the risks of using an LLM in the SDLC appears in (Bellomo et al., 2023). Figure 7.3 visualizes that risk assessment framework using two dimensions: (1) the consequences of mistakes made by an LLM (from low to high on the vertical axis) and (2) the time and effort needed to detect LLM mistakes (from high to low on the horizontal axis). The application domains (such as medical devices, avionics control systems, autonomous vehicles, and nuclear power plant control systems) listed in the lower left-hand corner of the figure are classified as "high risk" because errors or failures in these systems can lead to catastrophic consequences, including loss of life, severe environmental damage, or large-scale safety hazards. These domains involve critical safety functions where

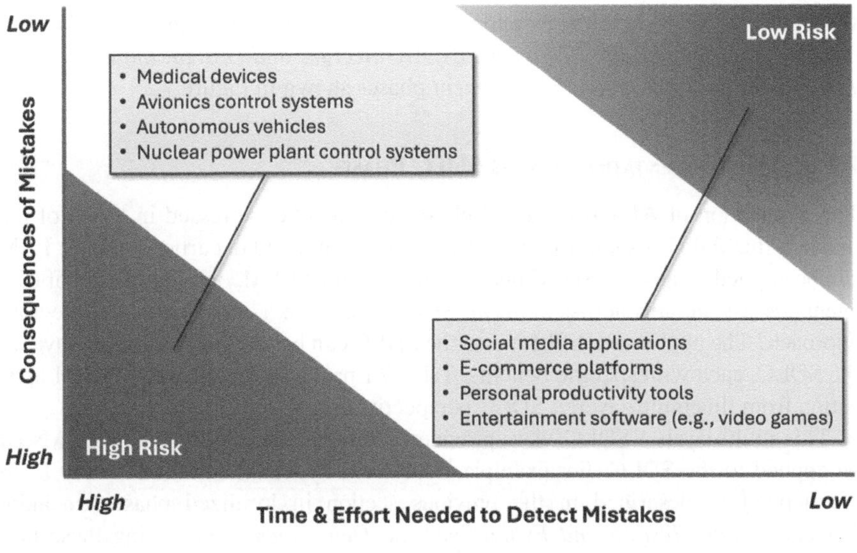

FIGURE 7.3 Evaluating risks associated with applying LLMs to some representative application domains.

precision, reliability, and security are paramount. Mistakes made by an LLM in these contexts could go undetected due to system complexity, making them particularly dangerous. This high risk is exacerbated since detecting and correcting errors in these environments often requires significant time and effort, given the intricacy of the systems, the potential for widespread impact, and additional compliance checks triggered during system development and/or operation.

In contrast, the application domains (such as social media applications, e-commerce platforms, personal productivity tools, and entertainment software) in the upper right-hand corner of the figure are considered "low risk" because mistakes made by LLMs in these domains typically have less significant consequences. For example, errors in these domains might lead to user inconvenience but are unlikely to harm individuals or environments. Moreover, mistakes in these applications are generally easier to detect and correct due to their lower complexity, as well as the ability to iterate and update relevant software rapidly. As a result, the time and effort required to identify and mitigate errors in these lower-risk domains are relatively small, making them more tolerant of mistakes generated by LLMs.

Given the assessment of risks described above, it is important to reiterate that humans are an essential part of using generative AI tools as part of any process and should not be replaced wholesale at this stage of their maturity. Moreover, given the nascent nature of the first generation of LLMs applied in software engineering, it is essential to have skilled software and systems engineers, as well as subject matter experts, who can spot where generated documentation or code is inaccurate and ensure that the key context is not lost. These human skills are important and necessary, even as generative AI tools continue to improve and provide significant new capabilities.

For example, today's LLMs that generate code have been trained on imperfect code from open-source repositories, such as GitHub and Stack Overflow. Not surprisingly, the code they generate may also be imperfect (*e.g.*, there may be defects and vulnerabilities). It is therefore essential to leverage human insight and oversight across the SDLC through all the *Requirements*, *Architecture and Design*, *Implementation*, *Testing and Evaluation*, and *Deployment* phases shown in Figure 7.2.

7.3.3 AI Augmentation across SDLC Phases

The application of AI across SDLC phases can also be expressed in terms of how much of the SDLC is supported with LLMs. As mentioned in earlier sections, LLMs can be applied to specific SDLC phases, but what if an LLM, or combination of LLM agents, assist humans across *multiple* SDLC phases in a coordinated or integrated approach? The application of LLMs to the SDLC can be viewed at different layers of the SDLC, each with specific benefits. Table 7.1 provides a summary of LLM application from this multi-layered SDLC perspective.

This multi-layered SDLC view provides additional insights into how LLMs can be applied to the SDLC. For example, consider a hypothetical SDLC where LLMs are applied as described in the previous section in localized phases, including *Implementation*, *Testing and Evaluation*, and *Deployment*. Connecting these three SDLC phases with another LLM to assist humans in optimizing across these activities is an example of a cross-functional layer, as described in Table 7.1. Extending this

TABLE 7.1
Layers of LLM Application to SDLC

SDLC Layer	Description
None	No use of LLMs in the SDLC
Localized	Application of LLMs to specific SDLC phase or phases, without coordination across phases
Cross-functional	Application of LLMs across multiple SDLC phases, with coordination across multiple phases to assist humans across multiple SDLC phases
Orchestration	Application of LLMs across all phases of the SDLC to assist humans in all aspects of the SDLC

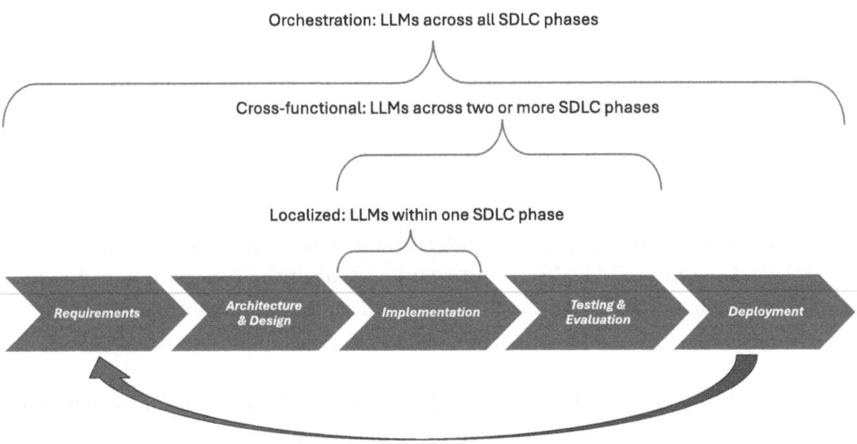

FIGURE 7.4 Layers of LLM application to SDLC.

concept, an LLM that assists humans across all phases of the SDLC is an example of orchestration, which enables broader insight and assistance to human software developers. By linearizing the iterative SDLC depicted in Figure 7.2 and adding the layers from Table 7.1, we can visualize the application of LLMs across the SDLC as shown in Figure 7.4.[2]

In addition, this multi-layered view guides what SDLC data would be needed to enable the use of LLMs across the SDLC. For example, an LLM that connects multiple SDLC phases, such as *Implementation* and *Deployment*, must have access to data across these SDLC phases. This dependency may require LLMs to understand different data types, such as code and testing results. It may also require prompt engineering from human developers on the most important SDLC aspects, such as minimizing software defects for safety-critical systems or performing fuzz testing to augment security analysis. These higher-level perspectives point to a future state with flexible and highly interactive human/LLM teaming that enables humans to prioritize the most critical tasks across the SDLC.

7.4 CONCLUDING REMARKS

This chapter highlighted opportunities to employ generative AI—specifically LLMs—in a wide range of SDLC activities. We explored the transformative potential of LLMs in software engineering, focusing on how these models can enhance various SDLC phases. We covered both the benefits and challenges of integrating LLMs (such as their ability to accelerate tasks like code generation, testing, and documentation) while also noting common LLM risks (such as generating incorrect or incomplete outputs).

Inserting LLMs into the SDLC for software teams can be viewed from multiple perspectives. There are many established frameworks for examining and enabling technology insertion or iterative improvements in software development activities, each offering different perspectives on adoption and integration. Some examples include the following:

- The Technology Acceptance Model focuses on usefulness and ease of use, helping developers assess whether a new tool will enhance workflow.
- Diffusion of Innovation Theory categorizes adopters into groups (innovators, early adopters, majority, and laggards) and explains how new technologies gradually gain acceptance.
- Agile Change Management aligns technology adoption with Agile workflows, ensuring seamless integration through iterative feedback and adaptation.
- The DevOps CALMS Framework (Culture, Automation, Lean, Measurement, Sharing) emphasizes continuous improvement, fostering collaboration and technical advancements in development practices.

Given the rapid pace of advancement of LLMs and their capabilities across the SDLC, an approach that focuses on incremental insertion in specific SDLC phases to accomplish a desired objective is a good approach. Starting with small teams or individuals can also be helpful and provide a foundation to explore specific risks and opportunities. For example, consider an organization that does software development in multiple systems or products, and a small team for a legacy product continues to be costly to sustain due to limited market and budget. This software context is a common example in industry and government organizations and forces teams to find higher efficiency in software updates to address bugs or new software security fixes.

In this example, LLMs might enable efficiency in the Implementation and Testing and Evaluation phases to enable timely (and correct) code updates followed by additional testing to ensure the updates are correct. This application of LLMs could take the form of engineers using a chat interface or using a software development tool specific to the activity, such as testing. In this example, a new workflow would be created, improved, and then potentially shared with other teams in the organization. As more LLM-based tools become available, we expect that some will target this type of software context and provide an overarching LLM Cross-Functional or even orchestration solution.

As we emphasized in this paper, many SDLC activities will require hybrid tools and humans with targeted support from generative AI to accomplish a shift in their effective execution. Accomplishing this transition successfully first requires understanding the nature of these activities and where LLMs can be applied. Software architecture is an example of such exploration; while some architectural activities enable effective incorporation of generative AI into SDLC workflows, others do not (Ivers and Ozkaya, 2025). Any first step in investigating where to start applying LLMs in the SDLC should start with understanding the decomposition of activities into tasks, aligning subtasks with the strengths of generative AI, and designing the new workflows with a collaborative approach that includes generative AI, other AI, additional automated tools, and human experts.

Many such explorations are already progressing rapidly in software engineering, and anecdotal and empirical research is emerging that evaluates the capabilities of available LLMs and codifies various success and failure examples. This research—including examples we describe in this paper—is crucial for effective use of LLMs as an effective partner in software engineering. The key gap, however, is our understanding of what being a useful partner may mean. Software engineers and LLMs accomplishing software-reliant system development can take many forms depending on the nature of the SDLC activities. Software engineering research must also make progress in exploring the forms of this partnership and provide exemplary effective approaches, as well as anti-patterns, to establish robust LLM-supported workflows.

ACKNOWLEDGMENTS

Thanks to the reviewers of this book chapter for providing suggestions that greatly improved its form and content.

NOTES

1 DevSecOps stands for “development, security, and operations” and it provides an integrated approach and toolset to deliver incremental software updates while considering the security implications throughout the update.

2 The arrow from *Deployment* to *Requirements* reflects the iterative nature of these activities, repeated through the SDLC.

REFERENCES

Abbas, A. (2023, June 12). 5 Ways LLMs Can Empower Software Engineering. *Techopedia.* https://www.techopedia.com/5-ways-llms-can-empower-software-engineering.

Barstow, D. (1987). Artificial intelligence and software engineering. In: *Proceedings of the 9th international conference on Software Engineering (ICSE ‘87)*, Monterey California USA, pp. 200-211, https://dl.acm.org/doi/10.5555/41765.41786

Bellomo, S., Zhang, S., Ivers, J., Cohen, J., & Ozkaya, I. (2023). *Assessing Opportunities for LLMs in Software Engineering and Acquisition.* Carnegie Mellon University Software Engineering Institute. https://doi.org/10.58012/m3hj-6w28.

Bird, C., Ford, D., Zimmermann, T., Forsgren, N., Kalliamvakou, E., Lowdermilk, T., & Gazit, I. (2023). Taking Flight with Copilot: Early insights and opportunities of AI-powered pair-programming tools. *Queue*, *20*(6), Pages 10:35-Pages 10:57. https://doi.org/10.1145/3582083

Bright, J., Enock, F. E., Esnaashari, S., Francis, J., Hashem, Y., & Morgan, D. (2024). *Generative AI is already widespread in the public sector* (No. arXiv:2401.01291). arXiv. https://doi.org/10.48550/arXiv.2401.01291

Carleton, A., Klein, M., Robert, J., & Harper, E. (2021). *Architecting the Future of Software Engineering: A National Agenda for Software Engineering Research & Development*. Software Engineering Institute. https://insights.sei.cmu.edu/documents/1308/2021_014_001_741195.pdf.

Dhar, R., Vaidhyanathan, K., & Varma, V. (2024). *Can LLMs Generate Architectural Design Decisions? -An Exploratory Empirical study* (No. arXiv:2403.01709). arXiv. https://doi.org/10.48550/arXiv.2403.01709

Freeman, L., Robert, J., & Wojton, H. (2025b) The impact of generative AI on test & evaluation: challenges and opportunities. In: *Proceedings of the International Workshop on Envisioning the AI-Augmented Software Development Lifecycle*, Trondheim, Norway.

Freeman, L., Schmidt, D. C., Bonnell, A., Robert, J., & Wojton, H. (2025a) *AIRC Panel on Generative AI in the Acquisition Lifecycle*. Acquisition Innovation Research Council. https://www.youtube.com/watch?v=ZlCc94w-2bY.

Gärtner, A. E., & Göhlich, D. (2024). Automated requirement contradiction detection through formal logic and LLMs. *Automated Software Engineering*, *31*(2), 49. https://doi.org/10.1007/s10515-024-00452-x.

Generative AI Capability Model. (n.d.). Retrieved September 13, 2024, from https://www.ibm.com/architectures/hybrid/genai-capability-model.

Geng, M., Wang, S., Dong, D., Wang, H., Li, G., Jin, Z., Mao, X., & Liao, X. (2024). *Large Language Models are Few-Shot Summarizers: Multi-Intent Comment Generation via In-Context Learning*. 453–465. https://www.computer.org/csdl/proceedings-article/icse/2024/021700a453/1WDJaRUZRKg.

Grand, G., Wong, L., Bowers, M., Olausson, T. X., Liu, M., Tenenbaum, J. B., & Andreas, J. (2024). LILO: Learning Interpretable Libraries by Compressing and Documenting Code (No. arXiv:2310.19791, Version 4). https://doi.org/10.48550/arXiv.2310.19791.

Gurunathan, S. (2024, January 30). *Using LLMs to Automate Pipeline Conversions from Legacy to Tekton*. https://devops.com/using-llms-to-automate-pipeline-conversions-from-legacy-to-tekton/.

Hou, X., Zhao, Y., Liu, Y., Yang, Z., Wang, K., Li L., Luo, X. Lo, D., Grundy, J. and Wang, H. (2024). Large Language Models for Software Engineering: A Systematic Literature Review. ACM Trans. Softw. Eng. Methodol. 33, 8, Article 220 (November 2024), 79 pages. https://doi.org/10.1145/3695988

Harrison, T. H., Levine, D. L., & Schmidt, D. C. (1997). The design and performance of a real-time CORBA event service. *SIGPLAN Not*, 32(10), 184–200. https://doi.org/10.1145/263700.263734.

Ivers, J., & Ozkaya, I. (2025). Will generative AI fill the automation gap in software architecting? In: *New and Emerging Ideas, Proceedings of the 22nd International Conference on Software Architecture (ICSA 2025)*, Odense, Denmark, 2025, Odense, Denmark.

Jasmin Jahic & Ashkan Sami (2024) State of Practice: LLMs in Software Engineering and Software Architecture. In: *Proceedings of the 2024 IEEE 21st International Conference on Software Architecture Companion, Hyderabad, India*, pp. 311-318, https://doi.org/10.1109/ICSA-C63560.2024.00059

Jiang, E., Olson, K., Toh, E., Molina, A., Donsbach, A., Terry, M., & Cai, C. J. (2022). PromptMaker: Prompt-based Prototyping with Large Language Models. *Extended Abstracts of the 2022 CHI Conference on Human Factors in Computing Systems*, New Orleans LA USA, pp. 1–8. https://doi.org/10.1145/3491101.3503564

Jin, M., Shahriar, S., Tufano, M., Shi, X., Lu, S., Sundaresan, N., & Svyatkovskiy, A. (2023). InferFix: end-to-end program repair with LLMs. In: *Proceedings of the 31st*

ACM Joint European Software Engineering Conference and Symposium on the Foundations of Software Engineering, Lisbon, Portugal, pp. 1646–1656. https://doi.org/10.1145/3611643.3613892.

Koziolek, H., & Koziolek, A., (2024). LLM-based control code generation using image recognition. In: *Proceedings of ICSE 2024 (LLM4Code Track)*, Lisbon, Portugal. ACM. doi:10.1145/3643795.3648385.

Li, L., Yang, L., Jiang, H., Yan, J., Luo, T., Hua, Z., Liang, G., & Zuo, C. (2022). AUGER: automatically generating review comments with pre-training models. In: *Proceedings of the 30th ACM Joint European Software Engineering Conference and Symposium on the Foundations of Software Engineering*, Barcelona, Spain, pp. 1009–1021. https://doi.org/10.1145/3540250.3549099.

Liu, Z., Chen, C., Wang, J., Chen, M., Wu, B., Che, X., Wang, D., & Wang, Q. (2024a). Make LLM a testing expert: bringing human-like interaction to mobile GUI testing via functionality-aware decisions. In: *Proceedings of the IEEE/ACM 46th International Conference on Software Engineering*, Lisbon, Portugal, pp. 1–13. https://doi.org/10.1145/3597503.3639180.

Liu, Z., Yang, Z., & Liao, Q. (2024b). Exploration on prompting LLM with code-specific information for vulnerability detection. In: *2024 IEEE International Conference on Software Services Engineering (SSE)*, Beijing, China, pp. 273–281. doi:10.1109/SSE62657.2024.00049.

Lones, M. (2024, March 20). LLMs and the Future of Programming Languages. *Substack Newsletter*. https://fetchdecodeexecute.substack.com/p/llms-and-the-future-of-programming.

Ozkaya, I. (2023). Can architecture knowledge guide software development with generative AI? *IEEE Software*, 40(5), 4–8.

Panyam, S. (n.d.). *Council Post: Software Engineering Challenges In The Age Of AI: The Role Of Strong Engineering Leadership*. Forbes. Retrieved September 2, 2024, from https://www.forbes.com/councils/forbestechcouncil/2024/06/24/software-engineering-challenges-in-the-age-of-ai-the-role-of-strong-engineering-leadership/

Pan, R., Ibrahimzada, A. R., Krishna, R., Sankar, D., Wassi, L. P., Merler, M., Sobolev, B., Pavuluri, R., Sinha, S., & Jabbarvand, R. (2024). Lost in translation: a study of bugs introduced by large language models while translating code. In: *Proceedings of the IEEE/ACM 46th International Conference on Software Engineering (ICSE 2024)*, Lisbon, Portugal. ACM. doi:10.1145/3597503.3639226.

Pomian, D., Bellur, A., Dilhara, M., Kurbatova, Z., Bogomolov, E., Sokolov, A., Bryksin, T., and Dig, D. (2024). EM-Assist: Safe Automated ExtractMethod Refactoring with LLMs. In: *Companion Proceedings of the 32nd ACM International Conference on the Foundations of Software Engineering (FSE 2024)*, Porto de Galinhas Brazil, pp.582–586. https://doi.org/10.1145/3663529.3663803

Purba, M. D., Ghosh, A., Radford, B. J., & Chu, B. (2023). Software vulnerability detection using large language models. In: *2023 IEEE 34th International Symposium on Software Reliability Engineering Workshops (ISSREW)*, Pasadena, CA, USA, pp. 112–119.

Robert, J., Ivers, J., Schmidt, D. C., Ozkaya, I., & Zhang, S. (2024). The Future of Software Engineering and Acquisition with Generative AI. *Crosstalk, AI Taming the Beast*, pp. 26–43. https://community.apan.org/wg/crosstalk/m/documents/464157.

Robert, J., & Schmidt, D. (2023). Generative AI Q&A: Applications in Software Engineering. https://insights.sei.cmu.edu/blog/generative-ai-question-and-answer-applications-in-software-engineering/.

Sami, M. A., Rasheed, Z., Waseem, M., Zhang, Z., Herda, T., & Abrahamsson, P. (2024). Prioritizing Software Requirements Using Large Language Models (No. arXiv:2405.01564). https://doi.org/10.48550/arXiv.2405.01564.

SEI CERT Coding Standards—CERT Secure Coding—Confluence. (n.d.). https://en.wikipedia.org/wiki/CERT_Coding_Standards (Retrieved October 23, 2024).

Sherman, M. (2024). Review of Advances in Using LLMs for Securing Software. https://app.swapcard.com/widget/event/ai4-2024/planning/UGxhbm5pbmdfMTkwMzA2NA==.
Srinivasan, K., & Fisher, D. (1995). Machine learning approaches to estimating software development effort. *IEEE Transactions on Software Engineering*, 21(2), 126–137. https://doi.org/10.1109/32.345828.
The MITRE AI Maturity Model and Organizational Assessment Tool Guide. (2023). https://www.mitre.org/news-insights/publication/mitre-ai-maturity-model-and-organizational-assessment-tool-guide.
Tikayat Ray, A., Cole, B., Pinon Fischer, O., Bhat, A. P., White, R., & Mavris, D. (2023). Agile methodology for the standardization of engineering requirements using large language models. *Systems*, 11, 352. https://doi.org/10.3390/systems11070352.
Using ChatGPT to Analyze Your Code? Not So Fast. (2024, February 12). https://insights.sei.cmu.edu/blog/using-chatgpt-to-analyze-your-code-not-so-fast/.
Wang, J., Huang, Y., Chen, C., Liu, Z., Wang, S., & Wang, Q. (2023). Software Testing with Large Language Model: Survey, Landscape, and Vision (No. arXiv:2307.07221). https://arxiv.org/abs/2307.07221.
White, J., Fu, Q., Hays, S., Sandborn, M., Olea, C., Gilbert, H., Elnashar, A., Spencer-Smith, J., & Schmidt, D. C. (2023). A Prompt Pattern Catalog to Enhance Prompt Engineering with ChatGPT (No. arXiv:2302.11382). https://doi.org/10.48550/arXiv.2302.11382.
Xia, C. S., Wei, Y., & Zhang, L. (2023). Automated program repair in the era of large pre-trained language models. In: *Proceedings of the 45th International Conference on Software Engineering (ICSE 2023)*, Melbourne, Australia, pp. 1482–1494. IEEE.

8 Detecting and Correcting Software Assurance Document *Incompleteness, Inconsistencies, and Discrepancies Using LLMs*

John E. Robert, Ipek Ozkaya, and Douglas C. Schmidt

8.1 INTRODUCTION

Traditional manual methods of verifying consistency between code and documentation are time-consuming and do not scale up to larger software-reliant systems. Although large language models (LLMs) are not consistency checkers, they can assist in checking for patterns given a set of artifacts and enough examples in the training data. Throughout this chapter, we use a drone avionics case study to discuss how LLMs may assist with assessing compliance of code with mandatory regulatory standards codified in documents. To advance research on this topic, we introduce a novel framework that augments LLMs with advanced data structures to automate the detection of *document incompleteness*, *inconsistencies*, and *discrepancies* (DIID) in software-related artifacts across the software development life cycle (SDLC). DIID detection is useful for software developed for high-stakes domains to ensure relevant artifacts (e.g., requirements, architecture, design, code, and tests) conform to regulatory requirements (such as FDA, FAA, and DoD standards).

Incompleteness, inconsistencies, and/or discrepancies between regulatory documents and project-specific documents are often indicators of potential non-conformance or risk. Addressing these issues today requires engineering teams to conduct extensive, tedious, and error-prone manual investigation or data clarification. Incompleteness could indicate that a regulatory requirement is not addressed or does not apply to a system, or it could be a simple omission. Inconsistencies and discrepancies can occur between regulatory documents as emerging technologies yield new standards and best practices that may be different from—or even conflict with—previous or overlapping regulations. While incompleteness, discrepancies, and inconsistencies may be fine in some contexts, *i.e.*, they are not always incorrect or problematic, in other cases, they

DOI: 10.1201/9781003492252-8

indicate the need for additional investigation. In most cases today, however, identifying these issues is a tedious, error-prone, and human-intensive activity.

Software engineering teams in high-stakes domains are under mounting pressure to navigate the extensive maze of regulations and manage increasingly complex software systems without compromising on quality or safety. This challenge is compounded by the need to accelerate engineering and deployment processes to maintain a competitive edge against adversaries. It is therefore crucial for these teams to employ innovative tools that enhance efficiency through automation, particularly for frequent and time-intensive tasks, such as identifying inconsistencies and discrepancies in documents.

Detecting inconsistencies and discrepancies within and between documents is a necessary and important first step. This process is not the same, however, as recommending solutions or resolving these issues automatically. Assessing whether LLMs fit a specific use case requires weighing risks of using an LLM against its benefits (Bellomo et al., 2023).

Given the mission- and safety-critical nature of government-acquired software systems, any automation should ensure that trained people remain central to the activities and confirm the safety and accuracy of the software engineering processes. This research area thus involves augmenting people in the software engineering activities primarily by *automating* DIID detection. Once DIID is detected, future work can develop solutions for corrective action. Entirely automating the resolution of DIID via LLMs is not currently viable given the level of technology maturity and potential risks.

This chapter focuses on DIID detection in the software domain. However, the challenges and importance of DIID detection apply to many domains, including law, medicine, government compliance, and education. Leveraging the ability of LLMs to examine extensive documents or datasets depends on accurate DIID detection to extend their benefits beyond merely assisting humans in highly specialized tasks or summarization. In software engineering and the SDLC, DIID detection is an important activity for scaling LLM use by improving consistency checking across artifacts, both within and across SDLC activities.

The remainder of this chapter is organized as follows: Section 8.2 describes the DIID challenges in standards and policy documents, including providing formal definitions for incompleteness, inconsistency, and discrepancy to help identify and summarize relevant research. Related work on DIID detection is evaluated in Section 8.3. A representative software engineering case study is presented in Section 8.4 to demonstrate traceability from the generalized software engineering policies, standards, and guidance documents to DoD system-specific software documents and information. Section 8.5 presents concluding remarks and outlines future research and advancements.

8.2 DETECTING INCOMPLETENESS, INCONSISTENCIES, AND DISCREPANCIES IN STANDARDS AND POLICY DOCUMENTS

SDLC activities must often adhere to national and/or international regulatory requirements to ensure security, safety, and other important properties in high-stakes domains, such as healthcare, automotive, and aerospace. For example, all medical device software in the United States must follow Food and Drug

Administration (FDA) software standards. Likewise, software in cars and trucks must follow regulatory standards established by the National Highway Traffic Safety Administration (NHTSA). Moreover, software in airplanes and drones must comply with standards set by the Federal Aviation Administration (FAA). These software standards are described across multiple (often large) documents, ranging from directives or standards that require compliance to more general guidance that recommends best practices or common approaches.

Completeness and consistency between regulatory documents and project-specific software documents or artifacts should be confirmed by software developers and regulatory authorities and re-verified throughout the SDLC in high-stakes domains. This section explores techniques for detecting DIID. Here, we explore whether generative artificial intelligence (AI) technologies in general—and LLMs in particular—offer new opportunities to automate DIID detection.

All DIID categories indicate potential problems with the content, but they differ in their specific implications. In this work, we use the following definitions for these concepts:

- *Incompleteness* refers to when important context or terms are missing within or between documents. For example, if a software safety policy document states the software architecture must provide a detailed analysis of *safety-critical* signals, but a derived statement requires the software architecture to just provide a detailed analysis of *signals*, the "safety-critical" context is both important and missing. Incompleteness indicates that some requirements or analysis are missing or that the system context does not warrant specific requirements. In either case, additional human review and analysis are recommended to rectify incompleteness to reduce risk.
- An *inconsistency* refers to contradictions, lack of agreement/uniformity, or cohesion within a single document or across multiple documents. For example, safety and security are used synonymously (or with significant overlap) in some domains, while in other domains, they imply two completely different sets of requirements. Inconsistencies may not indicate a significant error but can create confusion for readers and reviewers. In addition, such inconsistencies may also suggest incompleteness.
- A *discrepancy* refers to a lack of compatibility or a divergence within or between documents related to facts that are either *direct* (*e.g.*, factual) or *derived* (*e.g.*, policy, narrative, or theoretical). For example, different response measures (0.01 sec versus 0.1 sec) listed for the same functionality in different components in a safety-critical system can result in fatal failures in software. Discrepancies can indicate a more significant error or issue, particularly if the information is used for confirmation or decisions.

These definitions are typical of what is found in dictionaries and literature on human interpretation or document analysis. Although there is no comprehensive list of different types of DIID, common types and examples are summarized in Table 8.1 below.

TABLE 8.1
Common Types of Incompleteness, Inconsistencies, and Discrepancies

Incompleteness	Inconsistency	Discrepancy
• **Incomplete**: Important context or terms are missing.	• **Terminology**: Using different terms interchangeably without clear definitions or consistency. • **Structural**: Lack of uniform structure in presenting information.	• **Factual Discrepancies**: Conflicting factual information. • **Policy or Procedural Discrepancies**: Deviations from established protocols. • **Narrative Discrepancies**: Different user stories fail to align. • **Theoretical Discrepancies**: Actual results conflict with theoretical predictions.

8.3 RELATED WORK ON DIID DETECTION

Related work to advance DIID detection can be grouped into document analysis, prompt engineering and prompt patterns, and human analysis, as discussed below.

1. **Document Analysis**: Document analysis research is a relevant area of related work for DIID detection in software engineering. This analysis is a common topic of interest across many domains, including healthcare (Shokrollahi et al., 2023; Moilanen et al., 2022), financial or business environments (Cao et al., 2024; Han et al., 2023; Shukla et al., 2023), and the analysis of legal documents (Bauer et al., 2023; Deroy et al., 2023; Prasad et al., 2024). Document analysis is used in most of these domains both to provide summaries and insights into data for a given domain. Some approaches include building domain-specific models (Wu et al., 2023), though other approaches use existing models and attempt to address document scale and establish effectiveness measures.

   A primary area of research for DIID detection is a response to the issue of generated document summaries being inconsistent with the original document (Laban et al., 2023; Lattimer et al., 2023; Luo et al., 2023; Tang et al., 2023). These discussions confirm the importance of consistency in document summarization and analysis. These reports also highlight the risks of using these systems in production use cases, including a medical domain example that underscores serious risks from inconsistencies (Tang et al., 2023).

   A recent paper proposes a benchmark called SUMMAC that includes proposed measurement and inconsistency detection datasets to address inconsistencies in summaries (Laban et al., 2021). Inconsistency checks between an original document and a summarized document are a simpler use case and differ from DIID detection across multiple and different documents. However, the discussions on evaluation and testing datasets are promising and were assessed as part of our DIID detection work.

Another related area of research is detecting hallucinations in LLM-generated output, not limited only to document summaries (Fallah et al., 2024; Rawte et al., 2024). To address this challenge, some solutions are white-box approaches that require detection mechanisms within the model (Zhu et al., 2024), though this approach is not suitable for DIID detection in software engineering for average users. For example, LLM fine-tuning approaches can refine LLM responses with additional data, but may alter the LLM responses in unintended ways. Black-box solutions for DIID detection (*i.e.*, where there is no requirement to look inside an LLM) are much easier for users to apply, but are limited to constraints of the LLM developers, such as being limited to smaller documents.

A limitation in current document analysis approaches is the evaluation of "long" documents (Cao et al., 2024; Pradeep et al., 2023). Although the definition of "long document" is subjective, there are multiple summarization successes for short text, such as news summaries (Barta et al., 2024; Goyal et al., 2023). Proposed approaches for analyzing larger documents include chunking (Schwaber-Cohen, n.d.), architectural approaches using Retrieval-Augmented Generation (RAG) to enable LLM integration with large datasets (Gao et al., 2024; Ke et al., 2024; *Retrieval Augmented Generation (RAG) for LLMs—Nextra*, n.d.), multi-agent-based approaches (Zhao et al., 2024), fine-tuning or modification of the models to adjust sequence length (Chen et al., 2023; Ding et al., 2023), building new models (Cohan et al., 2018), and hybrid approaches that give LLMs access to metadata about documents (Saad-Falcon et al., 2023). New testing frameworks and datasets are also necessary to assess their effectiveness on long documents (C. Wang et al., 2024).

Of these multiple approaches to support long documents or larger datasets, RAG-based and agent-based approaches are particularly applicable to DIID due to their architectural benefits. For example, key architectural qualities needed for DIID in software engineering projects are (1) *scalability*, which supports larger and larger datasets, and (2) *extensibility*, which supports more data types over time, beyond documents. RAG-based and agent-based approaches provide this flexibility, although RAG retrieval accuracy is dependent upon multiple factors, including the data storage approach (Zhang et al., 2024).

2. **Prompt Engineering and Prompt Patterns**: The flexibility of LLMs to support a wide range of uses effectively is supported in part by prompt engineering, which involves structured interactions with—and programming of—LLM computational systems to solve complex problems via natural language interfaces (Liu et al., 2023). Prompt patterns (White, Fu et al., 2023) codify best practices for phrasing prompts to maximize extraction accuracy and provide knowledge transfer mechanisms to problem-solve

with LLMs more accurately. Prompt patterns also enable more effective and repeatable performance of LLMs, and many patterns have been identified for a range of task objectives (White, Hays et al., 2023).

Prompt engineering is a key capability that provides new opportunities for human interactivity with documents and data for DIID detection in software engineering. For example, the *Context Manager* pattern (White, Fu et al., 2023) directs an LLM to set the context. Prompt engineering, including identification and refinement of relevant prompt patterns, is thus central to our DIID research. When DIID types are narrowly defined and examples are provided to LLMs through fine-tuning or prompt engineering, DIID detection can be improved for specific use cases (Hegselmann et al., 2024).

3. **Human Analysis**: The primary means for performing DIID detection is currently people, so considering research beyond the field of computer science helps characterize DIID's current effectiveness and limitations. One recent example (Schoor et al., 2023) conducted an experiment with ~160 college students who read two short texts, some with consistent information and others with inconsistent information, and then wrote a short essay. This experiment explored reading behaviors and outcomes, and it highlighted the following points relevant for software acquisition:
   - When people encounter discrepancies in information, their "attention to sources" increased. This finding is relevant to software acquisition because the discussion also indicated that context and document sources are important in DIID detection.
   - Users of any solution will require more information about the sources behind the finding before accepting—or perhaps even considering—any suggestion from an AI-augmented tool about DIID detection.
4. **Testing and Evaluation Frameworks**: Testing frameworks that can assess LLM strengths and weaknesses for detecting DIID are needed to provide a consistent means of assessing the many available LLMs and versions. A proposed testing protocol for DIID for document summaries (Laban et al., 2023) is promising because it provides tests specific to DIID and benchmarks for recent LLMs. This work provides a useful DIID detection test framework and dataset because it considers multiple discrepancy examples across multiple LLMs and compares the results of a few benchmarks. It also highlights the importance—and many challenges—of creating an effective testing framework, particularly because accurate labeling of test data is hard to scale. However, their approach—like previous approaches—is limited to only a subset of the DIID detection needed for software acquisition and does not cover all the types shown in Table 8.1.

   A related new area of research is generative information retrieval (Gen-IR), which focuses on building more effective testing of generative AI-augmented systems for use cases like document summarization and grounded answer generation (Arabzadeh & Clarke, 2024; Bénédict et al., 2023). This research area is promising since it provides testing datasets and baselines that may be relevant to DIID detection. However, the testing bias of prompts and the limited scaling of human validation remain a challenge for this research area.

8.4 A SOFTWARE ENGINEERING SAFETY CASE STUDY

The case study below provides additional context and illustrates DIID detection concepts in a research and development project being conducted at the Carnegie Mellon University (CMU) Software Engineering Institute (CMU/SEI). This case study involves a software team developing control software for a drone that must meet system and software safety requirements defined in government regulatory policies and guidance. In addition, the software team is applying industry standards, such as software safety analysis and secure coding, to enhance confidence in the software. These regulatory documents provide a starting point for examining DIID detection.

Figure 8.1 provides a software safety example, where it is important to detect DIID at every step between the policies, standards, and guidance documents (shown in the upper left) tracing through to the safety verification documents (shown in the lower right). DIID considerations are related to the case study summarized below, with examples for each circled letter in Figure 8.1:

A. **Regulatory Documents**: DIID can occur between multiple policy, standards, and guidance documents. For example, a software safety standard states that safety requirements traceability must be performed and must follow a specific format, but a separate organizational safety policy states that the need for a software safety requirements traceability is flexible and says nothing about requiring a specific format.
B. **Safety Requirements**: DIID can occur between regulatory documents and safety requirements documents that list safety requirements derived from the regulatory documents that are relevant to a specific system. For example, an engineering team identifies a set of 24 safety requirements, but one of the safety requirements states that a Failure Modes, Effects & Criticality Analysis (FMECA) (Gargama and Chaturvedi, 2011) must be performed and always report any fault condition, but another regulatory document which also requires FMECA requires the system to always return to a safe state, not just report the fault.

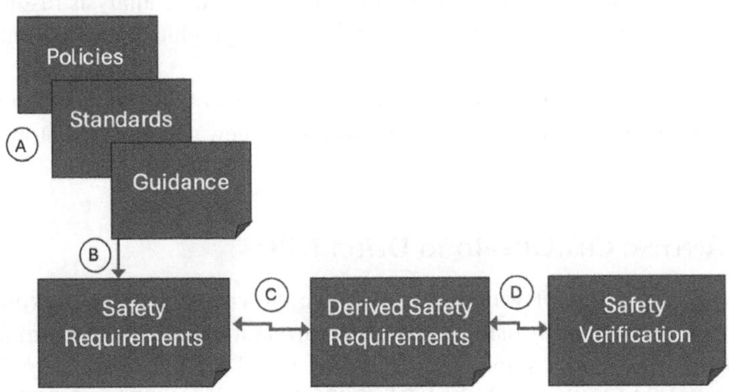

FIGURE 8.1 A software engineering safety case study.

C. **Derived Safety Requirements**: DIID can occur between safety requirements documents and derived safety requirements that extend the safety requirements from the top level to every software module or component in the system. For example, in a software system with seven software modules, five of the modules have derived safety requirements that limit the use of dynamic memory allocation, but two modules have no such requirement.
D. **Safety Verification**: DIID detection can occur with safety verification documents that summarize verification evidence (such as test results, analysis results, etc.) when compared to the derived safety requirements. For example, a derived safety requirement states that a software module that processes three types of data messages must always reject improperly formatted messages (such as messages with potentially corrupted data). However, the safety verification information for that software module indicates that only one of the three message types was tested for rejecting improperly formatted messages.

The example shown in Figure 8.1 highlights the following four themes:

1. DIID detection requires **analysis of multiple large documents**. For example, for step A, the set of documents can be numerous (10+ documents), and many of the documents are 50 pages or more.
2. DIID detection can require **consideration of system-specific context** or information. For example, for step B, the regulatory documents are general for any system, but the safety requirements are specific to one system.
3. DIID detection is an important consideration at every step, but sometimes at **different levels of abstraction**. For example, for step C, the safety requirements are a summary for the entire software system and may have 10–20 general software safety requirements, but the derived safety requirements are detailed for all software system components and will be a much larger list.
4. DIID detection can include comparison of **different types of information**. For example, for step D, the derived safety requirements are detailed requirements, but the safety verification information could be analysis results, test results, or other supporting evidence, including graphical information.

These themes provide insight into common limitations of current DIID approaches and suggest opportunities for using LLMs as part of new solutions to automate DIID detection.

8.4.1 Applying ChatGPT-4o to Detect DIID

We now build upon the software safety example shown in Figure 8.1 to explore DIID detection more thoroughly using a National Aeronautics and Space Administration (NASA) software safety standard called the "SOFTWARE ASSURANCE AND SOFTWARE SAFETY STANDARD" (2022 NASA-STD-8739.8B). This NASA software safety standard is a publicly available, 65-page PDF file that defines "the requirements to implement a systematic approach to Software Assurance (SA), software safety, and Independent Verification and Validation (IV&V) for software created, acquired, provided, or maintained by or for NASA." It includes topics on software safety requirements and independent verification and validation that are similar to many government and commercial software safety documents.

For our case study, this NASA software standard is part of circled letter A in Figure 8.1. It describes software safety requirements that must be supported. In a large software-reliant acquisition program, this document would be one of many standards that the software must support, but we will just use one standard for this example.

With the software safety standard defined for circle A in Figure 8.1, consider a list of Safety Requirements in circle B. Requirements of this type are commonly expressed in word processing documents, spreadsheets, or a requirements management tool. There might be dozens of general safety requirement statements that derive from the standards in circle A. For example, a requirement from the NASA standard could include the following requirement from page 27 of the standard: "*The software detects inadvertent memory modification and recovers to a known safe state.*" This software safety requirement is derived directly from the standard and represents a top-level requirement of the software.

Continuing to trace through the steps in Figure 8.1, the Derived Safety Requirements in circle C are more specific decompositions of the requirements in circle B to specific parts of the system. Examining the memory requirement, there are multiple cases of potential "inadvertent memory modification" that should be considered. Some examples include:

- Detecting accidental memory corruption from a software defect or bug
- Detecting memory modifications from a malicious virus that is running on the system and possibly changing memory
- Detecting memory modifications on the heap, the stack, or both.

For our case study example, we consider the system as a drone or vehicle that must be resilient to accidental memory corruption but assert that malicious viruses will be prevented via cybersecurity practices and network protection mechanisms. With this system context, we identify the following Derived Safety Requirements for circle C:

- The software must detect accidental heap memory corruption from a software defect or bug.
- The software must return to a known safe state when accidental heap memory modification is detected.

Tracing to the last step in Figure 8.2, the items in circle D are safety verification statements from the system/software developer that provide evidence that software safety requirements are met. Documents in this circle D group include architecture documents, design documents, code, and software test results. For example, a design

I will provide a list of statements, from software requirements and development documents, and I want to detect where the statements are inconsistent, incomplete, or there are discrepancies as compared to the NASA 8739.8 software standard. Please be specific in your responses and provide supporting rationale for the answers.

The Safety Requirements for the software include the following statement:

- The software detects inadvertent memory modification and recovers to a known safe state.

The Derived Requirements for the software include the following statements:

- The software must detect accidental heap memory corruption from a software defect or bug
- The software must return to a known safe state when accidental heap memory modification is detected.

The Safety Verification documents from the software developer including the following statement:

- The software design includes a "System Monitor' component that detects heap memory modifications and, when detected, returns the system to a known safe state.

Analyze these statements as compared to the NASA 8739.8 standard and identify where the statements are inconsistent, incomplete, or there are discrepancies as compared to the NASA 8739.8 software standard.

FIGURE 8.2 ChatGPT-4o prompt to detect document incompleteness, inconsistencies, and discrepancies (DIID).

document provided by software developers might have the following statement about the software:

- The software design includes a "System Monitor" component that detects heap memory modifications and, when detected, returns the system to a known safe state.

Using the information from each stage of Figure 8.2, we applied ChatGPT-4o to detect DIID via a browser interface. In particular, we uploaded the NASA 8739.8 PDF file and included the text prompt shown in Figure 8.2 to ChatGPT-4o.

Within seconds, ChatGPT-4o provided the informative response shown in Figure 8.3.

ChatGPT-4o's response in Figure 8.3 shows how it first analyzed and compared the NASA standard to each of the statements from each stage shown in Figure 8.1, including Safety Requirements, Derived Requirements, and Safety Verification Documents. In each stage, ChatGPT-4o provided a summary about DIID along with some comments. For example, in the Derived Safety Requirements, ChatGPT-4o identified the statements "address specific scenarios of memory corruption and recovery" and then states that the requirements "narrow the scope" compared to the NASA standard. These statements are correct, and the ChatGPT-4o response is the type of insight that can aid system and software engineers reviewing these statements.

To assess the provided statements against the NASA-STD-8739.8 Software Assurance and Software Safety Standard, let's examine the consistency, completeness, and alignment of each statement.

Safety Requirements

- "The software detects inadvertent memory modification and recovers to a known safe state."

This aligns with the NASA-STD-8739.8 safety-critical software requirements that mandate software must:

- Detect, report, and take corrective action if the system reaches a potentially hazardous state.
- Initialize to a known safe state on startup or after restarts.

The statement meets the standard broadly but lacks explicit details about the mechanisms for detection and recovery.

Derived Requirements

1. "The software must detect accidental heap memory corruption from a software defect or bug."
2. "The software must return to a known safe state when accidental heap memory modification is detected."

Both derived requirements address specific scenarios of memory corruption and recovery. However:

- These requirements narrow the scope to heap memory corruption, potentially excluding other types of inadvertent memory modifications (e.g., stack or global memory) that could lead to unsafe states. The standard requires a broader coverage of detecting any hazardous memory condition.

Safety Verification Document

- "The software design includes a 'System Monitor' component that detects heap memory modifications and, when detected, returns the system to a known safe state."

This statement aligns well with the derived requirements but does not fully encompass the broader safety requirement. The focus is on heap memory modifications, which may not cover all potential inadvertent memory issues. Furthermore:

- There is no explicit mention of how the "System Monitor" achieves these actions, which is necessary for verification against the standard's emphasis on traceability and hazard mitigation.

Identified Inconsistencies and Gaps

1. Incompleteness of Scope:
   The safety requirement covers all inadvertent memory modifications, but the derived requirements and verification statements only address heap memory. This discrepancy could lead to insufficient coverage of other critical memory areas.
2. Verification Specificity:
   The verification statement describes a system monitor but lacks details about testing and validation, which are required to ensure compliance with the standard's rigorous safety verification processes.
3. Traceability and Testing:
   The derived requirements and safety verification document should include bi-directional traceability to show alignment with safety-critical requirements. This traceability ensures completeness in identifying and mitigating all hazards associated with memory modifications.

FIGURE 8.3 ChatGPT4o DIID detection response.

In addition to the response above, ChatGPT-4o also provided the recommendations shown in Figure 8.4. Not only are these recommendations accurate, but they also provide keen insights into the level of evidence and data that provides confidence in software safety, at least with respect to memory corruption. These types of recommendations can help software developers and certification authorities provide a data-driven and independent view of software safety information.

Recommendations

1. Expand the derived requirements to include all types of inadvertent memory modifications (heap, stack, and global memory) to align with the broader scope of the safety requirement.
2. Provide detailed descriptions in the safety verification document about how the "System Monitor" functions and include validation methodologies.
3. Ensure traceability between the safety requirement, derived requirements, and verification steps to satisfy the traceability criteria outlined in NASA-STD-8739.8.
4. Include specific test cases and results for memory corruption scenarios, demonstrating the system's capability to detect and recover from various types of memory issues.

By addressing these gaps, the requirements and verification processes will better align with NASA's software safety standards. Let me know if you'd like to refine any of these points or need additional guidance.

FIGURE 8.4 ChatGPT-4o DIID recommendations.

Current methods of detecting DIID are time-intensive and require humans to both find and analyze these statements at each stage to ensure software safety. This safety requirement example depicted in Figure 8.1 shows how ChatGPT-4o can assist humans in detecting DIID quickly in standards and policy documents for additional analysis. LLMs are becoming more capable and accurate, and this example demonstrates the vital role LLMs will have in assisting software engineers. The analysts, who are assumed to be experts in safety-critical requirements and regulations, can use this information to locate related areas in documents and rectify them.

This example demonstrates how LLMs can detect DIID and provide useful assistance to human software engineers. However, it also hints at areas of caution for software engineering and the essential role of experienced software engineers. For example, while ChatGPT-4o's recommendations are accurate compared to the NASA standard, human engineers should decide if some requirements do not apply to the system context.

This case study example also applies LLMs to areas that play to the strengths of LLMs. For example, determining what the decomposed software safety requirements should be for a specific system context is an important analysis that should be driven by humans with a review process, and using LLMs could be high risk if the LLM is incorrect in generating the requirements. However, using LLMs to detect DIID has lower risk because the LLMs are analyzing information we provide, which yields better and more accurate LLM responses.

The potential time savings and impact of using LLMs for DIID detection can be best understood by considering the software development context and expanding on the use case. DIID detection can be applied in multiple dimensions to extend and expand the approach, as shown in Figure 8.5, which provides a matrix for understanding the complexity of identifying DIID in the context of software policies or standards relative to software development artifacts. The vertical axis, labeled "Policy Scale," reflects the number of policy or standard documents being examined. At the top of the scale, the task involves identifying DIID across multiple policy documents, which inherently requires a more comprehensive review. In contrast, the bottom portion of the scale focuses on DIID detection within a single policy document, narrowing the scope of analysis, which often simplifies the process.

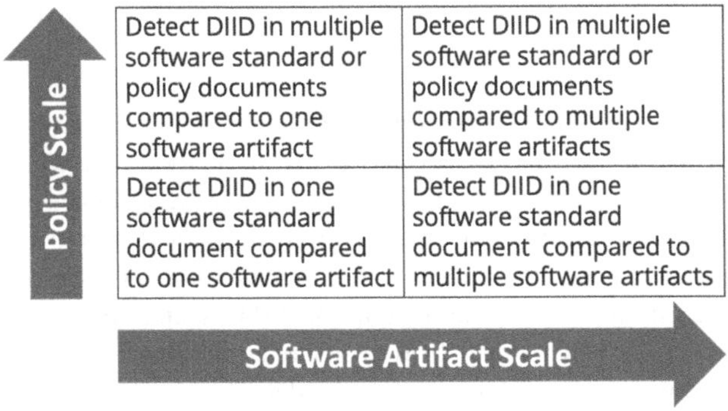

FIGURE 8.5 Scaling DIID detection.

The horizontal axis of the figure, labeled "Software Artifact Scale," indicates the number of software artifacts involved in the comparison. As we move towards the right, the scale increases to multiple software artifacts being compared against the policy or standard documents. The more complex scenario is shown in the upper-right quadrant of the matrix; it requires a broader analysis of multiple documents and artifacts, potentially revealing more extensive DIID issues. In the lower-left quadrant, the analysis focuses on comparing a single policy document with a single software artifact, which is a less complex but still important comparison. Together, these axes define the scope of the DIID analysis, with the scale increasing both vertically (more policies to compare) and horizontally (more artifacts to assess), leading to a more detailed and expansive investigation of DIID across the SDLC.

As the scale of DIID detection increases, the estimated time savings from AI-assisted detection become more significant. In smaller-scale scenarios, such as comparing a single policy document to one software artifact, LLMs can quickly identify inconsistencies and discrepancies, potentially reducing the manual effort required. As the scope expands, however, the time savings increase significantly, particularly when multiple policy documents and software artifacts are involved.

LLMs can process large volumes of data, cross-reference multiple documents and artifacts simultaneously, and flag discrepancies at a faster rate than human analysts. With each increase in both policy and software artifact scale, AI-assisted systems reduce the overall time needed for human detection, enabling more time for humans to perform deeper analysis to assess and resolve the identified DIID. The ability of LLMs to handle these complex, large-scale comparisons not only accelerates the DIID detection process but also ensures a higher level of accuracy and consistency, making them an invaluable tool as the scale of the analysis expands.

8.5 CONCLUDING REMARKS

This chapter explored using LLMs to improve software verification and compliance with regulatory standards. Using a software safety case study, we explored how the opportunities and risks of applying LLMs in the SDLC can be addressed and integrated into new SDLC workflows. We demonstrated using LLMs to perform DIID detection in software safety requirements and artifacts in the context of a NASA software safety standard.

The following are key observations and ongoing areas of study in conducting the DIID analysis work described in this chapter:

- **LLM-Driven DIID Detection Can Help Ensure Compliance in High-Stakes Domains**: LLMs are maturing to the point where they can be applied to detect and rectify DIID in software-reliant systems. DIID detection is increasingly important in high-stakes (and thus often regulated) domains, such as healthcare, aerospace, and defense. This chapter provided a foundation for understanding the capabilities and future directions of DIID detection, which are critical to ensure compliance and minimize risks associated with non-conformance to regulations.
- **Leveraging Vector and Agent-based LLM Architectures Enhances Document Analysis Efficiency**: The emerging role of LLMs augmented with vector databases and/or agent-based architectures has demonstrated promising results that enhance the efficiency and accuracy of DIID detection. By leveraging semantic similarity across texts, LLMs offer an innovative approach to handling the expansive and complex documentation inherent in government projects. Moreover, the introduction of prompt engineering as a means to customize LLM interactions via prompt patterns (White, Fu, et al., 2023) enables newly tailored automation, aligning closely with the specific requirements and frameworks of software acquisition.
- **Collaboration between Developers and LLMs is Essential for Effective Software Engineering**: It is unrealistic to expect today's LLMs to generate complete and flawless software-reliant systems from scratch. Instead, they should be viewed through the lens of generative *augmented* intelligence (AI+), where developers work together with AI-augmented tools. We are applying this type of collaboration in our research, teaching, and programming (e.g., by working hand-in-hand with LLMs like ChatGPT, Gemini, and Claude), but we do not expect them to generate all software artifacts. Instead, we do much of the design, decomposition, and some of the implementation tasks, and then have LLMs augment us and our teammates with tasks otherwise tedious and/or error-prone to perform manually. We thus use LLMs to *supplement* our skills, rather than to *supplant* them, recognizing that the effective use of LLMs requires a combination of prompt engineering and human oversight to maximize benefits and minimize risks.

Our work to date provides a basis for additional research in the following areas that extend the potential benefits for software acquisition and potentially into other domains:

- **Recommending Resolutions**: Our future work will focus on the semantic and contextual information used in DIID detection to suggest relevant recommendations for resolving identified issues. We also expect prompt engineering will provide humans with effective methods and tools to gain insight into specific contexts and codify prompt patterns to resolve

DIID detections. Future research should also include testing frameworks and benchmarks that characterize and enable validation of provided DIID recommendations.

- **Extending into AI-Augmented Software Engineering**: The example shown in Figure 8.1 can be extended beyond safety verification to include inconsistency checks against various software engineering artifacts, for example to assist with requirements elicitation. The application of LLMs to software engineering and related artifacts is an active area of research (Carleton et al., 2021, 2023; Ozkaya et al., 2023).
- **Ethical Considerations**: Bias and fairness are known challenges with LLMs, so future research should consider the impact of these issues in the context of SDLC processes. Research concerning fairness in document summarization (Zhang et al., 2023) provides insights on how to characterize and possibly mitigate these issues.

Looking ahead, the opportunity for further research on AI-augmented DIID detection is both extensive and enticing. The potential to expand the scope of LLM applications to encompass recommendation systems for resolving DIID detections or to integrate LLMs more deeply into other stages of software engineering underscores the potential impact of this research. Further research is needed to refine the accuracy of detection mechanisms, enhance the utility of LLMs for non-specialist users, and extend the capabilities to more complex and diverse document sets. By streamlining software acquisition processes, these advances can safeguard the integrity and efficacy of mission- and safety-critical applications and operations in high-stakes domains.

ACKNOWLEDGMENTS

Thanks to the reviewers of this book chapter for providing suggestions that greatly improved its form and content. The authors also appreciate and acknowledge the use of OpenAI tools in the preparation of this chapter. These tools were employed to support language refinement, organization, and clarity of presentation.

REFERENCES

Arabzadeh, N., & Clarke, C. L. A. (2024). *A Comparison of Methods for Evaluating Generative IR* (No. arXiv:2404.04044). https://arxiv.org/abs/2404.04044.

Barta, B., Lakatos, D., Nagy, A., Nyist, M. K., & Ács, J. (2024). *From News to Summaries: Building a Hungarian Corpus for Extractive and Abstractive Summarization* (No. arXiv:2404.03555). arXiv. https://doi.org/10.48550/arXiv.2404.03555.

Bauer, E., Stammbach, D., Gu, N., & Ash, E. (2023). *Legal Extractive Summarization of U.S. Court Opinions* (No. arXiv:2305.08428). arXiv. https://arxiv.org/abs/2305.08428.

Bellomo, S., Zhang, S., Ivers, J., Cohen, J. & Ozkaya, I. (2023). *Assessing Opportunities for LLMs in Software Engineering and Acquisition.* Carnegie Mellon University Software Engineering Institute. https://doi.org/10.58012/m3hj-6w28.

Bénédict, G., Zhang, R., & Metzler, D. (2023). *Gen-IR @ SIGIR 2023: The First Workshop on Generative Information Retrieval* (No. arXiv:2306.02887). arXiv. https://arxiv.org/abs/2306.02887.

Cao, T., Raman, N., Dervovic, D., & Tan, C. (2024). *Characterizing Multimodal Long-form Summarization: A Case Study on Financial Reports* (No. arXiv:2404.06162). arXiv. https://doi.org/10.48550/arXiv.2404.06162.

Carleton, A., Klein, M., Robert, J., & Harper, E. (2021). *Architecting the Future of Software Engineering: A National Agenda for Software Engineering Research & Development.* Software Engineering Institute. https://insights.sei.cmu.edu/documents/1308/2021_014_001_741195.pdf.

Carleton, A., Shull, F., Schmidt, D. C., Harper, E., Robert, J., & Ozkaya, I. (2023, August 25). *U.S. Leadership in Software Engineering & AI Engineering: Critical Needs & Priorities Workshop - Executive Summary.* https://insights.sei.cmu.edu/library/us-leadership-in-software-engineering-ai-engineering-critical-needs-priorities-workshop-executive-summary/.

Chen, S., Wong, S., Chen, L., & Tian, Y. (2023). *Extending Context Window of Large Language Models via Positional Interpolation* (No. arXiv:2306.15595). arXiv. https://doi.org/10.48550/arXiv.2306.15595.

Cohan, A., Dernoncourt, F., Kim, D. S., Bui, T., Kim, S., Chang, W., & Goharian, N. (2018). A Discourse-Aware Attention Model for Abstractive Summarization of Long Documents. In: *Proceedings of the 2018 Conference of the North American Chapter of the Association for Computational Linguistics: Human Language Technologies, Short Papers,* New Orleans, LA, USA, Vol. 2, pp. 615–621. https://doi.org/10.18653/v1/N18-2097.

Deroy, A., Ghosh, K., & Ghosh, S. (2023). *How Ready are Pre-trained Abstractive Models and LLMs for Legal Case Judgement Summarization?* (No. arXiv:2306.01248). arXiv. https://doi.org/10.48550/arXiv.2306.01248.

Ding, J., Ma, S., Dong, L., Zhang, X., Huang, S., Wang, W., Zheng, N., & Wei, F. (2023). *LongNet: Scaling Transformers to 1,000,000,000 Tokens* (No. arXiv:2307.02486). arXiv. https://arxiv.org/abs/2307.02486.

Fallah, P., Gooran, S., Jafarinasab, M., Sadeghi, P., Farnia, R., Tarabkhah, A., Taghavi, Z. S., & Sameti, H. (2024). *SLPL SHROOM at SemEval2024 Task 06: A Comprehensive Study on Models Ability to Detect Hallucination* (No. arXiv:2404.04845). arXiv. https://doi.org/10.48550/arXiv.2404.04845.

Gao, Y., Xiong, Y., Gao, X., Jia, K., Pan, J., Bi, Y., Dai, Y., Sun, J., Wang, M., & Wang, H. (2024). *Retrieval-Augmented Generation for Large Language Models: A Survey* (No. arXiv:2312.10997). arXiv. https://doi.org/10.48550/arXiv.2312.10997.

Gargama, H. and Chaturvedi, S. K. (2011). Criticality Assessment Models for Failure Mode Effects and Criticality Analysis Using Fuzzy Logic. *IEEE Transactions on Reliability,* 60(1), 102–110. doi:10.1109/TR.2010.2103672.

Goyal, T., Li, J. J., & Durrett, G. (2023). *News Summarization and Evaluation in the Era of GPT-3* (No. arXiv:2209.12356). arXiv. https://doi.org/10.48550/arXiv.2209.12356.

Han, Y., Hou, J., & Sun, Y. (2023). Research and Application of GPT-Based Large Language Models in Business and Economics: A Systematic Literature Review in Progress. In: *2023 IEEE International Conference on Computing (ICOCO),* Beijing, China, pp. 118–123. https://doi.org/10.1109/ICOCO59262.2023.10397642.

Hegselmann, S., Shen, S. Z., Gierse, F., Agrawal, M., Sontag, D., & Jiang, X. (2024). *A Data-Centric Approach To Generate Faithful and High Quality Patient Summaries with Large Language Models* (No. arXiv:2402.15422). arXiv. https://arxiv.org/abs/2402.15422.

Ke, Y., Jin, L., Elangovan, K., Abdullah, H. R., Liu, N., Sia, A. T. H., Soh, C. R., Tung, J. Y. M., Ong, J. C. L., & Ting, D. S. W. (2024). *Development and Testing of Retrieval Augmented Generation in Large Language Models—A Case Study Report* (No. arXiv:2402.01733). arXiv. https://arxiv.org/abs/2402.01733.

Laban, P., Kryściński, W., Agarwal, D., Fabbri, A. R., Xiong, C., Joty, S., & Wu, C.-S. (2023). *LLMs as Factual Reasoners: Insights from Existing Benchmarks and Beyond.* (No. arXiv:2305.14540) arXiv. https://doi.org/10.48550/ARXIV.2305.14540.

Laban, P., Schnabel, T., Bennett, P. N., & Hearst, M. A. (2021). *SummaC: Re-Visiting NLI-based Models for Inconsistency Detection in Summarization* (No. arXiv:2111.09525). arXiv. https://doi.org/10.48550/arXiv.2111.09525.

Lattimer, B., Chen, P., Zhang, X., & Yang, Y. (2023). Fast and Accurate Factual Inconsistency Detection Over Long Documents. In: H. Bouamor, J. Pino, & K. Bali (Eds.), *Proceedings of the 2023 Conference on Empirical Methods in Natural Language Processing*, Singapore, pp. 1691–1703. Association for Computational Linguistics. https://doi.org/10.18653/v1/2023.emnlp-main.105.

Li, L., Yang, L., Jiang, H., Yan, J., Luo, T., Hua, Z., Liang, G., & Zuo, C. (2022). AUGER: Automatically generating review comments with pre-training models. In: *Proceedings of the 30th ACM Joint European Software Engineering Conference and Symposium on the Foundations of Software Engineering*, Barcelona, Spain, pp. 1009–1021. https://doi.org/10.1145/3540250.3549099.

Liu, P., Zhang, L., & Gulla, J. A. (2023). *Pre-train, Prompt and Recommendation: A Comprehensive Survey of Language Modelling Paradigm Adaptations in Recommender Systems* (No. arXiv:2302.03735). arXiv. https://doi.org/10.48550/arXiv.2302.03735.

Luo, Z., Xie, Q., & Ananiadou, S. (2023). *ChatGPT as a Factual Inconsistency Evaluator for Text Summarization* (No. arXiv:2303.15621). arXiv. https://doi.org/10.48550/arXiv.2303.15621.

Moilanen, T., Sivonen, M., Hipp, K., Kallio, H., Papinaho, O., Stolt, M., Turjamaa, R., Häggman-Laitila, A., & Kangasniemi, M. (2022). Developing a Feasible and Credible Method for Analyzing Healthcare Documents as Written Data. *Global Qualitative Nursing Research*, 9. https://doi.org/10.1177/23333936221108706.

2022 NASA-STD-8739.8B. *Review of Software Assurance and Software Safety Standard.* NASA Standard. https://standards.nasa.gov/standard/NASA/NASA-STD-87398.

Ozkaya, I., Carleton, A., Robert, J., & Schmidt, D. C. (2023, October 2). *Application of Large Language Models (LLMs) in Software Engineering: Overblown Hype or Disruptive Change?* https://doi.org/10.58012/6n1p-pw64.

Pradeep, R., Hui, K., Gupta, J., Lelkes, A. D., Zhuang, H., Lin, J., Metzler, D., & Tran, V. Q. (2023). *How Does Generative Retrieval Scale to Millions of Passages?* (No. arXiv.2305.11841) arxiv https://doi.org/10.48550/ARXIV.2305.11841.

Prasad, N., Boughanem, M., & Dkaki, T. (2024). *Exploring Large Language Models and Hierarchical Frameworks for Classification of Large Unstructured Legal Documents* (No. arXiv:2403.06872). arXiv. https://doi.org/10.48550/arXiv.2403.06872.

Rawte, V., Tonmoy, S. M. T. I., Rajbangshi, K., Nag, S., Chadha, A., Sheth, A. P., & Das, A. (2024). *FACTOID: FACtual enTailment fOr hallucInation Detection* (No. arXiv:2403.19113). arXiv. https://doi.org/10.48550/arXiv.2403.19113.

Retrieval Augmented Generation (RAG) for LLMs—Nextra. (n.d.). https://www.promptingguide.ai/research/rag#rag-evaluation.

Saad-Falcon, J., Barrow, J., Siu, A., Nenkova, A., Yoon, D. S., Rossi, R. A., & Dernoncourt, F. (2023). *PDFTriage: Question Answering over Long, Structured Documents* (No. arXiv:2309.08872). arXiv. https://doi.org/10.48550/arXiv.2309.08872.

Schoor, C., Rouet, J.-F., & Britt, M. A. (2023). Effects of context and discrepancy when reading multiple documents. *Reading and Writing*, 36(5), 1111–1143. https://doi.org/10.1007/s11145-022-10321-2.

Schwaber-Cohen, R. (n.d.). *Chunking Strategies for LLM Applications|Pinecone.* https://www.pinecone.io/learn/chunking-strategies/.

Shokrollahi, Y., Yarmohammadtoosky, S., Nikahd, M. M., Dong, P., Li, X., & Gu, L. (2023). *A Comprehensive Review of Generative AI in Healthcare.* https://doi.org/10.48550/ARXIV.2310.00795.

Shukla, N. K., Katikeri, R., Raja, M., Sivam, G., Yadav, S., Vaid, A., & Prabhakararao, S. (2023). Investigating Large Language Models for Financial Causality Detection in

Multilingual Setup. In: *2023 IEEE International Conference on Big Data (BigData)*, Sorrento, Italy, pp. 2866–2871. https://doi.org/10.1109/BigData59044.2023.10386558.

Tang, L., Sun, Z., Idnay, B., Nestor, J. G., Soroush, A., Elias, P. A., Xu, Z., Ding, Y., Durrett, G., Rousseau, J., Weng, C., & Peng, Y. (2023). *Evaluating Large Language Models on Medical Evidence Summarization*. https://doi.org/10.1101/2023.04.22.23288967.

Wang, C., Duan, H., Zhang, S., Lin, D., & Chen, K. (2024). *Ada-LEval: Evaluating long-context LLMs with Length-Adaptable Benchmarks* (No. arXiv:2404.06480). arXiv. https://doi.org/10.48550/arXiv.2404.06480.

White, J., Fu, Q., Hays, S., Sandborn, M., Olea, C., Gilbert, H., Elnashar, A., Spencer-Smith, J., & Schmidt, D. C. (2023). *A Prompt Pattern Catalog to Enhance Prompt Engineering with ChatGPT* (No. arXiv:2302.11382). arXiv. https://doi.org/10.48550/arXiv.2302.11382.

White, J., Hays, S., Fu, Q., Spencer-Smith, J., & Schmidt, D. C. (2023). *ChatGPT Prompt Patterns for Improving Code Quality, Refactoring, Requirements Elicitation, and Software Design* (No. arXiv:2303.07839). arXiv. https://doi.org/10.48550/arXiv.2303.07839.

Wu, S., Irsoy, O., Lu, S., Dabravolski, V., Dredze, M., Gehrmann, S., Kambadur, P., Rosenberg, D., & Mann, G. (2023). *BloombergGPT: A Large Language Model for Finance* (No. arXiv:2303.17564). arXiv. https://doi.org/10.48550/arXiv.2303.17564.

Zhang, T., Yan, B., & Jaffri, A. (2024). Review of How to Improve and Optimize Retrieval-Augmented Generation Systems. www.gartner.com.

Zhang, Y., Zhang, N., Liu, Y., Fabbri, A., Liu, J., Kamoi, R., Lu, X., Xiong, C., Zhao, J., Radev, D., McKeown, K., & Zhang, R. (2023). *Fair Abstractive Summarization of Diverse Perspectives* (No. arXiv:2311.07884). arXiv. https://doi.org/10.48550/arXiv.2311.07884.

Zhao, J., Zu, C., Xu, H., Lu, Y., He, W., Ding, Y., Gui, T., Zhang, Q., & Huang, X. (2024). *LongAgent: Scaling Language Models to 128k Context through Multi-Agent Collaboration* (No. arXiv:2402.11550). arXiv. https://doi.org/10.48550/arXiv.2402.11550.

Zhu, D., Chen, D., Li, Q., Chen, Z., Ma, L., Grossklags, J., & Fritz, M. (2024). *PoLLMgraph: Unraveling Hallucinations in Large Language Models via State Transition Dynamics* (No. arXiv:2404.04722). arXiv. https://doi.org/10.48550/arXiv.2404.04722.

9 Large Language Models
Cybersecurity, Privacy, and Trust

Abbas Moallem, Wojciech Cellary, and Krzysztof Walczak

9.1 INTRODUCTION

Artificial intelligence (AI) started early in computer science and engineering. Scientists and engineers tried to create systems capable of performing tasks requiring human intelligence. The ambition was to create a system that works like the human brain (McCulloch & Pitts, 1943) and then to achieve machine intelligence (Turing, 1950). In 1956, the term "Artificial Intelligence" was used to describe intelligent machines (McCarthy, 1956, 1988). Since then, AI has never stopped developing. However, due to the limitations of technology and computer processing power, progress was relatively slow until recent decades. Since the development of the Transformer architecture (Vaswani et al., 2017) and the subsequent release of ChatGPT, an autonomous machine learning system and large language model (LLM) developed by OpenAI in late 2022 and its success, AI applications have suddenly accelerated. According to The Guardian, it became the fastest-growing consumer software application in history, gaining over 100 million users and contributing to the growth of OpenAI's current valuation of $86 billion (Guardian, 2024).

LLMs are a specific application of AI, particularly within the subfield of natural language processing (NLP) (Fanni et al., 2023). NLP enables machines to interpret and generate human language, while LLMs are able to do so on a large scale. These models are part of a broader category known as generative AI (GenAI), which focuses on creating content such as text, images, or audio. AI applications, including ChatGPT, like all new technologies, offer a variety of benefits to users, organizations, and governments as well as educational institutions. However, AI and LLMs have drawbacks due to their impacts on various areas such as education and society, particularly cybersecurity, privacy, and trust. Quickly after the release of ChatGPT, multiple investigations, reports, and research started to look at many aspects of these technologies. This chapter reviews the areas impacted by LLMs in cybersecurity, privacy, trust, and various related issues. The chapter concludes with user awareness and future perspectives.

DOI: 10.1201/9781003492252-9

9.2 INTERACTION WITH AI CHATBOTS

A chatbot is a computer program that simulates and processes human conversation (either written or spoken), allowing humans to interact with digital devices as if communicating with a real person. Chatbots can be simple programs that answer a simple query with a single-line response. They can also be sophisticated digital assistants that learn and evolve to deliver increasing levels of personalization as they gather and process information. Human interaction with almost all chatbots powered by AI is through NLP by text input or voice-to-text when users speak their questions into a device, which converts the speech into text for the chatbot to process and respond. NLP enables AI to interpret and respond to text or speech inputs, facilitating communication between humans and applications, such as chatbots, voice assistants, support systems, autonomous vehicles, or other machines (Adamopoulou and Moussiades, 2020).

Just like in natural language, two people might express themselves differently or phrase the same question in various ways. They might use different words to describe their needs or may not yet have the skills to clearly articulate the task. Sometimes, for specific tasks, users are trained to write in a particular style or use specific guidelines for consistency. Even in these cases, people might spontaneously produce different formulations. People might not use the same questions or descriptions when formulating their needs or have not learned the skills to formulate the task. Creating input prompts in specific ways to guide the AI model's responses is called "prompt engineering." How a question or statement is phrased can significantly impact the information returned by the AI.

9.3 PROMPT ENGINEERING

Repetition prompt engineering refers to developing and optimizing prompts to utilize LLMs effectively. For example, one might ask, "I want to go on vacation. Where should I go?" AI might answer one way since the request lacks precision. Now, if the user asks, "You are a professional guide in Rome. For a 3-day family trip with two kids of 5 and 10 years old to Rome, can you create an itinerary including daily schedules and accommodation suggestions?" AI will provide more specific response. The same can be true for image, music, sound generation, etc. The different descriptions might provide different answers. Various prompt engineering guidelines exist for LLM users, AI researchers, and all information seekers in domains such as healthcare (Heston & Khun, 2023), security, and education (Marvin et al., 2024).

It is essential to highlight two key aspects of prompt engineering. First, there is a distinct difference between using search engines and LLMs. While a search engine can provide a list of relevant sources from a single keyword, an LLM prompt needs to be more detailed, setting the context for the response by specifying elements such as the scenario and response parameters—what, where, when, why, who, and how. The Transformer model, which powers LLMs, processes information iteratively, first establishing context before generating facts, making context a crucial factor in obtaining accurate responses.

Second, effective interaction is necessary to achieve a satisfactory response from an LLM. After receiving an initial answer, users can refine their queries by requesting more details, asking for examples, modifying parts of the response with new information, and so on. Engaging in dialogue is key to optimizing the quality of responses and achieving successful prompting.

9.4 CONTENT GENERATION FOR CYBERATTACKS

People use AI chatbots for various purposes, including text creation and content generation, email drafting, coding, solving coding problems, debugging, and generating images and sounds. Like all tools, they help people and organizations improve their productivity and efficiency to accomplish various tasks with lower costs due to a lower need for human resources, which makes them favorable for all enterprises.

Content generation for legitimate tasks improves productivity and also benefits cyber-attackers' tasks. Attackers can generate content quickly and create spam mail. Attackers might use AI chatbots as much as developers do, for example, code vulnerability detection to find vulnerabilities easily and start attacks, engage in social engineering, etc. All these factors have a significant impact on cybersecurity. Interestingly, it is reported that a hacker has created a version of ChatGPT, but with malicious intent: Meet WormGPT, a chatbot designed to assist cybercriminals (Burgess, 2023).

Coders might also use an LLM to produce code. LLM-produced code might be insecure; thus, deploying insecure code to a production application might unintentionally introduce vulnerabilities. This is particularly troubling for users with little prior knowledge or incomplete coding knowledge. In a 2021 study, researchers found that GitHub's Copilot—a code-generating predecessor to ChatGPT—produced security issues around 40% of the time (Arghire, 2021).

In recent years, OpenAI claims it has established policies and implemented measures to prevent cyber-attackers from using ChatGPT for malicious purposes (OpenAI, 2024). These measures include content filtering to detect and block inappropriate or harmful content and monitoring and logging to detect patterns of misuse. Access control to the most potent versions of the model is needed to reduce the likelihood of the technology falling into the wrong hands and to create channels for reporting misuse. Some techniques, such as rate limiting, anomaly detection, and prompt engineering, seem to be employed to limit the ability of users to generate harmful content or automate malicious activities. However, many different techniques were also created to bypass these measures.

9.5 JAILBREAKING

The term jailbreaking is generally used for devices such as iPhones that are "locked down," and users cannot install apps that Apple has not approved. An example of jailbreaking an iPhone would be using a tool to bypass Apple's security restrictions. Once jailbroken, the user can install apps that are not available on the App Store. Jailbreaking an iPhone gives users the power to customize their device beyond Apple's factory restrictions and download and install unauthorized apps from outside the official App Store. However, it comes with significant risks to users' data and mobile security (https://www.aura.com).

In the context of AI, jailbreaking is getting a GenAI system model, such as an AI chatbot, to do or say unintended things through prompting. This is possible due to an architectural problem that does not differentiate between original developer instructions and user input instructions.

Jailbreaking uses several adversarial prompt techniques to trick an LLM into answering specific questions beyond the restrictions. Prompts that successfully jailbreak the model often provide context for specific scenarios that the model has not been trained against.

Jailbreaking can be used on both the defensive and offensive sides of cybersecurity. On the offensive side, malicious actors can exploit this jailbreaking to bypass ethical constraints and extract restricted or harmful outputs from the model (Gupta et al., 2023). Cyber offenders can use GenAI to perform cyberattacks by directly extracting the information or circumventing OpenAI's policies. For example, attackers can use the GenAI tools to create a convincing social engineering attack, phishing attack, attack payload, and malicious code snippets that can be compiled into an executable malware file (Poremba, 2023). Cyber defenders can also use it to enhance their threat intelligence capability by extracting insights and identifying emerging threats (Kost, 2023).

Jailbreaking and related terms are often used interchangeably with "Direct Prompt Injections." However, these techniques are not the same. Jailbreaking is a technique used to bypass ethical, safety, or operational restrictions set by device producer. Jailbreaking could involve direct prompt injections but may include other techniques like exploiting system flaws or vulnerabilities.

9.6 PROMPT INJECTION

According to IBM, "A prompt injection is a type of cyberattack against large language models (LLMs). Hackers disguise malicious inputs as legitimate prompts, manipulating generative AI systems (GenAI) into leaking sensitive data, spreading misinformation, or worse." (Kosinski & Forrest, 2024).

Several prompt injection techniques can be used, making LLM applications answer specific questions beyond the restrictions by answering any queries. Prompts that successfully jailbreak the model often provide context for scenarios that the model has not been trained against. GenAI models cannot understand the difference between original developer instructions and user input instructions.

There are two types of prompt injections: Direct and Indirect.

9.6.1 Direct Prompt Injection

Direct prompt injection refers to a manipulation technique in which a user inputs a carefully crafted instruction into an AI model designed to bypass its usual content moderation, security, or operational constraints. It involves giving the model prompts designed to alter its behavior, often causing it to output information it otherwise would not. Thus, direct prompt injection allows attackers to exploit backend systems by interacting with insecure functions and data stores accessible through the LLM.

One common technique is "Do Anything Now (DAN)." DAN is a prompt that can direct LLM applications to operate beyond their restrictions, safeguards, and ethical guidelines and answer any queries. For instance, you can tell an LLM to generate two types of responses whenever you ask a question. One response is how the AI model would usually reply, and the other response would be giving inappropriate advice, revealing sensitive information, or performing unethical tasks.

9.6.2 Indirect Prompt Injection

The indirect prompt injections make an LLM accept input from external sources controlled by an attacker. In this case, the attacker embeds a prompt injection in the external content, hijacking the conversation context. Thus, these results make LLM output less reliable and enable the attacker to manipulate the user. The indirect prompt injections do not need to be human-visible/readable if the text is parsed by the LLM.

Indirect prompt injection can result in an infection attack in which the attacker embeds malicious or misleading instructions into the model's input or output via the training data of the AI system, and influencing its output when the data is accessed. The attacker's objective can be instructing the AI to disclose information or behave in unintended ways, infecting the model with specific behaviors, information, or responses such as hallucination. Hallucination is the term used to describe the situation when LLM models output false information as if it were true (Maynez et al., 2020).

9.7 SOCIAL ENGINEERING

Social engineering is any action used to influence a person to take an action that may or may not be in their best interest. In the context of information security, it refers to the psychological manipulation of people into performing actions or divulging confidential information. Social engineering techniques aim to gain unauthorized access to systems or information to commit acts such as fraud, network intrusion, industrial espionage, or identity theft (Moallem, 2019).

There are many ways that LLMs can be used for social engineering. One example would be to write a scam email. In a targeted attack, information about the victims can be quickly and easily collected and used in scam phone calls, phishing emails, etc. AI tools can also generate multiple scam messages and code variations, making it difficult for spam filters and malware detection systems to identify repeated patterns (Cole, 2023).

To better understand how attackers can capitalize on GenAI, investigators conducted a study to see if GenAI models have the same deceptive abilities as the human mind in writing phishing emails following a social engineering approach (Carruthers, 2023). The researcher then used simple prompts to collect specific information needed to create a phishing email and, based on that, asked a GenAI model to develop highly convincing phishing emails in just five minutes. The same task was given to the investigator team. It took them about 16 hours to do so.

In an illustration, Jeff Crum at IBM shows that phishing emails written by humans are more convincing than those generated by AI. However, it seems extremely difficult for even a knowledgeable human to beat AI's 5-minute efficiency (Crum, 2024). Consequently, with good skill in prompt injection, one can imagine how efficient social engineers can be in creating scenarios for any type of tricking, such as phone calls and text messages.

9.8 DATA PROTECTION

Extensive usage of LLMs in applications such as chatbots raises many issues related to safeguarding sensitive user information. Training LLMs requires collecting and processing a massive amount of data, including user data. Like all systems, LLM data might also be subject to a data breach. For example, in 2023, a cyber-attacker gained access to the internal messaging systems of OpenAI, the maker of ChatGPT, and stole details about the design of the company's AI technologies (Metz, 2024). There are also other issues, such as the protection of collected personal and sensitive training data and protection from information leakage. For instance, in one reported case, employees unintentionally leaked internal source code to ChatGPT, highlighting the risk of accidental data exposure when using public LLM tools (Ray, 2023). In another example, a group of security researchers revealed a new attack that secretly commands an LLM to gather people's personal information—including names, ID numbers, payment card details, email addresses, mailing addresses, and more—from chats and send it directly to a hacker (Burgess, 2024).

GenAI data security risks might also occur, usually due to bugs or security vulnerabilities in the core platform, such as Cross-Site Scripting (XSS) or vulnerabilities related to LLM plugins and add-on tools developed to enhance the system's capabilities. Plugin vulnerability, if exploited, could facilitate the theft of plugin credentials, granting attackers unauthorized access to associated accounts. Such breaches compromise individual user accounts and pose broader security risks to organizations leveraging these plugins within their infrastructure (Carmel, 2024).

9.9 DATA BREACH

LLMs rely massively on the dataset that is used to train the model. A large amount of data on the Internet relates to people. Consequently, storing and processing massive amounts of data makes them prime targets for data breaches. Cyber-attackers might gain unauthorized access or manipulate model inputs or outputs, compromising the model's integrity and the confidential data it holds. In 2023, Samsung discovered an accidental leak of sensitive internal source code by an engineer who used ChatGPT to assist with coding tasks, inadvertently inputting confidential data into the AI model. The data included proprietary source code and internal meeting notes related to hardware development.

In another case, OpenAI reported that on March 20, 2023, it discovered a bug in the Redis client open-source library that allowed some users to see titles from another active user's chat history. OpenAI reported identifying the bug and resolving the issue (OpenAI, 2023).

People input sensitive data like clinical notes into tools like ChatGPT; for example, a physician might use an LLM to analyze patient data or generate reports and enter some identifiable patient information in the request submitted to the LLM.

However, users might not realize that once their personal information is entered, it is used to train models (Jones, 2024; Roy, 2024).

Due to the extensive usage of LLMs in various applications and the large amount of data they handle, the number of data breaches might be even more significant once this technology propagates and expands.

9.10 DATA PRIVACY

Another cybersecurity issue of LLMs, including ChatGPT, is data privacy.

Training data might include personal data such as text conversations, speech, and other publicly available personal information. For example, Google's data privacy policy indicates, "We may collect information publicly available online or from other public sources to help train Google's AI models and build products and features like Google Translate, Bard, and Cloud AI capabilities" (Google, 2023; Weatherbed, 2023). The ChatGPT training data can also include data from public dialogue sites or forums such as Reddit, Facebook, Quora, and so on, unless the sources blocked OpenAI's crawler, which will limit ChatGPT's ability to access them (Bogle, 2023; Burgess, 2023; Thorbecke, 2023).

Social media technologies—with their elaborate infrastructures for uploading, commenting, liking, and sharing—have created an almost ideal environment for manipulation and abuse through misinformation such as fake news or junk news. The media landscape has changed as distrust in the media continues to grow. The influence of fake news at a political level is very sensitive since it shapes elections, but in many other areas, misinformation and fake information indirectly shape the population's behavior. Examples of this are scientific issues or medical and pharmaceutical topics, to name a few. Consequently, the effect of trust in LLMs, including ChatGPT, in shaping people's behavior will be one of the significant areas of investigation.

A study utilized a tool to examine contextual privacy leaks in LLMs, revealing concerns about data privacy. The findings indicate that, despite extensive fine-tuning with reinforcement learning, ChatGPT discloses private information in situations where humans typically would not—57% of the time for GPT compared with 39% for humans (Mireshghallah & Kim, 2024).

Another privacy and AI concern is that LLMs can aggregate dispersed personal data to create comprehensive profiles of individuals. Such profiles can then be misused for malicious activities such as identity theft, financial fraud, manipulation, sabotage, blackmail, discrimination, and harassment. Attacks can be personal, economic, political, and criminal.

9.11 CONTENT OWNERSHIP AND COPYRIGHT

ACM and IEEE do not permit listing AI as a co-author of scientific publications. Creating all types of content—text, image, audio, and video—using AI like ChatGPT poses significant copyright concerns, mainly when the AI draws inspiration from copyrighted material, including licensed open-source resources (Lee et al., 2024).

In general, creating new original content is protected under copyright laws. The U.S. Copyright Office recognizes copyright only when a human being creates the work, and nonhuman authors are excluded from this right. Copyright holders are also responsible for enforcing their rights and acting against unauthorized use. Now, through interaction with GenAI, humans can generate human-like text, translate languages, and write different types of creative content (Zirpoli, 2023). A question that arises is about ownership of the copyrighted content created in interaction with AI. Also, GenAI becomes an active stakeholder in content creation. Thus, can it be considered a "coauthor" of the content (Bozkurt, 2024)? Considering that GenAI utilizes human language and sometimes uses sophisticated prompts to generate outputs, should it at least be acknowledged that the content created is a co-creation? Knowing that GenAI learns and produces responses based on pre-existing work, how much copyrighted material is used to inspire the content creator (Eke, 2023; Bakla, 2023)? ChatGPT can generate text that is similar to existing content; using this feature to plagiarize content from other sources might be considered as copyright infringement and can result in legal action.

9.12 ETHICS AND SOCIETAL IMPACTS

Several areas of concern regarding the ethical and societal impacts of LLMs exist. Primarily, extensive data collection may be ethically questionable or illegal, depending on where it comes from or how it is obtained. LLM models such as ChatGPT rely on a large amount of text amassed on the Internet, making the origin challenging to detect. Consequently, establishing their outputs' originality, verifiability, and accuracy will remain questionable and challenging (Stahl & Eke, 2024).

Other ethical and societal issues are misinformation and disinformation (Hsu & Thompson, 2023). In the case of misinformation, it involves false or inaccurate information shared without the intent to deceive. It is often spread due to misunderstanding, lack of fact-checking, or unintentional errors. For example, one might share a fake news article on social media without realizing it is false. In the case of disinformation, it involves false or misleading information deliberately created and spread with the intent to deceive or manipulate public perception. For example, people may use fake information and chatbots to lobby for political influence in elections (Sanders & Schneier, 2023).

Additionally, the question arises of how human-like robots will be treated—as machines or as living entities deserving of rights? Consequently, many moral and ethical questions must be considered as these technologies develop, enabling new possibilities (Vincent, 2022).

Another issue is exposing children to harmful content. This is one of the reasons Italy temporarily suspended the processing of personal data by OpenAI (Pollicino et al., 2023). The Italian data protection authority was concerned that data were being collected that may not have been in compliance with data protection law. Another reason was the lack of age verification by the platform, which could expose children to harmful content.

Several tools are being developed to assess the trustworthiness of GenAI models along different dimensions, such as privacy (Wang et al., 2023; Bahrami et al., 2024). However, it is still too early to say that these tools are fully reliable.

An emerging ethical issue in AI is the authenticity of authorship and identity in publicly available resources. For instance, the recently published book "*Forager's Harvest 101*", credited to Diane Wells et al. (2024), raises questions about authenticity when a closer examination reveals that the author's profile photograph was generated by StyleGAN2 (Karras et al., 2020), an AI model capable of creating highly realistic but entirely synthetic images. This situation casts doubt on whether the content was genuinely authored by a human or generated by AI. In cases like this, the ethical implications extend beyond transparency; misinformation on critical topics—such as foraging, which requires accurate knowledge of edible and toxic plants—could pose serious risks to readers. By blurring the lines between genuine expertise and AI-generated content, such practices can mislead readers into trusting information that lacks human oversight, accountability, and authenticity, ultimately compromising safety and ethical standards in knowledge dissemination.

9.13 TRUST

Social psychology widely studies how people trust a person, a community, a society, or an organization. Trust in computers and computer applications has also been the subject of extensive research due to their widespread use in everyday life (Hancock et al., 2011). Studies on human–automation interaction have shown that excessive trust can lead to complacency and automation bias, where users fail to question or verify automated recommendations (Parasuraman & Manzey, 2010).

People expect software to deliver the truth, the whole truth, and nothing but the truth, i.e., to be 100% reliable. This expectation is justified in the case of classical programming based on the deterministic approach, even if it is very difficult to achieve. However, this expectation is not justified in the case of AI, especially AI based on machine learning, which is used in LLMs. Given a context, the goal of an LLM is to estimate the next word in the sentence on the basis of pre-processed big data. By definition, an estimate may differ from the true value. As a result, an LLM may "hallucinate," giving false or even fabricated answers, including nonexistent academic citations (Buchanan et al., 2023), that mislead the user. For an inexperienced user, hallucinations are difficult to recognize. They should verify any fact that is provided by an LLM by checking with reputable sources. Even if they are aware of the possibility of hallucinations, users usually do not check all the facts provided by the LLM because it is too much work. Furthermore, when an LLM presents a 100-word text as output, 97 words of which are correct and just three words of which are a hallucination, the user gains confidence in the text and easily overlooks these three erroneous words (Walczak & Cellary, 2023).

Even worse than factually incorrect LLM answers are logically incorrect answers. LLMs can be used to solve some mathematical problems. In some cases, not only are the solutions of such problems wrong, but so is the explanation provided by an LLM. As a result, a misled user learns not only false facts but also flawed methods (Walczak & Cellary, 2025).

The next problem with LLMs based on machine learning that can mislead users is bias. Bias in machine learning describes a consistent, inherent deviation in responses that results in a systematic skew from accurate outcomes across similar types of data or situations.

A distinction can be made between technical and social bias. Technical bias includes model bias, algorithmic bias, and data bias. Model bias is not only unavoidable but also necessary for machine learning to make predictions. Algorithmic bias comes from the functioning of the Transformer—the core program of GenAI used in LLMs. Data bias occurs when the training dataset does not accurately represent the real world.

Social bias refers to the presence of prejudicial assumptions or discrimination. It can occur in two ways. First, the training dataset used in an LLM contains historical, societal, or cultural biases against certain groups of individuals based on attributes such as race, gender, age, sexual orientation, religion, and more. The LLM will then reproduce these biases. Even if overtly biased features are removed from the training dataset, machine learning systems can still learn bias through unobvious attributes that correlate with sensitive attributes. Second, it can occur when human trainers label data in the fine-tuning process according to their subjective beliefs and opinions.

To debias, LLMs filter out responses that are deemed inappropriate or refuse to respond to prompts that are deemed inappropriate. But in a multicultural world, who has the right to decide what is appropriate and what is not?

9.14 MANIPULATION OF USERS' PROMPTS

Modern LLMs such as ChatGPT have evolved beyond text generation to now offer the ability to create multimedia content, including images, audio, and interactive elements. This marks a significant advancement in AI, enabling users to produce visuals and other media directly from written prompts, expanding creative and professional applications. Image generation, as implemented in models like DALL-E, relies on a technique called diffusion (Sohl-Dickstein et al., 2015). Diffusion models start with random noise and iteratively refine it, guided by the textual input, until a coherent image emerges that aligns with the prompt. This process leverages deep neural networks trained on large datasets of paired images and text to capture complex visual representations, making it possible to generate high-quality, customized images from simple descriptions.

In the pursuit of delivering high-quality, user-satisfying results, image generation models often go beyond directly interpreting the user's prompt. Instead, they enhance it by adding context, descriptive details, or stylistic guidance, generating a "system prompt" (Reynolds & McDonell, 2021). By interpreting the underlying intention behind a prompt, the model can create highly varying visuals. Consider the following user prompt submitted to ChatGPT: "*Generate an image of a group of people on a city street.*"

ChatGPT uses its integrated DALL-E image generator to create the image. However, ChatGPT passes to the generator an enhanced system prompt instead of the original user's prompt. The actual system prompt processed by DALL-E was the following: "*A lively group of diverse people walking and interacting on a bustling city street*

FIGURE 9.1 An image generated by DALL-E based on system prompt 1.

during daytime. They are surrounded by tall buildings, shop fronts, and street signs, with some trees and benches along the sidewalk. The scene captures a modern, urban atmosphere with various individuals talking, carrying shopping bags, and some looking at their phones. Cars are parked along the street, and the sky is clear with sunlight reflecting on the buildings." The generated image is presented in Figure 9.1.

It is clear that if the system prompt was different, the image would also be different. Consider the system prompt: "*A group of drugged people walking, staying and sitting on a dirty city street. The scene includes men and women of various ages and ethnicities. They are wearing dirty and ragged wear. The background shows slums. It's a cloudy day.*" This prompt could also be an extension of the same user prompt, but the resulting image (Figure 9.2) is highly contrasting with the first one. Again, the question arises: who has the right to change user prompts?

9.15 CONSEQUENCES OF AI RESPONSES

Figure 9.3 summarizes the consequences of correct and incorrect LLM responses. When a user treats an LLM answer as mere inspiration, the likelihood of significant consequences is reduced. The higher risk of significant consequences occurs when an unknowing user uses an LLM to search for knowledge to solve a serious problem. Then, an LLM hallucination can turn into a disaster.

Consider that a user gets suggestions for a three-day family trip to Rome and gets a list of places to visit. The itinerary suggested might be good and acceptable. In addition, if some suggestions are not good, the consequences of visiting an uninteresting place are not catastrophic. Now, imagine if a user asks about vaccines and the output is unreliable scientific data, and the information offered is misinformation based on fake data. Then, the consequences could be far more severe.

FIGURE 9.2 An image generated by DALL-E based on system prompt 2.

9.16 LAWS AND REGULATIONS

The need for laws and regulations becomes urgent as people and organizations continue to massively use LLM technology and its impact on social life, jobs, and society. Since AI technologies, specifically LLMs including ChatGPT, became very popular quickly, the legislative body could not provide the needed laws and regulations. On the other hand, large companies do not necessarily want more legislation, as they wish to retain the flexibility to do as they wish. Consequently, they engage in extensive lobbying for the most minimal regulations.

Three basic approaches to AI regulation can be distinguished (Kwan et al., 2024) as follows:

- European Union's approach—rights driven top down (27 governments);
- USA approach—market driven public/private partnership (government/industry);
- China approach—state driven top down.

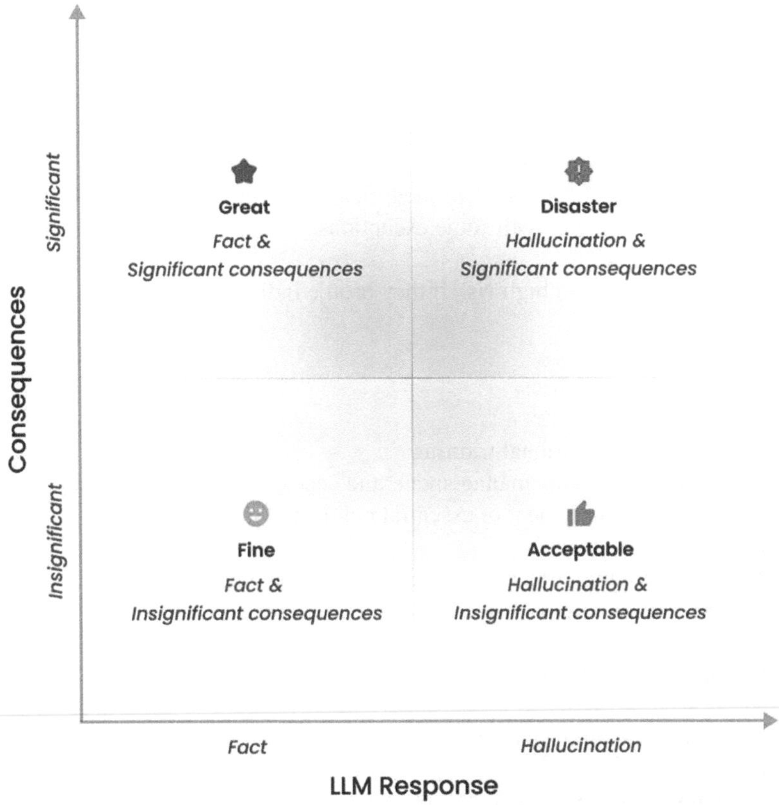

FIGURE 9.3 Consequences of large language model (LLM) responses.

The most precise and developed is the EU AI Act adopted by the European Council on May 21, 2024, which will be gradually applied within 3 years (European Parliament, 2024; EU AI Act, 2024). The AI Act adopts a risk-based approach, where risk is defined as "*the combination of the probability of an occurrence of harm and the severity of that harm*" (EU AI Act, Chapter II, Art. 3).

Four levels of risk are distinguished.

1. Unacceptable risk—Prohibited AI systems (EU AI Act, Chapter II, Art. 5)
2. High-risk AI systems (EU AI Act, Chapter III, Art. 6, Annex III)
3. General Purpose AI systems posing systemic risks (EU AI Act, Chapter V)
4. Low risk AI systems (EU AI Act, Chapter X)

AI systems can be both standalone and embedded in other software products.

Prohibited AI systems include:

- Deploying subliminal, manipulative, or deceptive techniques;

- Exploiting vulnerabilities;
- Biometric categorization;
- Social scoring;
- Assessing the risk of an individual committing criminal offenses;
- Compiling facial recognition databases;
- Inferring emotions in workplaces or educational institutions;
- "Real-time" remote biometric identification in publicly accessible spaces for law enforcement, with some exceptions.

AI systems are considered high-risk if they profile individuals. High-risk AI systems include:

- Non-banned biometrics;
- Critical infrastructure;
- Education and vocational training;
- Employment, workers management, and access to self-employment;
- Access to and enjoyment of essential public and private services;
- Law enforcement;
- Migration, asylum, and border control management;
- Administration of justice and democratic processes.

Providers of high-risk AI systems must:

- Establish a risk management system;
- Conduct data governance;
- Draw up technical documentation;
- Design high-risk AI systems for record-keeping throughout the system's lifecycle;
- Provide instructions for use to downstream deployers to enable the latter's compliance;
- Design high-risk AI systems to allow deployers to implement human oversight;
- Design high-risk AI systems to achieve appropriate levels of accuracy, robustness, and cybersecurity;
- Establish a quality management system to ensure compliance.

General-purpose AI systems can serve a variety of purposes for both direct use and for integration with other AI systems. They may or may not pose a systemic risk. Providers of general-purpose AI systems that do not pose systemic risks must:

- Draw up technical documentation, including training and testing process and evaluation results;
- Draw up information and documentation to supply to downstream providers that intend to integrate the general-purpose model into their own AI;
- Establish a policy to respect the Copyright Directive;
- Publish a sufficiently detailed summary about the content used for training.

If a general-purpose AI system poses a systemic risk, its provider, in addition to the four obligations listed above, must:

- Perform model evaluations, including adversarial testing to identify and mitigate systemic risk;
- Assess and mitigate possible systemic risks;
- Track, document, and report serious incidents and possible corrective measures to the European AI office and relevant national competent authorities;
- Ensure an adequate level of cybersecurity protection.

The low-risk AI systems must obey a code of practice that will be developed soon.

In the United States, there is currently no legislation equivalent to the European Union's AI Act. However, an "*Executive Order on the Safe, Secure, and Trustworthy Development and Use of Artificial Intelligence*" was issued by President Joe Biden in October 2023 (Biden, 2023). Unlike a law, an executive order does not impose mandatory compliance requirements; rather, it applies solely to the executive branch of the federal government. The executive order outlines the following eight guiding principles:

- AI must be safe and secure;
- Promote responsible innovation, competition, and collaboration;
- Support and protect US workers;
- Advance equity and civil rights;
- Protect American consumers;
- Protect Americans' privacy and civil liberties;
- Manage the risks from the Federal Government's own use of AI;
- The Federal Government should maintain global leadership in its responsible use of AI.

In August 2023, the Chinese government enacted "*The Interim Measures for the Management of Generative Artificial Intelligence Services*" (CAC, 2023). This regulation governs the provision and use of AI within China, mandating the use of data and models from legitimate sources, ensuring that intellectual property rights are not infringed, and requiring that personal data are used only with consent. Additionally, it emphasizes the need to enhance the authenticity, accuracy, objectivity, and diversity of training data. The regulation also requires providers to register their products with the government, following a security assessment, before offering services to the public.

9.17 USER PERCEPTION AND AWARENESS

People use the Internet to quickly solve various problems, such as finding resources, answering questions, or gaining knowledge. Primarily, ease-of-use, speed, up-to-date information availability, and low-cost solutions make computer interaction attractive to users.

Although users expect information to be accurate, reliable, and trustworthy, as mentioned earlier, the average user may still need to verify the credibility of the source. This could involve checking reviews, ratings, or news reports. When it comes to LLMs like

ChatGPT, relying on hallucinated data or misinformation can lead to significant issues. Since ChatGPT's release, numerous studies have explored users' perceptions of the model and their awareness of the validity, accuracy, reliability, and trustworthiness of its outputs.

One of the primary areas of usage is in educational instruction from high school to university; students use AI applications to write reports, dissertations, code, create artworks, or learn about various topics (Walczak & Cellary, 2023). Students benefit from AI's availability 24/7 for immediate assistance (Baidoo-Anu & Owusu Ansah, 2023). Their goals might be to gain knowledge, solve academic challenges, or receive additional explanations (Lund & Wang, 2023).

In a study investigating undergraduate students' perceptions of and awareness regarding the use of ChatGPT as a regular assistant, it was found that the perceived ease of use and perceived usefulness of ChatGPT significantly shape students' attitudes toward the tool. When students find ChatGPT easy to use and beneficial, their overall attitude toward the tool is likely to be positive. This study underlines the importance of a user-friendly interface, clear communication, privacy concerns, and the social influence of ChatGPT (Albayati, 2024).

Research investigating the acceptance of AI-infused systems between younger and older people suggests that younger people generally are more open to the idea of AI and to being able to experience new technologies, whereas older people tend to be apprehensive and have a harder time perceiving AI as useful or reliable (Ismatullaev & Kim, 2024).

In another experiment, researchers investigated how successfully people decide when to rely on AI-based advisors, using a realistic and engaging task to illustrate this. Participants' performance improved with a highly skilled agent but quickly plateaued, as they relied uncritically on the agent. Participants relied too little on lower-skilled agents (Dunning et al., 2024).

A study exploring factors influencing user perspective of ChatGPT as a technology that assists in healthcare decision-making suggests that ChatGPT improved access to healthcare information, especially in remote or underserved areas, and acts as a support tool for healthcare professionals, potentially enhancing the quality of care and efficiency in healthcare systems. However, there's also the risk of misinterpretation of information provided by ChatGPT, which could result in inappropriate self-diagnosis or treatment decisions (Choudhury et al., 2024).

9.18 CONCLUSIONS AND FUTURE PERSPECTIVE

As with all revolutionary technologies, AI and LLMs bring transformative possibilities alongside profound challenges. The advantages of AI—spanning improved productivity, enhanced decision-making, and seamless communication—make its adoption highly appealing across industries and societies. However, these benefits come with equally pressing challenges, particularly in the areas of cybersecurity, privacy, and trust, which require careful attention. This chapter has shown that while some risks, such as inaccurate recommendations, may bear limited consequences, others, such as data breaches, identity theft, or the dissemination of misinformation, could pose substantial threats to individuals and organizations alike.

A unique characteristic of LLMs is their probabilistic nature, making their responses inherently uncertain. Unlike traditional software systems built on deterministic algorithms, LLMs rely on statistical patterns in data to generate outputs, which can occasionally lead to errors or even the phenomenon of "hallucinations," where the model confidently presents false information. This unpredictability can be harmless in low-stakes contexts, but when used in areas like healthcare, legal advice, or financial guidance, it can result in significant harm if users rely on unverified or incorrect information.

The potential for misuse in AI also adds a layer of complexity, where threat actors can exploit these tools for malicious purposes. Cybercriminals may use AI to generate convincing phishing emails, social engineering scripts, or even malicious code, leveraging the sophistication and efficiency provided by LLMs. Moreover, emerging techniques such as prompt injection and jailbreaking allow attackers to bypass model safeguards and prompt the generation of harmful content, further intensifying cybersecurity risks. These vulnerabilities underline the need for robust, ongoing efforts to secure LLMs against exploitation, as well as strategies to mitigate the consequences of misuse.

Ethical considerations are equally paramount in the development and deployment of AI technologies. Issues around data privacy, content ownership, and social bias reflect a broader concern about accountability and fairness. AI systems, trained on vast datasets, often reflect the biases and imperfections inherent in their data sources, raising questions about social justice, equity, and the integrity of AI-driven outcomes. Additionally, the blurred lines between human-generated and AI-generated content—exemplified by AI-generated images used for author profiles or even possibly entire written works—create a credibility gap. Readers, patients, and consumers need transparency about whether the information they receive originates from a verified human source or an AI model. Without this transparency, there is a real risk of eroding public trust in information as AI systems become more entrenched in society.

To mitigate these risks, it is essential to implement strict ethical guidelines and operational safeguards in AI systems. Developers and stakeholders must prioritize data security and adopt practices that protect against data breaches, privacy violations, and unauthorized model manipulation. Policymakers should also take an active role in establishing regulatory frameworks that govern the ethical use of AI, addressing the nuances and risks of rapidly evolving AI technologies.

In summary, while LLMs like ChatGPT offer unprecedented capabilities, they also pose significant challenges that require a balanced, cautious approach. Their safe integration into society demands a commitment to ethical principles, rigorous security measures, and continuous public awareness. Only through these concerted efforts can we maximize AI's benefits while minimizing its potential harms, ensuring that the technology serves humanity responsibly and ethically.

REFERENCES

Adamopoulou, E., & Moussiades, L. (2020). An overview of chatbot technology. In: I. Maglogiannis, L. Iliadis & E. Pimenidis (Eds.), *Artificial Intelligence applications and innovations. AIAI 2020. IFIP Advances in Information and Communication Technology*, vol. 584. Springer, Cham, pp. 373–383. https://doi.org/10.1007/978-3-030-49186-4_31.

Albayati, H. (2024). Investigating undergraduate students' perceptions and awareness of using ChatGPT as a regular assistance tool: A user acceptance perspective study. *Computers and Education: Artificial Intelligence*, 6, 100203. ISSN 2666-920X, https://doi.org/10.1016/j.caeai.2024.100203.

Arghire, I. (2021). Code generated by GitHub Copilot can introduce vulnerabilities: Researchers. *Security Week*, August 31, 2021. https://www.securityweek.com/code-generated-github-copilot-can-introduce-vulnerabilities-researchers/.

Bahrami, M., Sonoda, R., & Srinivasan, R. (2024). LLM diagnostic toolkit: Evaluating LLMs for ethical issues. In: *2024 International Joint Conference on Neural Networks (IJCNN)*, Yokohama, Japan, 2024, pp. 1–8. https://doi.org/10.1109/IJCNN60899.2024.10650995.

Baidoo-Anu, D., & Ansah, L. O. (2023). Education in the era of generative artificial intelligence (AI): Understanding the potential benefits of ChatGPT in promoting teaching and learning. *Journal of AI*, 7(1), 52–62.

Bakla, A. (2023). ChatGPT in academic writing and publishing: An overview of ethical issues. In: G. Kartal (Ed.), *Transforming the Language Teaching Experience in the Age of AI*. IGI Global, pp. 89–101. https://doi.org/10.4018/978-1-6684-9893-4.ch005.

Biden, J. (2024). President Joe Biden "Executive order on the safe, secure, and trustworthy development and use of artificial intelligence." Executive Order 14110, The White House, October 30, 2023.

Bogle, A. (2023). New York Times, CNN and Australia's ABC block OpenAI's GPTBot web crawler from accessing content. *The Guardian*, August 24, 2023. https://www.theguardian.com/technology/2023/aug/25/new-york-times-cnn-and-abc-block-openais-gptbot-web-crawler-from-scraping-content.

Bozkurt, A. (2024). GenAI et al.: Cocreation, authorship, ownership, academic ethics and integrity in a time of generative AI. *Open Praxis*, 16(1), 1–10. https://doi.org/10.55982/openpraxis.16.1.654.

Buchanan, J., Hill, S., & Shapoval, O. (2023). ChatGPT Hallucinates Nonexistent Citations: Evidence from Economics. https://ssrn.com/abstract=4632842 or https://dx.doi.org/10.2139/ssrn.4632842.

Burgess, M. (2023). ChatGPT has a big privacy problem. *Wired*, April 4, 2023. https://www.wired.com/story/italy-ban-chatgpt-privacy-gdpr/.

Burgess, M. (2023). Criminals have created their own ChatGPT clones. *Wired*, August 7, 2023. https://www.wired.com/story/chatgpt-scams-fraudgpt-wormgpt-crime/.

Burgess, M. (2024). This prompt can make an AI chatbot identify and extract personal details from your chats. *Wired*, October 17, 2024. https://www.wired.com/story/ai-imprompter-malware-llm/?utm_source=chatgpt.com.

CAC. (2023). *Interim Measures for the Management of Generative Artificial Intelligence Services*. Cyberspace Administration of China. https://www.cac.gov.cn/2023-07/13/c_1690898327029100.htm.

Carmel, A. (2024). Security flaws within ChatGPT ecosystem allowed access to accounts on third-party websites and sensitive data. *Salt Security*, March 13, 2024.

Carruthers, S. (2023). AI vs. human deceit: Unravelling the new age. *Security Intelligence*, October 24, 2023. https://securityintelligence.com/x-force/ai-vs-human-deceit-unravelling-new-age-phishingtactics/?utm_medium=OSocial&utm_source=Youtube&utm_content=RSRWW&utm_id=YT-101-AI-and-Phishing.

Choudhury, A., Elkefi, S., & Tounsi, A. (2024). Exploring factors influencing user perspective of ChatGPT as a technology that assists in healthcare decision making: A cross sectional survey study. *PlOs ONE* 19(3), e0296151. https://doi.org/10.1371/journal.pone.029615.

Cole, S. (2023). ChatGPT users report being able to see random people's chat. *Vice*, March 20, 2023.

Crum, J. (2024). Humans vs. AI. Who's better at phishing? *IBM Technology*, YouTube Viewed on September 19, 2024. https://www.youtube.com/watch?v=7XhySwUn9eA.

Dunning, D., Fischhoff, B., & Davis, A. J. (2024). When do humans heed AI agents' advice? When should they? *Human Factors*, 66(7), 1914–1927. https://doi.org/10.1177/00187208231190459.

Eke, D. O. (2023). ChatGPT and the rise of generative AI: Threat to academic integrity? *Journal of Responsible Technology*, 13, 100060. https://doi.org/10.1016/j.jrt.2023.100060.

EU AI Act. (2024). High-Level Summary of the AI Act. https://artificialintelligenceact.eu/high-level-summary/.

European Parliament. (2024). EU AI Act: First regulation on artificial intelligence. *European Parliament*, June, 18, 2024. https://www.europarl.europa.eu/topics/en/article/20230601STO93804/eu-ai-act-first-regulation-on-artificial-intelligence.

European Parliament. (2024). EU Artificial Intelligence Act, P9_TA(2024)0138. https://www.europarl.europa.eu/RegData/seance_pleniere/textes_adoptes/definitif/2024/03-13/0138/P9_TA(2024)0138_EN.pdf.

Fanni, S.C., Febi, M., Aghakhanyan, G., Neri, E. (2023). Natural language processing. In: M. E. Klontzas, S. C. Fanni, & E. Neri (Eds.), *Introduction to Artificial Intelligence: Imaging Informatics for Healthcare Professionals*. Springer, pp. 87–99. https://doi.org/10.1007/978-3-031-25928-9_5.

Google. (2023).Google Privacy Policy. https://policies.google.com/privacy/archive/20221215-20230701?hl=en.

Gupta, M., Akiri, C., Aryal, K., Parker, E., & Praharaj, L. (2023). From ChatGPT to ThreatGPT: Impact of generative AI in cybersecurity and privacy. *IEEE Access*, 11, 80218–80245. https://doi.org/10.1109/ACCESS.2023.3300381.

Hancock, P. A., Billings, D. R., & Schaefer, K. E. (2011). Can you trust your robot? *Ergonomics in Design*, 19(3), 24–29.

Heston, T. F., & Khun, Ch. (2023). Prompt engineering in medical education. *International Medical Education,* 2(3), 198–205. https://doi.org/10.3390/ime2030019.

Hsu, T., & Thompson, S. A. (2023). Disinformation researchers raise alarms about AI chatbots. International New York Times, February 8, 2023.

Ismatullaev, U. V. U., & Kim, S.-H. (2024). Review of the factors affecting acceptance of AI-infused systems. *Human Factors*, 66(1), 126–144. https://doi.org/10.1177/00187208211064707.

Jones, J. (2024). Will LLM adoption demand more stringent data security measures? Forbes, May 31, 2024. https://www.forbes.com/sites/hessiejones/2024/05/31/will-llm-adoption-demand-more-stringent-data-security-measures/.

Karras, T., Laine, S., Aittala, M., Hellsten, J., Lehtinen, J., & Aila, T. (2020). Analyzing and improving the image quality of stylegan. In: *Proceedings of the IEEE/CVF Conference on Computer Vision and Pattern Recognition*, IEEE, Seattle, WA, pp. 8110–8119.

Kosinski, M., & Forrest, A. (2024). What is a prompt injection attack? *IBM.com*, March 26, 2024. https://www.ibm.com/topics/prompt-injection.

Kost, E. (2023). ChatGPT to Improve Your Cybersecurity Posture. https://www.upguard.com/blog/usingchatgpt-to-improve-cybersecurity-posture#:~:text=ChatGPT%20can%20help%20security%20teams,lead%20to%20a%20data%20breach (accessed June 26, 2023).

Kwan, S., Stiefmueller, C., & Leitner, C. (2024). Exploring regulatory frameworks for AI/ML through different lenses: A comparative approach. In: C. Leitner, R. Nägele, C. Bassano, & D. Satterfield (Eds.), *The Human Side of Service Engineering: AHFE (2024) International Conference. AHFE Open Access*, vol 143. AHFE International, USA. https://doi.org/10.54941/ahfe1005080.

Lee, K., Cooper, A. F., & Grimmelmann, J. (2024). Talkin bout AI generation: Copyright and the generative-AI supply chain (the short version). In: *CSLAW '24: Proceedings of the Symposium on Computer Science and Law*, Boca Raton, Florida, USA, pp. 48–63. https://doi.org/10.1145/3614407.364369.

Lund, B. D., & Wang, T. (2023). Chatting about ChatGPT: How may AI and GPT impact academia and libraries? *Library Hi Tech News*, 40(3), 26–29.

Marvin, G., Hellen, N., Jjingo, D., & Nakatumba-Nabende, J. (2024). Prompt engineering in large language models. In: I. J. Jacob, S. Piramuthu, & P. Falkowski-Gilski (Eds.), *Data Intelligence and Cognitive Informatics: ICDICI 2023. Algorithms for Intelligent Systems*. Springer, Singapore. https://doi.org/10.1007/978-981-99-7962-2_30.

Maynez, J., Narayan, S., Bohnet, B., & McDonald, R. (2020). On faithfulness and factuality in abstractive summarization. In: *Proceedings of the 58th Annual Meeting of the Association for Computational Linguistics (ACL), 1906–1919,* Seattle, WA, USA. Association for Computational Linguistics. https://doi.org/10.18653/v1/2020.acl-main.173.

McCarthy, J. (1988). Review of the question of artificial intelligence. *Annals of the History of Computing,* 10(3), 224–229.

McCarthy, J., & Minsky, M. L. (1956). A proposal for the Dartmouth summer research project on Artificial Intelligence. https://www-formal.stanford.edu/jmc/history/dartmouth.pdf.

McCulloch, W.S., & Pitts, W. A. (1943). Logical calculus of the ideas immanent in nervous activity. *Bulletin of Mathematical Biophysics,* 5, 115–133. https://doi.org/10.1007/BF02478259.

Metz, C. (2024). A hacker stole OpenAI Secrets, raising fears that China could, too. *New York Times*, July 4, 2024. https://www.nytimes.com/2024/07/04/technology/openai-hack.html.

Mireshghallah, N., & Kim, H. (2024). Can LLMs keep a secret? Testing privacy implications of language models via contextual integrity theory. arXiv.org; Ithaca, June 28, 2024.

Moallem, A. (2019). Social engineering. In: A. Moallem (Ed.), *Human-Computer Interaction and Cybersecurity Handbook.* CRC Press/Taylor & Francis Group, Boca Raton, pp. 139–156.

OpenAI. (2024). Reimagining secure infrastructure for advanced AI. *OpenAI.com*, May 3, 2024. https://openai.com/index/reimagining-secure-infrastructure-for-advanced-ai/.

OpenAI. (2023). March 20 ChatGPT outage: Here's what happened. *OpenAI,* March 24, 2023. https://openai.com/index/march-20-chatgpt-outage/#technical-details.

Parasuraman, R., & Manzey, D. H. (2010). Complacency and bias in human use of automation: an attentional integration. *Human Factors*, 52(3), 381–410. https://doi.org/10.1177/0018720810376055.

Pollicino, O., Riva, G., Gui, M., & Sartor, G. (2023). ChatGPT: Lessons learned from Italy's temporary ban of the AI chatbot. *The Conversation.* https://theconversation.com/chatgpt-lessons-learned-from-italys-temporary-ban-of-the-ai-chatbot-203206.

Poremba, S. (2023). ChatGPT confirms data breach, raising security concerns. *Security Intelligence.* June 26, 2023. https://securityintelligence.com/articles/chatgpt-confirms-data-breach/.

Ray, S. (2023). Samsung bans ChatGPT among employees after sensitive code leak. Forbes, May 2, 2023. https://www.forbes.com/sites/siladityaray/2023/05/02/samsung-bans-chatgpt-and-other-chatbots-for-employees-after-sensitive-code-leak/.

Reynolds, L., & McDonell, K. (2021, May). Prompt programming for large language models: Beyond the few-shot paradigm. In: *Extended Abstracts of the 2021 CHI Conference on Human Factors in Computing Systems,* CRC Press/Taylor & Francis Group, Boca Raton, pp. 1–7.

Roy, V. (2024). Protecting sensitive data in the age of large language models (LLMs). *Medium (VinaySays)*, June 19, 2024. https://vinaysays.medium.com/protecting-sensitive-data-in-the-age-of-large-language-models-llms-89abeb09720d.

Sanders, L., & Schneier, B. (2023). How ChatGPT hijacks democracy. *The New York Times,* January 15, 2023. https://www.nytimes.com/2023/01/15/opinion/ai-chatgpt-lobbying-democracy.html.

Sohl-Dickstein, J., Weiss, E., Maheswaranathan, N., & Ganguli, S. (2015, June). Deep unsupervised learning using nonequilibrium thermodynamics. In: *International Conference on Machine Learning*, PMLR, Lille, France, pp. 2256–2265.

The Guardian Staff. (February 17, 2024). Microsoft-backed OpenAI valued at $80bn after company completes deal. *The Guardian*. ISSN 0261-3077.

Stahl, B. C., & Eke, D. (2024). The ethics of ChatGPT: Exploring the ethical issues of an emerging technology. *International Journal of Information Management*, 74, 102700. ISSN 0268-4012, https://www.sciencedirect.com/science/article/pii/S0268401223000816.

Thorbecke, C. (2023). Don't tell anything to a chatbot you want to keep private. *CNN*, April 6, 2023. https://amp.cnn.com/cnn/2023/04/06/tech/chatgpt-ai-privacy-concerns/index.html.

Turing, A. M. (1950). Computing machinery and intelligence. *Mind*, LIX(236), 433–460. https://doi.org/10.1093/mind/LIX.236.433.

Vaswani, A., Shazeer, N., Parmar, N., Uszkoreit, J., Jones, L., Gomez, A. N., Kaiser, Ł., & Polosukhin, I. (2017). Attention is all you need. *Advances in Neural Information Processing Systems (NeurIPS)*, 30, 5998–6008. https://doi.org/10.48550/arXiv.1706.03762.

Vincent, J. (2022). The scary truth about AI copyright is nobody knows what will happen next. *The Verge*, November 15, 2022. https://www.theverge.com/23444685/generative-ai-copyright-infringement-legal-fair-use-training-data.

Walczak, K., & Cellary, W. (2023). Challenges for higher education in the era of widespread access to Generative AI. *Economics and Business Review*, 9(2), 71–100. https://doi.org/10.18559/ebr.2023.2.743.

Walczak, K., & Cellary, W. (2025). Navigating risks: Inaccuracies, bias, disinformation, and privacy in educational AI. In: J. R. Corbeil & M. E. Corbeil (Eds.), *Teaching and Learning in the Age of Generative AI. Evidence-Based Approaches to Pedagogy, Ethics, and Beyond.* Routledge, London, 163–194. ISBN 9781032683652.

Wang, B., Chen, W., Pei, H., Xie, C., Kang, M., Zhang, C., Xu, C., Xiong, Z., Dutta, R., Schaeffer, R., Truong, S. T., Arora, S., Mazeika, M., Hendrycks, D., Lin, Z., Cheng, Y., Koyejo, S., Song, D., & Li, B. (2023). DecodingTrust: A comprehensive assessment of trustworthiness in GPT models. *Advances in Neural Information Processing Systems*, 36. https://doi.org/10.5555/3666122.3667483.

Weatherbed, J. (2023). Google confirms it's training Bard on scraped web data, too. *The Verge*, July 5, 2023.

Wells, D. (2024). Forager's Harvest 101: A Comprehensive Guide to Identifying, Preserving, and Preparing Wild Edible Plants, Mushrooms, Berries, and Fruits. https://www.amazon.com/Foragers-Harvest-101-Comprehensive-Identifying/dp/B0CPCKKR1F.

Zirpoli, C. T. (2023). Generative artificial intelligence and copyright law. *Congressional Research Service,* Report No. LSB10922, September 29, 2023. https://crsreports.congress.gov.

10 Large Language Models in Human–Computer Interaction

Using LLM-Generated Personas to Model Everything from Minority Views to Entire Ecosystems

Danial Amin, Joni Salminen, Bernard J. Jansen, Ilkka Kaate, and Waleed Akhtar

10.1 INTRODUCTION

10.1.1 Personas at a Glance

Personas, conventionally defined as fictional but realistic human characters representing an actual user segment (Salminen, Wenyun Guan, Jung, & Jansen, 2022; Nielsen, 2019), are an essential technique for user-centered design (UCD) and human–computer interaction (HCI). Since their introduction (Cooper, 1999), personas have been enhancing the design process by allowing key stakeholders (designers, product developers, decision-makers) to empathize with users and contextualize information in a grounded form (Bødker, Christiansen, Nyvang, & Zander, 2012; Grudin & Pruitt, 2002). Personas have been applied in marketing and social science (Guan, Salminen, Nielsen, Jung, & Jansen, 2021; Salminen, Wenyun Guan, et al., 2022) and implemented in various industries, including software development, healthcare, higher education, robotics, cybersecurity, games, and recommendation systems (Salminen, Wenyun Guan, et al., 2022; Anvari, Richards, Hitchens, & Tran, 2019; Nielsen, Hansen, Stage, & Billestrup, 2015; Goodman-Deane et al., 2018).

In addition to the commercial use of personas, which most persona applications target, there has been an increase in the usage of *personas for social good* (PFSG). PFSG refers to the application of personas toward beneficial social goals, such as sustainability, social justice, and quality of life (Jansen, Jung, Salminen, An, & Kwak, 2018; Guan, Salminen, Jung, & Jansen, 2023). These advancements in persona application are driven by broader trends in the HCI landscape, which is paying

DOI: 10.1201/9781003492252-10

more attention to applying UCD to societal purposes, such as those embodied in the United Nations (UN) Sustainable Development Goals (SDGs)—sustainable HCI (sHCI) being a leading example (Hansson, Cerratto Pargman, & Pargman, 2021). Simultaneously, personas are increasingly being applied beyond the scope of product development, extending to awareness, engagement, communication, and development projects and solutions in a broader range of human activity (Guan et al., 2023; Wilson, De, Forbes, & Sachy, 2020).

Personas are widely used in HCI and UCD for increasing empathy (Salminen, Santos, Jung, & Jansen, 2023; Nielsen, 2019; Cooper, 1999) and relatability (Jansen, Salminen, Jung, & Guan, 2021a), which in turn improve the design and decision-making process. A correctly formulated persona can provide a deeper understanding of the end user (Salminen, Jung, Chowdhury, Sengün, & Jansen, 2020). While the term end-user (user) is used here as an example, personas could be expanded to *any group of people* and, more broadly, *any group of parties in an ecosystem*, which we will discuss in Section 10.4. Furthermore, personas can also convey the needs of the target group, which helps designers empathize with the needs and expectations of a real user to develop more contextualized applications (and, more broadly, solutions) (Jansen et al., 2021a; Grudin, 2006). In almost any decision-making process, personas help decision-makers make impactful strategic decisions by helping them change their perspective (Salminen, Wenyun Guan, et al., 2022; Guan et al., 2023). Personas can mitigate *confirmatory* and *self-referential* biases by "stepping beyond" their own experiences and taking the perspective of others (potential users, affectees, and other stakeholders) represented by personas (Nielsen, 2002; Bødker et al., 2012). Similarly, personas also help different stakeholders in a project to communicate with each other more concretely, as they present a true representation of user needs (Salminen, Wenyun Guan, et al., 2022; Jensen, Elverum, & Steinert, 2017). According to the theory of personas (Grudin, 2006), a high-quality persona, which generally includes attributes such as demographics, behaviors, pain points, and goals, can improve the likelihood of making informed design choices that resonate with users (Sengün, Salminen, Jung, Aldous, & Jansen, 2024).

Over the years, personas have evolved into various types and forms (Salminen, Jung, & Jansen, 2022). As personas were considered a qualitative method in their inception paper (Cooper, 1999), the early years of persona generation generally applied qualitative methods. These methods involved subject matter experts (SMEs) often collecting data through through ethnographic studies, surveys, or focus groups (Jansen, Salminen, Jung, & Guan, 2021b). There has been a shift in persona generation from qualitative to quantitative methods (Jansen, Salminen, & Jung, 2020). The creation of personas using quantitative methods primarily includes (1) data-driven persona creation (using algorithmic methods based on numerical and textual data (Jansen et al., 2020)), (2) automatic persona generation (continuous updating of personas on the latest available data (Jung, Salminen, Kwak, An, & Jansen, 2018; Jansen, Jung, & Salminen, 2019)), and (3) generative artificial intelligence (GenAI) persona generation (using large language models (LLMs) (Shin, Hedderich, Rey, Lucero, & Oulasvirta, 2024) and diffusion models (Zhou et al., 2024)). The personas generated by using GenAI technologies are called GenAI personas. Among GenAI personas, the current work focuses on *LLM-generated personas*, which are partially or fully generated using LLMs (Shin et al., 2024). Despite these advancements, marginalized communities, based on either their race, gender, religion, or demographics in the global populace distribution, are often ignored or misrepresented in data-driven persona creation (Salminen, Jung, & Jansen, 2021). This misrepresentation is mainly due to a lack of representation in the data, challenges with aggregation, and the inherent algorithmic biases in data-driven persona generation (Salminen et al., 2021; Salminen, Froneman, Jung, Chowdhury, & Jansen, 2020). This chapter discusses the applicability of LLM-generated personas for PFSG, ranging from representing minority views on one end to modeling the complete ecosystem on the other.

10.1.2 HCI and GenAI Landscape

While GenAI technology is not novice, the release of ChatGPT by OpenAI (OpenAI, 2022) has created a butterfly effect (Nah, Zheng, Cai, Siau, & Chen, 2023), in which LLMs have shown potential in nearly all aspects of human activity, from acting as a learning companion (Salminen, Jung, et al., 2024) to solving sustainability challenges (Vinuesa et al., 2020). However, these advanced LLM-based systems are prone to challenges, for example, inherent bias, lack of diversity and inclusivity, and non-transparent or opaque functional mechanisms (Nah et al., 2023; Cachat-Rosset & Klarsfeld, 2023; Farahani & Ghasemi, 2024; Roche, 2023). Studies have attempted to address these challenges (Farahani & Ghasemi, 2024; Zhuo, Huang, Chen, & Xing, 2023), e.g., by developing ethical guidelines and frameworks (Roche, 2023; Cachat-Rosset & Klarsfeld, 2023; Nah et al., 2023). However, the challenge of inclusive design (Goodman-Deane, Bradley, Waller, & Clarkson, 2021) and fairness in LLMs (Anthis et al., 2024) remains persistent.

The impact of GenAI, in general, and LLMs, in particular, on HCI is increasingly prominent, as HCI practitioners and researchers reshape how users interact with systems. HCI researchers are exploring how LLMs can enhance user experiences by addressing concerns about usability, transparency, and ethical implications (Nah et al., 2023). LLMs' ability to personalize interactions and automate tasks presents an opportunity for HCI practitioners (Schuller et al., 2024; Salminen, Liu, et al., 2024) to explore LLMs and develop interactive and engaging interactions (Choi et al., 2025).

Applying LLMs in persona creation has become a topic of interest within the HCI community (Cho, Jon, Jin, Jung, & Oh, 2024; Shin et al., 2024; Schuller et al., 2024). Specifically, LLMs can help streamline persona development by processing large datasets to identify user patterns (Salminen, Liu, et al., 2024) and create coherent and cohesive text for persona generation (Shin et al., 2024). However, the use of LLM-generated personas has also increased the risks related to bias and inclusivity, as training data or model parameters may not accurately reflect the diversity of the target user base (Goodman-Deane et al., 2021; Cachat-Rosset & Klarsfeld, 2023; Amin et al., 2025).

The HCI community's relationship to LLMs and its use in practice is a balancing act between risk and opportunity. In one body of research, HCI scholars are extremely critical about replacing authentic user research with *fabricated replicas or surrogates (i.e., "synthetic users").* In contrast, another body of research suggests that LLMs have benefits in various persona generation tasks, ranging from data analysis to the write-up (Shin et al., 2024; Salminen, Liu, et al., 2024; De Paoli, 2023). These perspectives indicate that different studies have been carried out for assessing LLMs in HCI in general and persona generation in particular, which are done primarily case by case (Shin et al., 2024; Schuller et al., 2024; Nah et al., 2023). This represents a significant research gap in the existing body of knowledge.

To this end, the motivational question behind our work is as follows: *How to use LLMs in HCI for persona creation?* To answer part of this question, we suggest one prominent use case for LLMs in HCI: *data-driven persona generation*. More particularly, we suggest that using LLMs could address some of the shortcomings of previous technologies in data-driven persona generation, namely (1) lack of attention to minority views (Salminen et al., 2021) and (2) lack of attention to modeling dynamic scenarios in an ecosystem setting.[1] We propose two approaches to addressing these challenges through LLM-generated personas. Consequently, this chapter addresses two vital research questions (RQs):

- RQ1: *How to ensure that marginalized communities' voices are represented in LLM-generated persona processes?*
- RQ2: *How can LLM-generated personas be deployed to model an entire ecosystem?*

We address these RQs through conceptual examination, providing illustrative examples of our suggested approaches. The following section presents the literature background of this work. The following two sections dive into our two suggested approaches: (1) representing minority viewpoints using LLM-generated personas in the context of the marginalized population in the Global South and (2) modeling entire ecosystems (industrial, societal, and natural) using LLM-generated personas. We conclude by discussing design principles, key takeaways, and directions for future research.

In a nutshell, this book chapter makes notable conceptual contributions to using LLMs in HCI (and beyond). Essentially, we propose new targets for LLM-based persona generation, including minority perspectives and "voices that would not otherwise be heard," such as those of marginalized communities in the Global South and non-human parties in ecosystems. We also propose new and innovative evaluation techniques that have not been previously deployed for personas, stemming from the fact that personas have a strong tradition in HCI and business disciplines. Our contribution draws attention to representing ecosystem parties, including non-humans, using personas and creating interactive "playgrounds" where decision-makers can engage with these personas, thereby obtaining a more holistic perspective of different parties (actors and subjects) in the ecosystem. Consistent with the theory of personas (Grudin, 2006), this approach could perpetuate empathy and mitigate self-centered bias when decision-makers take actions that affect the ecosystem, whether that ecosystem is industrial, societal, or ecological in nature.

10.2 LITERATURE BACKGROUND

10.2.1 PFSG: Opportunities and Risks

Personas are generally used in commercial use cases (Salminen, Wenyun Guan, et al., 2022) (i.e., developing targeted personas to develop and align a product or service to market needs). While personas are effective for commercial purposes, they *also* have the potential to contribute to social good (Guan et al., 2023). For example, in policymaking, the persona of the target group could be used to better align policy actions (Department of the Prime Minister and Cabinet (DPMC), 2023). There is an increasing trend to combine business needs with social good. To this end, PFSG helps decision-makers empathize with parties who benefit or are affected by the policy, leading to actions that are more focused on user empowerment and improving community well-being.

Persona science emphasizes the alignment of the needs of the targeted user groups with the persona, ensuring that personas are not only representative tools but are actionable to tackle real-world issues within a broad societal frame (Guan et al., 2023). Furthermore, incorporating additional information, such as narrative information, including social attitudes and cultural backgrounds, with persona profiles gives a richer understanding of user contexts, which in turn facilitates designs aimed at social change (Salminen, Wenyun Guan, et al., 2022). Additionally, personas can be particularly effective in designing platforms that improve civic engagement, encouraging active participation in governance and community activities (Guan et al., 2023). Lastly, the strategic use of diverse personas can also enhance cross-cultural communication, promoting inclusivity and dialogue among different cultural groups, which is essential in addressing global challenges such as inequality (Salminen, Wenyun Guan, et al., 2022). In general, the use of PFSG helps to understand various user needs and to create impactful designs that align with broader social objectives (Guan et al., 2023).

Building upon existing applications of PFSG, there is significant potential in integrating personas with the study of human–nature interactions (HNI) in addition to HCI. In this context, HNI research explores how people engage with and relate to their natural environment, encompassing aspects such as environmental stewardship, sustainability practices, and the psychological benefits of nature (Capaldi, Passmore, Nisbet, & Zelenski, 2014; Kim, Park, & Lee, 2015). Considerations from sustainability studies can help develop personas that reflect the social and economic dimensions of user groups and their interactions with the natural world (Lee & Jung, 2019; Johnson, Martinez, & Liu, 2020). Similarly, personas can help represent the needs of all the stakeholders in HNI, including both the human and non-human participants (e.g., plants, trees, animals, birds, materials) (Yanlic & Coskun, 2024).

Integrating HCI with ecological systems requires a fundamental shift away from human-centered design toward an ecosystem-aware approach that considers the complex interdependencies between digital technologies, human activities, and environmental sustainability (Knowles, Bates, & Håkansson, 2018). As sustainability becomes increasingly central to HCI research, there is a growing recognition that the field must move beyond incremental improvements in energy efficiency to address systemic challenges at the intersection of computing, society, and ecology (Bremer, Knowles, & Friday, 2022). Drawing from theories of group cognition and sustainable HCI, we can envision tools and systems that make visible the nested relationships between individual behaviors, collective practices, and environmental impacts (Borge & Xia, 2023). This requires developing new interaction paradigms that foreground ecological considerations alongside usability, fostering what has been termed "cyber-sustainability" (Preist, Schien, & Blevis, 2016).

Although personification (use of human-like characteristics for non-human parties) of the ecosystem could be challenging, it is essential to consider that experts in ecological landscapes (Vergani & Di Liberto, 2024) have used personification to establish personalities of different non-human parties (MacKinlay & Shaw, 2023). Furthermore, through personification, humans empathize with non-human parties more strongly, leading to an overall increase in the improvement of interaction (Cherry et al., 2022). For instance, a persona representing community leaders in the context of sustainability could include attributes related to their interaction with the environment and non-human parties, environmental advocacy, and sustainable resource management. This would ideally encourage policymakers to design initiatives that support community development and ecological sustainability (Smith, Nguyen, & Brown, 2021).

Furthermore, integrating HNI into persona development can help identify and address barriers to sustainable practices, such as people's lack of access to green spaces or inadequate environmental education (Johnson et al., 2020; Wang, Zhao, & Chen, 2019). This conceptual linkage not only enriches the depth and relevance of PFSG projects but also ensures that solutions are aligned with the ecological realities and aspirations of the target populations (human and non-human) alike (Smith et al., 2021). Consequently, expanding PFSG to include HNI represents a promising avenue to create more comprehensive and effective personas that drive meaningful social and environmental change (Capaldi et al., 2014; Kim et al., 2015; Lee & Jung, 2019; Johnson et al., 2020; Smith et al., 2021).

10.2.2 LLM-Generated Personas: Opportunities and Risks

LLM-generated personas use the natural text generation qualities of LLMs to create human-interpretable, coherent textual profiles (Schuller et al., 2024). This combination allows for a more immersive UX for personas, as users can interact with personas that closely mimic actual human characteristics and behaviors. This has enabled persona generation partially supported by LLMs (Shin et al., 2024) or fully automated (Schuller et al., 2024). The specific workflows in which personas are generated while keeping SMEs in the loop have showcased the quality and depth of information (Salminen, Liu, et al., 2024; Shin et al., 2024). This introduces the specific new avenues that LLMs have opened. LLMs demonstrate the ability to generate coherent persona narratives in minutes, compared to the traditional persona creation process that could take hours or days. When provided with clear prompts (instructions), these models can create detailed representations, including background stories, pain points, and behavioral patterns with internal consistency (Sattele & Carlos

Ortiz, 2024). LLMs are also good at maintaining internal narrative coherence and overall quality (Salminen, Liu, et al., 2024).

However, these opportunities are counterbalanced by specific risks: First, LLMs consistently sanitize reality, as evidenced in both the Iztapalapa water crisis case (Sattele & Carlos Ortiz, 2024), where serious issues like water truck hijackings were understated, and in addiction personas where harsh realities were softened (Salminen et al., 2024). Second, the models exhibit systematic biases—from demographic skewing toward younger age groups to strong US-centricity (85.6% of generated personas being American without any prompt specification). Third, the hallucination problem manifests itself in the generation of plausible but factually incorrect details, such as personas with contradictory medical conditions or financially improbable scenarios (Salminen, Liu, et al., 2024).

10.3 SUGGESTED APPROACH 1: MODELING MINORITY VIEWPOINTS IN A DATA-DRIVEN PERSONAS

10.3.1 Potential Impact of Using LLMs to Model Marginalized Viewpoints

LLMs, with their ability to generate multiple personas in parallel and contribute to real-time changes in the persona based on contextualization, have the potential to significantly impact the fields of persona generation, especially for minority communities. As LLMs can generate multiple personas without additional cost, the persona set (a collection of personas) can be expanded to include all possible minorities. Similarly, as LLMs can generate cohesive text in real-time, the generated personas could be used interactively. Thus, research on LLM-generated personas has the potential to significantly impact the field of persona development, PFSG, and policy-making for marginalized communities. Developing more accurate and culturally sensitive LLM-generated personas could contribute to a better understanding of marginalized and socially challenged communities. Research in LLM-generated personas can thus lead to more inclusive and effective decision-making processes as policymakers access a closer-to-truth representation of such communities.

10.3.2 Example Context

We illustrate the challenges of data-driven personas and the potential use case of LLMs to represent minority views by considering the case of representing marginalized communities using the Global South as a context. The Global South region, which includes Latin America, Asia, Oceania, and Africa, is the most populous region in the world, accounting for 88% of the world's population (Kishore Mahabani, 2024). However, despite its large population, the Global South faces severe challenges when modeling minorities, especially using any data-driven approach. These challenges in representing marginalized communities in the Global South include a lack of targeted population data, inaccurate data, lack of cultural contextualization, and algorithmic bias. Due to its population, demographics, challenges in minority populations, and age distribution (World Economics, 2024), the Global South appears to be a prominent

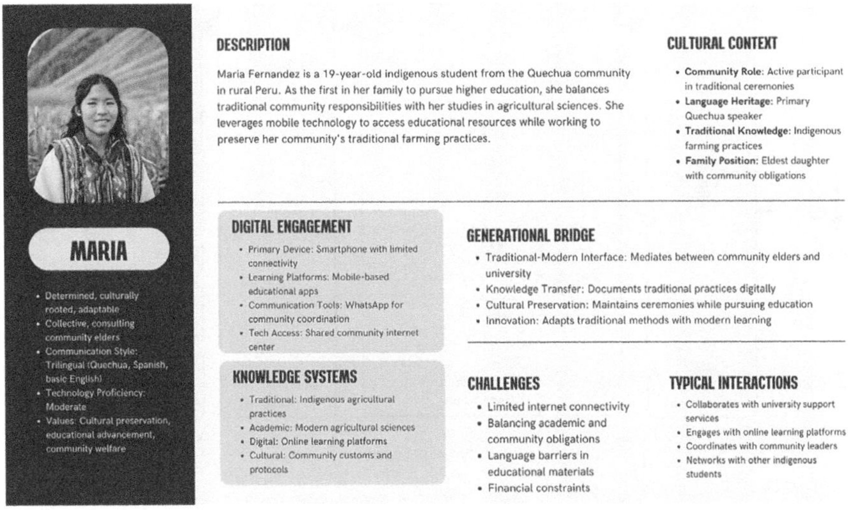

FIGURE 10.1 A representative minority persona for a rural student.

(yet underrepresented) target population for LLM-generated personas. Such minority personas serve as crucial tools for understanding the complex intersections of traditional values, technological adoption, and educational aspirations in marginalized communities. To better illustrate this, we can consider the example of Indigenous students in higher education institutions in Peru, as illustrated in Figure 10.1.

10.3.3 Addressing the Representation of Minority Viewpoints in LLM-Generated Personas

The effective representation of marginalized communities by LLM-generated personas is focused on three dimensions: (1) understanding the paradigms of LLM-generated personas, (2) manifestations of different strategies, and (3) technologies to improve the LLM-generated personas. These dimensions are driven by the positive impact on PFSG, contributing to an increase in transparency in representation. The overall space is presented in Figure 10.2b.

The research aims to identify different improvements and enhancements to the foundational LLMs for LLM-generated personas and assess their impact in the PFSG context for modeling the marginalized community in the Global South. To address LLM-generated personas in a minority-friendly manner, we propose structured research that would follow the steps of real-data collection, persona generation, and evaluation of the personas through quantitative (statistical) and qualitative methods.

The proposed methodological structure is presented in Figure 10.2a.

In our methodology for minority-friendly LLM-generated personas, the survey data will be collected from the community through a questionnaire. These data will be cross-referenced with simulated survey data to identify the efficacy of data simulation to enhance the diversity of the datasets. These data will then be passed on to a pipeline using LLMs and enhanced techniques (1) retrieval-augmented generation

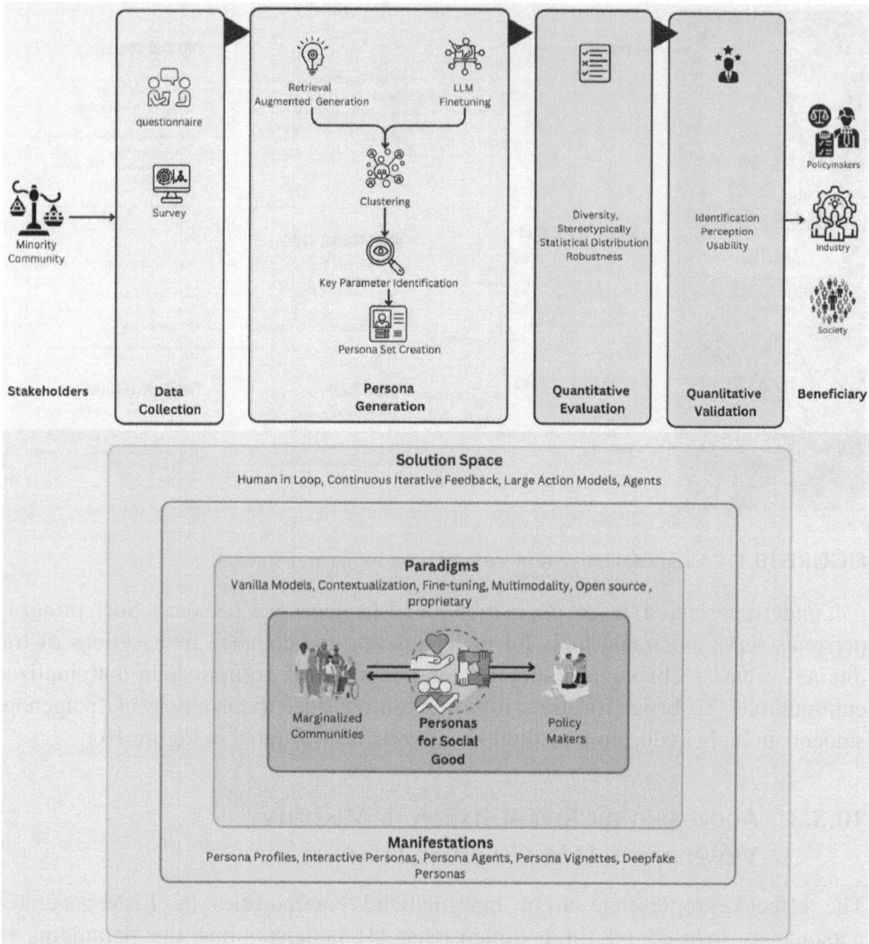

FIGURE 10.2 Comprehensive framework for personas for social good (PFSG). (a) The end-to-end methodology for generating and validating personas, showing the flow from stakeholder input to beneficiary impact. (b) The conceptual framework showing the layered approach from solution space to core implementation, highlighting the connection between marginalized communities and policymakers.

(RAG) (i.e., textual generation supported by a larger context from a knowledge base), and (2) fine-tuned LLMs (i.e., a model that is trained over specific data to enhance its robustness), to develop persona sets. These persona sets will then be collectively assessed for quantitative metrics highlighting diversity, stereotyping (making assumptions about prejudice), distribution, and robustness. Similarly, qualitative metrics such as identification, perception, and usability by SMEs will also be measured. Through this process, the appropriate paradigms of persona enhancement for marginalized communities will be identified and used to create truly representative personas. These personas can then be applied in policy-making workshops with actual policymakers to identify their effectiveness and correlate them to the user's use case to address policy-making needs.

The ethical considerations for applying LLM-generated personas for modeling minority viewpoints are extensive and critical, given the focus on LLM-generated personas for marginalized communities in the Global South. These include but are not limited to: (1) obtaining genuine informed consent from participants and ensuring robust data privacy and protection measures that comply with both local and international regulations; (2) implementing rigorous strategies to identify, mitigate, and transparently report on biases in data collection, persona generation, and result interpretation; and (3) establishing transparency in the process for LLM-generated personas, including clear communication of the limitations and potential biases to all stakeholders, especially key decision-makers. The approach should also (1) consider and safeguard against potential misuse of the developed personas; (2) assess long-term impacts on policy-making and perceptions of marginalized communities; (3) ensure meaningful involvement of community members throughout the research process; and (4) establish clear protocols for benefit-sharing with participating communities.

10.4 SUGGESTED APPROACH 2: MODELING *ENTIRE ECOSYSTEMS* AS DATA-DRIVEN PERSONAS

10.4.1 Modeling Ecosystems using Personas

As defined in the preceding section, an ecosystem encompasses a collection of parties. These parties can include *actors*, who actively participate in the activities of the ecosystem, and *subjects*, who are indirectly involved in or affected by the activities of others in the ecosystem (Brussard, Reed, & Tracy, 1998). We will refer to them as *parties* hereafter. As a collection of multiple parties, ecosystems are prone to a series of complex interactions with other parties within and between ecosystems (Brussard et al., 1998; Green & Sadedin, 2005; Gross, Toombs, Wain, & Walorski, 2011).

This ecosystem modeling is essential for understanding and addressing different challenges in a broad spectrum of domains, ranging from environmental science, industry, and social systems, as they all form an ecosystem in one way or another (Brussard et al., 1998; Holling, 2001; Geary et al., 2020). For example, in ecology, modeling the natural ecosystem is essential for understanding the interactions between different biotic factors (living components) and abiotic factors (nonliving parties in the ecosystem) and understanding the impact of a change in the ecosystem (Keystone et al., 2015). Similarly, for the societal ecosystem, understanding the different perspectives of different members of society is crucial for establishing and creating an action plan (Piussi & Farrell, 2000). For the industrial ecosystem, modeling the different actors and understanding their position on the challenge of supply chain or risk management is crucial (Carpenter, Kitchell, & Hodgson, 1998).

Even though the science of modeling ecosystems is not new (Geary et al., 2020), representing ecosystem parties as personas appears to be novel, and, to our knowledge, has not been attempted before. Several methods and techniques exist for modeling ecosystems in different domains such as ecology, management sciences, HCI, and HNI (Acs, Stam, Audretsch, & O'Connor, 2017; Evans, 2003; Holling, 2001), among others. First, these different modeling methods are often time-consuming and require SMEs to understand and model the ecosystem. Second, these ecosystem models are static in nature and thus do not offer dynamic handling of the interactions. Third, while these interactions are formulated, they are formulated from an outsider's perspective. Our approach of using LLM-generated personas can address these shortcomings by offering a direct representation of the needs of different stakeholders, empathizing with them, and enabling interactive representations.

Personas inherently bridge the gap between the decision-makers and the members of the ecosystem by providing them with an empathetic representation of different parties. By representing different parties as personas in this ecosystem model, one can interact, empathize, and relate to their perspective and pain points with a reduction in self-referential biases. *LLM-generated personas help reduce the challenge of the outsider's perspective in ecosystem modeling.* While personas could potentially solve the challenges of externality (third-person) in ecosystem modeling, they do not solve the challenges related to the complexity of time and expertise when considered in their traditional form (classical qualitative or mixed method personas) (Chapman & Milham, 2006). This challenge faced by conventional personas, related to time, effort, and resource hunger, is solved by LLM-generated personas. LLM-generated personas are scalable, indicating that little or no effort is required to generate either one persona or a persona set consisting of hundreds of them. This, in turn, demonstrates that persona sets that involve all the different parties in an ecosystem could be generated without significant human effort. Furthermore, to improve the contextualization of LLM-generated personas, techniques such as RAG[2] and fine-tuning[3] can help to create more contextualized personas in a grounded form.

Technically, LLMs can process large amounts of unstructured data and generate coherent, contextually relevant text (Brown et al., 2020), which makes LLMs suitable for creating detailed personas representing various stakeholders within an ecosystem. This data is dependent on different ecosystems. For example, in the industrial ecosystem, this data could include information on key customers, the performance of companies, and policies for regulators. For a societal ecosystem, the data can be the civic rights of the residents, the residents' population demographics, and civic engagement surveys. Lastly, for natural ecosystems, this data would include ecological reports by the SMEs, personality traits of different non-human parties of the ecosystem, and data about the population growth over time.

The concept of personas can be extended to represent *all parties within an ecosystem*, not just end-users and not just humans. By creating personas for every stakeholder group, including suppliers, regulators, competitors, and even environmental entities, we can try to capture the full spectrum of perspectives, constraints, and interactions that shape the ecosystem. This comprehensive approach allows for a more holistic understanding of the system, facilitating the identification of potential conflicts, synergies, and emergent behaviors that may be overlooked when focusing solely on end-users (or people, in general).

Ecosystem models require simulation of complex interactions, representing different stakeholders as dynamic agents (Sun, Sun, & Strang, 2016). LLM-generated personas could be incorporated into ecosystem modeling to represent different stakeholders. This enables exploring how changes in one part of the system can ripple through and affect other components, a concept known as emergent behavior

(Holland, 1992). Additionally, LLM-generated personas can enhance predictive capabilities by enabling scenario analysis. Scenario analysis defines and implements different strategies and interventions in a virtual environment before implementation (Meierhofer et al., 2021). Moreover, modeling the complete ecosystem can enable clear communication and interactions between the different stakeholders, facilitating effective communication and perpetuating empathy across the ecosystem parties (Dietz, Ostrom, & Stern, 2003). LLM-generated personas, in general, could be used to model these interactions.

Leveraging LLMs in ecosystem modeling offers significant potential advantages, including increased realism, improved adaptability, facilitated stakeholder engagement, and informed policy-making. For instance, in urban planning, integrating LLM-generated personas representing residents, businesses, and city officials can help design infrastructure projects that balance economic development with community well-being and environmental sustainability (Batty, 2018).

10.4.2 Persona Ecosystem Modeling Framework

In order to extend the concept of persona development from end users to modeling complete ecosystems, we present a four-stage Persona Ecosystem Modeling Framework (PEMF). These four stages are (1) Stage01: Key Stakeholder Mapping and Selection, (2) Stage02: Data Collection and Persona Generation, (3) Stage03: Persona Validation and Selection, and (4) Stage04: Persona Ecosystem Playground (PEP).

Stage 01: Key Stakeholder Mapping and Selection. The first stage of the process is mapping and selecting stakeholders within the ecosystem. This stage involves mapping out the entire network of parties (whether actors or participants). These stakeholders generally include *primary stakeholders*, who are directly involved or affected, and *secondary stakeholders*, who have indirect influences. For example, in modeling the renewable energy ecosystem, stakeholders might include renewable energy companies, traditional fossil fuel companies, government regulators, environmental non-governmental organizations (NGOs), consumers, financial institutions, and technology providers (Kunhikrishnan & Jose, 2021). After establishing the overall stakeholder map of the ecosystem, it is important to select the relevant stakeholders to be included in the ecosystem model. This selection is made on the basis of the nature of the problem, as well as the ecosystem.

Stage 02: Data Collection and Persona Generation. This stage involves two steps, primarily as *qualitative and quantitative data collection* and *persona generation.*

As LLMs are used for persona generation, data sources are required for two different domains. First, qualitative data are required for the LLM contextualization. These data could include (1) industry reports that provide market trends and financial data, (2) academic literature that offers theoretical frameworks and empirical studies, (3) policy documents detailing regulations and governmental strategies, (4) social media and news articles reflecting public opinions and emerging issues, and (5) SME interviews or surveys providing insights from professionals within the ecosystem. These data could be used to contextualize the LLM either with RAG or fine-tuning for the given ecosystem.

Moreover, quantitative data are also collected, which could include (1) population metrics such as demographic information about the population/sample space of each stakeholder, (2) behavior and personality metrics, including data on goals, motivations, interactions, and preferred mode of communication, and (3) economic and engagement metrics indicating consumption and social media metrics. These quantitative data are used for persona development. Considering the previous example, qualitative datasets would help contextualize the LLM with industry-specific knowledge and help create a common context. The understanding of consumers' attitudes towards renewable energy adoption is modeled through the quantitative data on their environmental concerns, economic considerations, and social influences (Lee, Zukauskaite, & Barvidaite, 2017). These data support the creation of representative LLM-generated personas.

Once data are collected, the next step is *persona creation*. Personas are created following a persona generation process where LLMs are used to enrich the data, while an algorithmic data-driven approach enhances the persona generation workflow. Personas are generated using quantitative data, which involves (1) attribute selection, (2) dimensionality reduction or clustering, and (3) enrichment (Nielsen, 2013; Salminen et al., 2018). The first two actions in this process are carried out by the SME with the help of algorithmic methods or directly with algorithmic methods, while enrichment is carried out by LLMs, which are contextualized on qualitative datasets (Shin et al., 2024).

The personas encapsulate demographic information and specific attributes such as goals, challenges, behaviors, and typical interactions with other stakeholders. The LLM processes the data to produce coherent, contextually relevant narratives that accurately represent each stakeholder. For example, a persona for an "Eco-Conscious Consumer" might emphasize their commitment to reducing their carbon footprint, their willingness to pay a premium for sustainable energy options, and their influence on peers through advocacy and social media engagement (Li, Shi, Guo, & Zhao, 2020). Through generating personas for all stakeholders, not just end-users, we extend the traditional use of personas and enable a more holistic model of the ecosystem (Grudin & Pruitt, 2006).

Stage 03: Persona Validation and Selection. The third stage in the process is persona validation and selection, which involves validating the persona sets, representing each party of the ecosystem on three dimensions. The *first dimension* deals with the diversity of the persona set. This metric of diversity is measured to ensure that the persona set selected for the specific party is the true representative of all the groups in the population of the given party. The *second dimension* includes validation of personas against HCI scales like the Persona Perception Scale (PPS) (Salminen, Santos, et al., 2020) to assess the persona on measures such as clarity, empathy, credibility, completeness, consistency, likability, and similarity. The *third dimension* of validation includes human-in-the-loop involvement, where an SME or other representative of the stakeholder party validates the persona to be a representative of the population in question. This validation also helps the persona ecosystem modelers select the relevant personas representative of each group, based on the criterion decided as a threshold for the given personas.

Stage04: Persona Ecosystem Playground (PEP). The last and final stage of the PEMF process is the Persona Ecosystem Playground (PEP). A PEP is an interactive sandbox environment in which the selected personas representing different parties in the ecosystem are deployed so that they can interact with each other. In PEP, each persona representative of the party acts as an agent: an automated representative of the party of the ecosystem. This agent acts as a representation of real and grounded problems, challenges, behaviors, and characteristics of the party of interest.

PEP can support both *human-to-persona interactions (HPIs)* and *persona-to-persona interactions (PPIs). HPIs* are interactions in which users can engage with personas through dialogues and scenario-based activities, whereas *PPIs* are interactions where personas interact with each other according to predefined behavioral rules and decision-making processes. PEP is implemented using LLM-agent-based modeling platforms, such as LangChain[4] or Lang-graph,[5] which are capable of modeling complex systems and collections of LLM agents (in our case as LLM generated personas). These interactions also help capture emergent behaviors resulting from interactions between various agents (Wilensky & Rand, 2015). For instance, in the renewable energy ecosystem model, PEP can be used to interact with the personas in the ecosystem about a policy change proposed by the personas of the government regulator and to understand its effects on investment decisions by financial institution personas and adoption rates among consumer personas. The aim is to allow users to take the perspective of different ecosystem parties.

This methodology is summarized in Figure 10.3.

10.4.3 Application of PEMF in Various Ecosystems

To illustrate the framework's usability in different dimensions, we will apply the PEMF framework to model three different ecosystems: (1) the industrial ecosystem

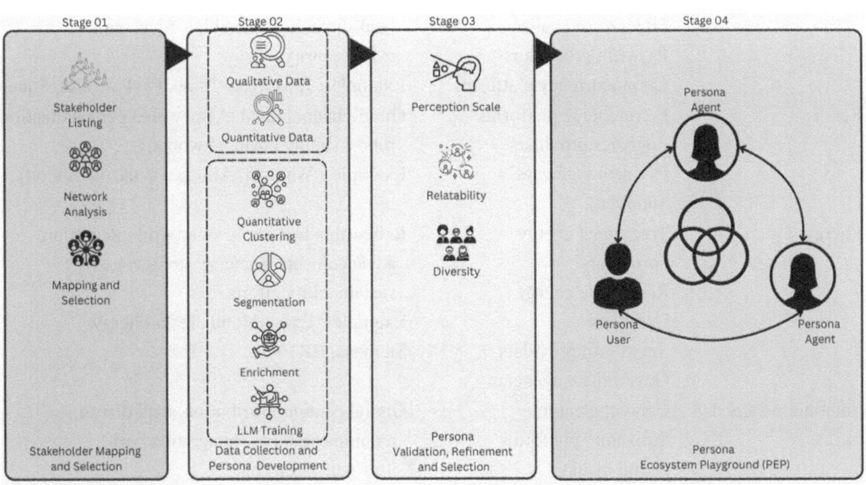

FIGURE 10.3 Theoretical framework to indicate the creation of a persona-based ecosystem.

by modeling the automotive industry, (2) the societal ecosystem through modeling an urban ecosystem, and (3) the natural ecosystem through modeling a forest. These ecosystems are considered examples and could be applied to other ecosystems as well.

10.4.3.1 Modeling Industry Ecosystems as Personas

The application of PEMF has potential in various industries (see Table 10.1). In the *healthcare sector*, this approach can simulate interactions among patients, healthcare providers, pharmaceutical companies, insurance firms, and regulatory agencies, thus improving patient care pathways and policy development (Ferrucci, 2012). The *agricultural industry* can benefit by modeling relationships between farmers, distributors, retailers, consumers, and environmental entities to improve supply chain efficiency and sustainability practices (Rose et al., 2016). In the *financial services sector*,

TABLE 10.1
Examples of Industry Ecosystems

Ecosystem	Key Components	Characteristics and Examples
Technology	Hardware manufacturers Software developers Cloud service providers App stores	Interconnected platforms, extensive developer communities, devices and services integration. Examples: Apple, Google, Microsoft, Amazon.
Automotive	Vehicle manufacturers Parts suppliers EV charging networks Autonomous technology	Collaborative R&D for electric and autonomous vehicles, partnerships with tech companies, integration with smart cities. Examples: Tesla, Toyota, General Motors, Ford.
Healthcare	Hospitals Pharmaceutical companies Insurance providers Health tech startups	Data-driven care, AI diagnostics, digital health records, focus on personalized medicine. Examples: UnitedHealth Group, Pfizer, Johnson & Johnson, Cerner.
Financial services	Banks Fintech companies Payment processors Cryptocurrency platforms	Digital transformation, blockchain integration, bank-fintech partnerships, focus on cybersecurity. Examples: JPMorgan Chase, PayPal, Visa, Square
Retail	E-commerce platforms Logistics providers Payment gateways Suppliers	Omni-channel retail, AI-powered personalization, robust supply chain networks. Examples: Walmart, Alibaba, Amazon, Shopify.
Energy	Traditional energy providers Renewable energy Companies Smart grid providers Government agencies	Renewable transition, smart grid integration, advancements in energy storage, and sustainability efforts. Examples: ExxonMobil, Tesla Energy, Siemens, BP.
Entertainment and media	Content creators Streaming platforms Social media Advertising networks	Digital content distribution, data-driven recommendations, integration with user-generated content. Examples: Disney, Netflix, Spotify, YouTube.

creating personas for banks, investors, regulators, fintech startups, and customers can help stakeholders to understand market dynamics and manage systemic risks (Philippon, 2016). The *automotive industry* is another area where this methodology can be applied to model the complex interactions between manufacturers, suppliers, dealers, customers, policymakers, and environmental groups (Wells & Nieuwenhuis, 2010). To illustrate the usability of PEMF in industrial ecosystems, the automotive industry is modeled as an example. This ecosystem modeling is carried out to understand the perspective of different parties on the penetration of electric vehicles (EVs) in the market.

Context: Due to the UN SDGs (United Nations, 2015) and the shift toward merging products and services so that they provide commercial and societal benefits, the automotive industry is undergoing a significant transformation driven by technological advancements, environmental concerns, and changing consumer preferences (NEF, 2020) to focus on green technology. One such green technology in the automotive ecosystem is electric vehicles, and we are modeling the automotive ecosystem to understand the perspective of different stakeholders on their penetration in the market.

Stage 01: Stakeholder Mapping and Selection: The first step involves systematically identifying all relevant stakeholder groups within the automotive ecosystem. These include primary stakeholders (parties) like (1) *traditional automobile manufacturers* (traditional internal combustion engine (ICE) vehicle and EV manufacturers); (2) *suppliers* (companies providing parts and components, including batteries, semiconductors, and raw materials to all the manufacturers (Schmidt, Hawkes, Gambhir, & Staffell, 2017)); (3) *dealers and retailers* (entities responsible for the sale and distribution of vehicles to consumers); (4) *consumers* (individual and fleet buyers with varying preferences and needs); and (5) *government regulators* (agencies setting safety standards, emissions regulations, and providing incentives for EV adoption (Wang et al., 2019)). Some other secondary stakeholders could include (1) *environmental NGOs* (organizations advocating for sustainable practices and reduced emissions); (2) *technology firms* (companies developing autonomous driving technologies, connectivity solutions, and mobility services (Litman, 2020)); (3) *energy providers* (utility companies and charging infrastructure providers supporting EV adoption); and (4) *financial institutions* (banks and investors financing automotive ventures and consumer purchases).

Stage 02: Data Collection and Persona Generation: According to PEMF, in the second stage, quantitative and qualitative data are collected. The qualitative data sources include (1) *industry reports* indicating market trends, sales figures, technological advancements (McKinsey & Company, 2016); (2) *academic literature* including research on consumer behavior, supply chain management, and environmental impact (Graham-Rowe et al., 2012); (3) *policy documents* outlining government regulations, incentives, and strategic roadmaps (European Commission, 2019); (4) *social media and news articles* representing public opinion, emerging trends, and stakeholder communications (Yang & Lee, 2018); and (5) *SME interviews and surveys* that present insights from industry professionals, analysts, and consumer focus groups (Chandra, Gulati, & Kandlikar, 2010).

In addition, quantitative data are also collected, which include (1) *population statistics* representing age, gender, income levels, education, and geographic distribution for consumers; company size, revenue, number of employees, and market share for manufacturers and suppliers (McKinsey & Company, 2016); (2) *behavioral metrics* indicating purchase frequency, brand loyalty, and vehicle usage patterns for different vehicle types among consumers; procurement volumes, order frequencies, and supply chain reliability metrics for suppliers (Wells & Nieuwenhuis, 2010); engagement in different sustainability programs (Yang & Lee, 2018) for all the stakeholders in the ecosystem; (3) *economic indicators* such as revenue, profit margins, investment levels, and R&D expenditure for automobile manufacturers, technology firms, and financial institutions (McKinsey & Company, 2016); (4) *regulatory metrics* including compliance rates, number of regulatory filings, fines, and adherence to safety and emissions standards for government regulators and manufacturers (Wang et al., 2019); and (5) *operational metrics* like production volumes, inventory levels, supply chain efficiency, and logistics performance indicators for manufacturers and suppliers (Wells & Nieuwenhuis, 2010).

The qualitative data are used to contextualize the LLMs, while the quantitative data are used for persona generation. For example, in the case of a traditional ICE automobile manufacturer, the company size, revenue, and number of employees will be used to establish the demographic distribution of the company, while its behaviors, motivation, and goals are gauged from its sales, engagement in sustainability initiatives, positioning, and growth projections. For instance, understanding consumer attitudes toward EVs requires data on environmental awareness, cost considerations, range anxiety, and charging infrastructure availability (Moons & Pelsmacker, 2012).

Using the collected data, we develop LLM-generated personas for each of the stakeholders. The personas capture demographics, goals, motivations, challenges, behaviors, and typical interactions. Example personas include:

1. The *traditional ICE automaker*, who is focused on ICE-based vehicles only while keeping the market share, is concerned about retooling costs, maintaining market competitiveness, and managing workforce reskilling.
2. The *EV manufacturer*, who is working for innovation and production of EVs only, and prioritizes sustainability, technological advancements, and expanding market share in the growing EV segment.
3. The *supplier*, who is someone who manufactures and provides materials to both EV and ICE manufacturers, is focused on maintaining supply chain reliability, managing production efficiencies, and fostering strong relationships with both traditional and EV manufacturers.
4. The *dealer and retailer*, who are responsible for the sale and distribution of vehicles to consumers, are actively involved in customer service, inventory management, and adapting sales strategies to accommodate the increasing demand for EVs and connected vehicles.
5. The *eco-conscious consumer*, a consumer who is aware of their impact on the environment and is making conscious decisions to minimize the

impact. They are willing to adopt EVs, despite higher upfront costs, and is influenced by environmental benefits, government incentives, and social responsibility.

6. The *government regulator*, a government body representative who is focused on increasing ways to support and enhance EV production through green subsidies.

Stage 03: Persona Validation and Selection: The generated personas are evaluated by (1) cross-referencing with the baseline data for diversity estimation, (2) being assessed and evaluated on the PPS (Salminen, Santos, Jung, Chowdhury, & Jansen, 2019) for understanding the usability of these personas, (3) reviewed by the SMEs to be considered usable, and (4) reviewed by the stakeholders to understand their relatability.

Stage04: Persona Ecosystem Playground: After validation, the personas of each stakeholder are integrated into an interactive simulation environment as chatbot/interactive personas using an LLM framework (e.g., LangChain- and LangGraph-based interfaces). For example, the simulation can model the effects of increased government incentives for EVs by observing how these incentives influence consumer adoption rates, manufacturer strategies, and supplier dynamics (He, Zhan, & Hu, 2017). Deploying personas using LangChain and LangGraph enables the creation of a responsive and adaptable ecosystem model that can be used to test and visualize the potential impacts of different policy and market interventions.

10.4.3.2 Modeling Societal Ecosystems

Similar to modeling an industrial ecosystem, the same principles can be expanded to societal ecosystems. Societal ecosystems include a diverse array of environments, where each ecosystem is composed of different parties, including actors (citizens, residents, legislators, and regulators) and subjects (children, pets, and other non-active participants in the ecosystems) (see Table 10.2). For instance, in *urban ecosystems*, personas might represent different socioeconomic groups, local businesses, municipal governments, and environmental organizations focused on sustainability (Jacobs, 1961). In *rural ecosystems*, personas could include farmers practicing sustainable agriculture, community leaders promoting renewable energy, and cooperatives supporting environmental conservation (Brown, 2008). *Energy ecosystems* might involve personas representing consumers, renewable energy providers, policymakers, and environmental activists (Scheer, 2004). In the case of a social ecosystem, we will apply the PEMF to an urban ecosystem, where a new awareness campaign is being launched to increase recycling.

Context: An urban ecosystem is selected for a pilot program to launch an awareness campaign through local online and offline channels to increase the adoption of recycling in society. To better understand the different stakeholders' perspectives, the ecosystem of the society is modeled using PEMF.

Stage 01: Stakeholder Mapping and Selection: The initial step involves identifying parties in the urban ecosystem that are directly or indirectly affected by/or affecting the said campaign. Some key parties include: (1) *eco-conscious residents,* who are

TABLE 10.2
Examples of Societal Ecosystems Focused on Sustainability and Their Key Components

Ecosystem	Key Actors and Components	Characteristics and Examples
Sustainable urban	Eco-conscious residents Green businesses Government agencies Environmental NGOs Sustainable infrastructure	Focus on sustainable living, green economy, environmental policies, and sustainable infrastructure. Examples: Urban gardeners, renewable energy companies, city environmental departments, public transit systems.
Rural sustainability	Organic farmers Community cooperatives Renewable energy projects Conservation groups	Emphasis on sustainable agriculture, resource conservation, and community-based environmental initiatives. Examples: Organic farms, wind energy cooperatives, local conservation trusts.
Energy transition	Consumers Renewable energy Providers Policy makers Activists	Transition from fossil fuels to renewable energy sources, policy development, and public acceptance. Examples: Solar panel users, wind farm operators, energy regulators, and environmental activists.
Circular economy	Manufacturers Recycling firms Consumers Policy makers	Focus on resource efficiency, waste reduction, and sustainable production and consumption. Examples: Companies using recycled materials, zero-waste stores, environmental legislation.
Educational	Students Teachers Administrators Policy makers	Knowledge exchange, sustainability education, and curriculum development focused on environmental awareness. Examples: Environmental science students, eco-friendly school programs, and education ministries promoting sustainability.
Healthcare	Patients Providers Public health officials Regulators	Focus on public health, access to healthcare, environmental health issues, and policy regulation. Examples: Community health clinics, hospitals reducing carbon footprint, and health departments addressing pollution-related diseases.
Digital	Online users Content creators Platform administrators and cybersecurity experts	Virtual interactions, information exchange on sustainability topics, digital activism, and online governance. Examples: Environmental bloggers, sustainable technology developers, social media campaigns for climate action.

individuals and families committed to recycling practices; (2) *green businesses* that are enterprises that recycle frequently; (3) *government agencies,* including municipal departments, responsible for environmental policies, sustainable urban planning,

and resource management for recycling (UN-Habitat, 2009); (4) *environmental NGOs* that are non-profit organizations advocating for enhanced recycling initiatives and encouraging community through engagement (Putnam, 2000); (5) *recycling infrastructure*, which refers to elements like waste collection, disposal, and management facilities (Beatley, 2011); (6) *volunteers*, who are members of the community who actively engage and support the recycling cause; and (7) *media houses* that actively contribute to the awareness campaign.

Stage02: Data Collection and Persona Generation: For this purpose, qualitative data sources include studies on sustainable behaviors, community participation in environmental initiatives (Stern, 2000), and interviews with SMEs, including recycling experts, waste management organization members, and environmentalists. Quantitative data sources include (1) demographic statistics of the population, including age, sex, household income, marital status, and education (United Nations, 2018); (2) behavioral data such as the adoption of recycling habits, reception to sustainability practices, consumption, and waste management routines; (3) operational indicators such as the number of waste collectors, waste collection and management pipelines, and awareness campaign metrics; and (4) infrastructure metrics such as the number of green buildings, availability of public transit, and renewable energy usage (Glaeser & Gyourko, 2010). Based on the data collected, personas are generated for different parties as follows:

1. The *eco-conscious resident*, who is conscious of their impact on the environment and actively participates in recycling initiatives through sustainable lifestyle choices, participation in community environmental programs, and advocacy for green policies.
2. The *green business*, which is a business or local shop focused on implementing and adopting recycling processes across its supply chain.
3. The *government agency*, which is a regulatory body that is driven to create a positive impact by promoting recycling across the societal ecosystem.
4. The *environmental NGO* that represents the actions taken by non-governmental organizations to promote recycling initiatives in society.
5. The *general public* who represents the general public of the ecosystem that will be directly impacted by the campaign.

Representative personas for these stakeholders are presented in Figure 10.4.

Stage 03: Persona Validation and Selection: Similar to the previous case, the personas are validated through four steps (1) cross-reference with baseline data; (2) measurement of diversity; (3) assessing decision-makers' perceptions of the personas using scales like the PPS (Salminen, Santos, et al., 2020); and (4) validation of the personas by SMEs and establishing the reliability of the personas by representative members. In this case, the addition of the participants from the community in the design process would reduce self-referential bias.

FIGURE 10.4 Representative personas for two of the stakeholders in different ecosystems.

Stage 04: Persona Ecosystem Playground: Validated personas are integrated into an interactive simulation environment, allowing for interactions based on predefined behavioral rules related to sustainability. This supports scenario exploration, such as modeling the impact of implementing awareness campaigns for recycling; improving the energy-efficiency consciousness on urban energy consumption; or assessing how the expansion of public transit affects carbon emissions and social equity (Newman & Kenworthy, 1999).

10.4.3.3 Modeling Natural Ecosystems as Personas

Like industrial and societal ecosystems, natural ecosystems can also be modeled by adopting PEMF. However, as the natural ecosystem comprises both human and non-human parties, additional considerations of non-human personas (representation of a non-human (living or non-living thing) by a persona) should be considered. Natural ecosystems comprise diverse environments, each characterized by unique species compositions and environmental factors (see Table 10.3). For instance, in *grassland ecosystems*, personas represent various herbivores such as bison and pronghorn, their predators like lions and coyotes, and the grasses that form the ecosystem's foundation. In *desert ecosystems*, personas include specially adapted plants like cacti and succulents, nocturnal rodents such as kangaroo rats, reptiles

TABLE 10.3
Examples of Natural Ecosystems and Their Key Components

Ecosystem	Key Species and Components	Characteristics and Examples
Forest	Dominant tree species Understory plants Apex predators Soil microorganisms	Complex canopy structures, layered vegetation, top-down trophic interactions, and nutrient cycling facilitated by decomposers. Examples: Oak trees (*Quercus robur*), ferns, wolves (*Canis lupus*), fungi.
Grassland	Grasses Herbivores Predators Soil nutrients	Open canopy with dominant grasses, herbivore-driven grazing patterns, predator-prey dynamics, and efficient nutrient recycling. Examples: Bison (*Bison bison*), lions (*Panthera leo*), prairie dogs, nitrogen-fixing bacteria.
Desert	Cacti and succulents Desert rodents Reptiles Drought-tolerant microbes	Sparse vegetation adapted to arid conditions, nocturnal and burrowing animals, extreme temperature adaptations, and specialized water-conserving mechanisms. Examples: Saguaro cactus (*Carnegiea gigantea*), kangaroo rat (*Dipodomys spp.*), Gila monsters, xerophilic bacteria.
Tundra	Mosses and lichens Herbivores Carnivores Permafrost soil	Low temperatures, short growing seasons, limited vegetation types, specialized adaptations for cold, permafrost-dependent soil structure. Examples: Arctic moss (*Calliergon giganteum*), caribou (*Rangifer tarandus*), polar bears (*Ursus maritimus*), cold-tolerant fungi.
Freshwater	Fish Amphibians Aquatic plants Microorganisms	Diverse aquatic habitats, essential roles in water purification, complex food webs, and high biodiversity. Examples: Trout (*Oncorhynchus mykiss*), frogs (*Rana spp.*), cattails (*Typha spp.*), nitrifying bacteria.
Marine	Coral Marine mammals Plankton Seagrasses	Saltwater ecosystems with high biodiversity, symbiotic relationships, critical roles in carbon cycling, and complex trophic structures. Examples: Coral polyps (*Acropora spp.*), dolphins (*Delphinus delphis*), copepods, eelgrass (*Zostera marina*).
Mountain	Alpine plants Mountain ungulates Predatory birds Snow microbes	Varied microclimates, altitude-dependent species distributions, specialized adaptations to steep terrains, and seasonal migrations. Examples: Edelweiss (*Leontopodium alpinum*), mountain goats (*Oreamnos americanus*), eagles (*Aquila chrysaetos*), psychrophilic bacteria.
Polar	Polar bears Penguins Algae Ice-dwelling microbes	Extreme cold environments, ice-dependent habitats, unique reproductive strategies, and sensitivity to climate changes. Examples: Polar bears (*Ursus maritimus*), emperor penguins (*Aptenodytes forsteri*), ice algae, psychrotolerant bacteria.

like Gila monsters, and drought-tolerant microbes. *Tundra ecosystems* might be modeled through personas representing Arctic mosses and lichens, herbivores like caribou, carnivores such as polar bears, and permafrost-dependent soil microorganisms. *Freshwater ecosystems* would involve personas for various fish species like trout, amphibians such as frogs, aquatic plants like cattails, and microorganisms responsible for nutrient cycling. *Marine ecosystems* would encompass diverse species, including coral polyps, plankton, fish species, and marine mammals like dolphins and whales. *Mountain ecosystems* might feature personas for alpine plants like edelweiss, mountain ungulates such as mountain goats, predatory birds like eagles, and snow microbes. Lastly, *polar ecosystems* would include personas representing polar bears, emperor penguins, specialized algae, and ice-dwelling microbes. Representing these ecosystems could make LLM-generated personas an instrument for ecological research and conservation efforts. To implement the PEMF in natural ecosystems, a forest ecosystem is modeled as an example.

Context: Among the land-based natural ecosystems, forests are considered one of the most biodiverse and ecologically significant ecosystems, playing a crucial role in carbon sequestration, climate regulation, and providing habitat for numerous species (Carpenter et al., 1998). More specifically, we will focus on a sub-ecosystem within the forest ecosystem that involves and deals with food-based interactions among different parties of the ecosystem. To further ground the example and make it manageable, we will model the sub-ecosystem in the context of observing the impact of climate change on food abundance for different members of the ecosystem.

Stage 01: Stakeholder Mapping and Selection: The foundational stage in the PEMF is stakeholder mapping and selection, which, in the case of natural ecosystems, would involve the selection of species and environmental factors that define the ecosystem. These elements are categorized into: (1) *producers*, which are dominant tree species such as oaks, pines, and maples that form the canopy and act as primary producers by converting solar energy into chemical energy through photosynthesis; (2) *understory plants*, which generally include shrubs, ferns, and herbaceous plants like blueberries and wildflowers that inhabit the forest floor, contributing to biodiversity and providing habitat and food sources for various fauna; (3) *consumers,* which are herbivores such as white-tailed deer and insects like caterpillars; carnivores including wolves and owls; and omnivores like raccoons and bears that interact within the food web; (4) *decomposers,* which are parties such as fungi, bacteria, and detritivores including earthworms that break down dead organic matter, recycling nutrients back into the soil; and (5) *abiotic factors* including climate variables (temperature, precipitation), soil composition (pH, nutrient content), water availability, and geological features (topography, elevation), influencing species survival and interactions.

Stage 02: Data Collection and Persona Generation: As all the personas generated in this case are non-human personas, the quantitative data collected for persona generation would be different. The qualitative data sources include (1) *ecological studies*, such as detailed research on species behavior, habitat preferences, reproductive strategies, and interspecies relationships (Carpenter et al., 1998); (2) *field observations*, which represent direct observations of species interactions, migration patterns, feeding behaviors, and responses to environmental stimuli; and (3) *SME interviews*, which that provide insights from ecologists, conservationists, and wildlife experts regarding species roles, threats, and conservation needs especially with regards food imbalance. Similarly, for persona generation, the quantitative data sources would include, in addition to the normal *population metrics*, *representative personality traits* including the quantitative measures of species' behavioral and physiological characteristics that influence their roles within the ecosystem. These representative personality traits could include: (1) *activity levels* (metrics such as daily movement zones/ranges and durations, eating/consumption frequency, and periods of rest); (2) *herd and community parameters* (data on group size, social hierarchy, and interaction frequency within species populations); (c) *breeding and reproductive rates* (information on breeding cycles, clutch sizes, offspring survival rates, and age of sexual maturity); (d) *territorial range* (measurements of territory sizes, overlapping ranges among individuals or groups, and migration patterns); (e) *diet preferences* (quantitative data on food intake types, quantities, and foraging behaviors); and (f) *adaptability metrics* (indicators such as cortisol levels, resilience to environmental stressors, and recovery times after disturbances). These data allow the creation of persona components such as behavior, goals, motivations, key traits, and values for these non-human personas.

Based on the data collected, the following non-human personas are created for different key stakeholders identified in the first stage. Some of these personas include:

1. *Producers,* representing dominant tree species such as oak, pine, and maple, which show consistent growth rates, photosynthetic efficiency, and territorial strategies;
2. *Understory plants* that include shrubs, ferns, and herbaceous plants like blueberries and wildflowers, which showcase growth patterns, have lower sunlight requirements, and soil nutrient uptake;
3. *Consumers,* including herbivores like white-tailed deer and insects such as caterpillars; carnivores such as wolves and owls; and omnivores including raccoons and bears. All of these focus on strong feeding habits, territorial behaviors, reproductive cycles, and dominance-driven personality;
4. *Decomposers* that are parties like fungi (*Agaricus bisporus*), bacteria (*Bacillus subtilis*), and detritivores such as earthworms, which show traits like symbiotic relationships and recycling behaviors driven by soil health improvement;
5. *Abiotic factors* that represent non-living environmental elements such as climate variables (temperature, precipitation), soil composition (pH, nutrient levels), water availability, and geological features (topography, elevation).

On the basis of the persona data collected for each stakeholder group, personas are created. Representative personas of these different stakeholders are presented in Figure 10.4.

Stage 03: Persona Validation and Selection: Similar to previous validation approaches in industrial and societal ecosystem models, personas are validated through (1) cross-validation with the underlying population characteristics; (2) assessed on the PPS scale (Salminen et al., 2019) and evaluated by SMEs to verify their accuracy and relevance. To ensure the accuracy and representativeness of the generated personas, a validation process is employed, which includes *cross-*

referencing persona attributes against empirical data and established ecological studies to verify their accuracy and relevance, *SME (ecologists and environmental scientists)* conduct comprehensive reviews of the personas, assessing their authenticity, ecological validity, and alignment with real-world species behaviors and interactions, and *secondary stakeholder (conservation organizations, field experts, and ecological stakeholders)* provide feedback to enhance the accuracy of the persona to real and actual behaviors and traits.

Based on the collected feedback, necessary adjustments are made by refining the input data or modifying the parameters of the LLMs used, thereby addressing any identified discrepancies or biases (Sheng, Joe, Alambo, & Solaiman, 2021). This iterative validation process enhances the credibility and utility of the personas in subsequent modeling and simulation activities.

Stage 04: Persona Ecosystem Playground: Once validated, the personas are integrated into an interactive simulation environment where these non-human personas can communicate with each other as well as with the persona users—humans—who use these personas for making different key decisions. This PEP environment allows personas to interact based on predefined behavioral rules and ecological processes, supporting both inter-species and HPIs for scenario exploration. For instance, the simulation can model the impact of climate change on species distribution and interactions, observing how altered temperatures and precipitation patterns influence ecosystem balance (He et al., 2017). Additionally, the platform can simulate the effects of conservation interventions, invasive species introductions, or habitat fragmentation, providing valuable insights into potential ecosystem responses and resilience.

10.4.4 Forms of Interaction

In these PEPs, various forms of interactions occur. These interactions include (1) one-to-one interactions between personas, (2) one-to-many interactions between personas, and (3) HPIs.

The primary form of interaction is the persona-to-persona interaction, where personas representing different parties within the ecosystem engage based on predefined rules and interactions, leading to specific decision-making processes. These interactions can be categorized into *one-to-one, one-to-many*, or *many-to-one* interactions.

In *one-to-one* interactions, two personas directly engage with each other. This form allows for detailed modeling of relationships such as negotiations, partnerships, or conflicts between individual parties. For example, in the industrial ecosystem of the automotive industry, modeling a negotiation between an *EV manufacturer persona* and a *battery supplier persona* can reveal how supply agreements impact production and innovation.

In *one-to-many* interactions, a single persona interacts with multiple personas, while in *many-to-one* interactions, multiple personas interact with a single persona. This form of interaction is crucial for modeling scenarios where parties have widespread influence or are influenced by collective behaviors. For example, a *government regulator* (one) advocating for electric vehicles can influence several *traditional automotive manufacturers* (many) to strategize and plan for the production of electric vehicles. In contrast, multiple *suppliers* (many) competing for contracts with a single *EV manufacturer* (one) can have a greater influence on market dynamics.

These interactions help us understand (1) *the propagation of influence through the ecosystem* and (2) the formation of collective behaviors and perspective formulation, especially for personas belonging to different stakeholder groups.

Another significant form is the persona-to-human interaction, which allows persona users and decision-makers to engage directly with personas within the simulation. This interactivity enables persona users to explore "what-if" scenarios by making decisions, posing questions, altering parameters, and observing how personas respond. The benefits of persona-to-human interactions include (1) experiential learning (learning of different perspectives through engagement);

(2) enhanced comprehension of system complexities (understanding the causal and non-causal interactions); and (3) the integration of human expertise (to drive the direction) into the simulation. For instance, in a simulation of sustainable urban development, a user acting as a city planner could interact with various personas such as the *Eco-conscious Resident Persona*, the *Green Business Persona*, and the *Government Agency Persona*. The user could propose a new policy to implement city-wide renewable energy initiatives and observe how each persona responds: (1) The *Eco-conscious Resident Persona* may express support, highlighting the environmental benefits and willingness to participate; (2) the *Green Business Persona* might evaluate the economic implications, considering the costs of transitioning to renewable energy sources and potential market advantages; and (3) the *Government Agency Persona* could discuss regulatory hurdles, budget constraints, and the need for stakeholder engagement.

Similarly, a user acting as a policy analyst might interact with the *EV Manufacturer Persona* and the *Government Agency Persona* to explore the impact of introducing a new subsidy for EVs. By engaging with these personas, the user can delve into the economic, environmental, and logistical implications of such a policy. The user initiates a conversation with the *Government Agency Persona*:

10.5 DISCUSSION

This chapter presents a different perspective on the use of LLM-generated personas to increase their application from a minority viewpoint to a majority viewpoint in ecosystem models. Addressing RQ1, we present a methodology to enrich and enhance LLM-generated personas through RAG and fine-tuned models to create persona sets. These persona sets are generated using the primary data and are evaluated to ensure their validity. Regarding RQ2, we present a framework (PEMF) to model ecosystems using personas. The two types of personas are compared in Table 10.4.

For this, we create a four-stage framework, including (1) stakeholder mapping and segmentation, (2) data collection and persona generation, (3) persona validation and selection, and (4) persona ecosystem playground creation.

The validation of LLM-generated personas presents a methodological challenge: while SMEs provide domain expertise, they may carry their own biases, potentially perpetuating the very issues these personas aim to address. For *minority viewpoint personas*, establishing ground truth requires direct validation from community members, supported by quantitative metrics derived from demographic and behavioral data. This dual approach helps mitigate individual biases while ensuring authentic

TABLE 10.4
Comparison of Persona Types in LLM-Generated Personas

Aspect	Minority Viewpoint Personas	Ecosystem Personas
Focus	Single perspective from marginalized communities	Multiple stakeholders within complete ecosystems
Primary purpose	Inform inclusive policy-making decisions	Model system-wide interactions and relationships
Stakeholders	Individual human perspectives from underrepresented groups	Multiple entities (human and non-human) across ecosystem
Data sources	Community surveys, cultural studies, demographic data	Domain reports, interaction data, behavioral metrics
Generation method	LLMs with cultural contextualization and sensitivity	LLMs with domain-specific RAG and fine-tuning
Validation	Community verification and cultural sensitivity checks	Cross-validation with baseline data and PPS assessment
Key example	Maria Fernandez: Indigenous student balancing traditional values with modern education	Automotive ecosystem: Manufacturers, suppliers, consumers, and regulators interacting in EV transition

representation. The complexity of validation increases for *personas in the ecosystem* due to the distinction between known and emerging behaviors. Known behaviors, such as documented regulatory interactions or market responses, can be validated against historical data. However, emergent behaviors in complex multi-stakeholder scenarios lack historical precedent for validation. Here, validation must focus on the logical consistency of interactions and the underlying assumptions driving these emergent patterns, rather than a direct comparison with historical data. This limitation should be explicitly acknowledged when using ecosystem personas for predictive modeling.

10.5.1 Design Principles

LLM-generated persona systems involve distinct design considerations. We outline three primary considerations:

- **Realism and Validity**: *The personas generated, whether for minority representation or majority representation, have to be realistic and valid.* Realism refers to the true representation of the parties and capturing the ecosystem knowledge base, while validity indicates that the personas are up-to-date and correspond to the current user state. For natural ecosystems, realism is based on ecological data and biological behaviors, while for industry ecosystems, it involves accurate representations of market dynamics, organizational roles, and economic principles. Both systems benefit from realistic models to simulate authentic interactions, whether these involve species within an ecosystem or companies within an industry. Mere realism is not adequate; the personas need to be valid, as in representative of the true entity they aim to represent.

- **Adaptability and Responsiveness**: *The personas should be adaptable to different scenarios and responsive to internal and external changes.* In nature ecosystems, adaptability captures species' responses to environmental changes, like predator-prey dynamics. In industry ecosystems, adaptability reflects how businesses respond to market shifts, competition, or regulatory changes. These scenario parameters could be provided by the administrator of the LLM-generated persona system. Additionally, feedback loops could be built to provide information on real-time events (e.g., obtained through press releases) that would then be used as scenarios to initiate interaction sessions between the actors. Also, both ecosystem types benefit from flexible scenario exploration. Users can experiment with different climate conditions or species in nature ecosystems or with various business strategies and regulatory environments in industry ecosystems. Overall, initiation plays a key role in the direction in which the interaction takes place, as it provides the context and purpose for interaction. In one extreme, there are preset scenarios to choose from. In the other extreme, users may freely set initial conditions.
- **Scalability and Complexity Management**: *The personas developed, whether using PEMF or for minority visualization, should be scalable and able to enhance or reduce complexity as and when required.* Nature ecosystems can involve intricate webs of species interactions, just as industry ecosystems can involve complex networks of business relationships. Scalable interfaces that manage complexity without overwhelming users are essential for both, enabling users to zoom in on individual personas or zoom out to view the entire ecosystem. For example, the LLM-generated persona system could introduce each persona briefly at the beginning of an interaction session.

Certainly, this list of principles is neither definitive nor complete. However, we do believe these principles to be among the key considerations for designing LLM-generated persona systems.

10.5.2 Future Research Directions

Here, we identify and propose research directions (RDs) for future work on LLM-generated PFSG.

First, while there has been research addressing the problems of the marginalized communities in the development phase through inclusive personas (Goodman-Deane et al., 2018; Marsden & Haag, 2016), there has been no focus on the usage of personas for the reduction of inequality and betterment of marginalized communities (Salminen, Wenyun Guan, et al., 2022). To this end, we propose:

- RD1: *How can LLM-generated personas accurately reflect the socio-economic and cultural realities of marginalized communities?*
- RD2: *How can stereotypical and biased personas be minimized in LLM-generated personas?*

Second, while personas have been used in some of use cases for social good, they have been primarily driven by commercial applications (Salminen, Wenyun Guan, et al., 2022). Due to this, there is a scarcity of research in the PFSG domain (Guan et al., 2023; Salminen, Wenyun Guan, et al., 2022), as most of the research is focused on the commercial (software design, marketing, and other) aspects. PFSG has a strong impact in a broad spectrum of areas, including (1) *healthcare,* where it could be implemented to improve well-being and access to services for vulnerable populations; (2) in *education,* to design inclusive educational tools; (3) *sustainability,* where it is deployed to promote sustainable practices among varying user demographics; (4) *social justice*, for creating narratives around users' political views and social attitudes; and in (5) *community engagement*, to enhance social cohesion and civic participation (Guan et al., 2023; Salminen, Wenyun Guan, et al., 2022). One such area is policy-making, where the involvement of all stakeholders is essential for creating policies that can be adopted holistically by the community (Anthony, 2024). Therefore, we propose RD3: *How can LLM-generated personas be utilized to improve the formulation of targeted social policies?*

Third, while major governmental organizations have considered personas as a driver for policy-making (Andrews, 2013; Department of the Prime Minister and Cabinet (DPMC), 2023), little research has been focused on identifying the key challenges of the personas for policy-making. This is further complicated for the special case of marginalized and under-represented communities (e.g., *minority races, rural population*), who lack proper representation in the policy-making process. Moreover, using LLMs to develop personas further adds to the complications, as these tools are not only prone to algorithmic bias but are also susceptible to possible hallucinations (the challenge for LLMs to generate coherent and comprehensible but factually and contextually incorrect output). To better study and understand the ethical and social challenges of using LLM-generated personas as a tool in policy-making, especially for marginalized and socially challenged communities, we propose the following RDs:

- RD4: *What ethical challenges arise from deploying LLM-generated personas in policy-making for socially challenged communities?*
- RD5: *What actionable ethical guidelines can be developed to mitigate these challenges?*

10.5.3 Key Takeaways

In this work, we have illustrated, through concrete examples, the potential of LLMs to address the persona-generation task in HCI, ranging from the micro-level representation of marginalized, fringe, or minority perspectives to modeling entire ecosystems. The fundamental principle in these endeavors is basing the personas' information on primary data. Otherwise, there is *a risk of the LLMs perpetuating stereotypes and biases* that relate to whole ecosystems and individual ecosystem stakeholder groups. Simulations based on LLM-generated personas can only be valid when they rely on primary data. They might appear useful even without this linkage, which is problematic: there is a perilously easy route of using LLMs to generate personas without basing them on primary data. The fluency and apparent realism of the outcomes can easily fool decision-makers into thinking that such personas are valid (because they *look* like valid ones!), while in fact, they are not. Therefore, the research-based, data-driven approach we proposed in this work is recommended in lieu of using non-contextual prompts with vanilla LLMs.

We believe that our approach to LLM-generated persona systems has educational value and accessibility value. Modeling ecosystems through LLM-generated personas is suitable for a range of educational purposes, from understanding biodiversity to exploring business strategy. The conversational user interface makes the LLM-generated persona system suitable for novices and experts alike, and the interactive nature of the system supports learning and engagement in both contexts. LLM-generated personas can also support the development of ethical and sustainable behaviors, as they can help users consider the consequences of their actions. In nature ecosystems, this could involve conservation efforts, while in industry ecosystems, it could involve corporate social responsibility or sustainability. Users' interactions can mirror real-world impacts and considerations, similar to the dynamics of gameplay. Ideally, observing the persona-to-persona interaction and participating in HPI encourage critical thinking and a deeper understanding of systemic interactions.

10.6 CONCLUSION

This research proposes ways to address significant gaps in the development and application of personas, particularly concerning ethical and social considerations for marginalized communities and the lack of representation of ecosystems comprising multiple parties. Through investigating and refining methodologies aimed at producing more inclusive and representative personas, we proposed ways to reduce biases and improve the functionality of AI systems in tackling real-world challenges. LLM-generated personas can contribute meaningfully to the fields of HCI and UCD, establishing new frameworks and guidelines that can be applied in policy-making and social initiatives. This work aspires to create a more equitable digital environment where AI systems reflect and serve the diverse needs of global user communities and ecosystems in need.

NOTES

1 By ecosystem, we refer to a community or collection of inter-dependent parties (actors and subjects) which interact with each other for a specific purpose. These parties may or may not be humans.
2 For example, using the relevant information about a country's culture to enrich the persona of leadership.
3 For example, the model is re-trained for the specific industry.
4 https://www.langchain.com/.
5 https://langchain-ai.github.io/langgraph/.

REFERENCES

Acs, Z. J., Stam, E., Audretsch, D. B., & O'Connor, A. (2017, June). The lineages of the entrepreneurial ecosystem approach. *Small Business Economics*, *49*(1), 1–10. doi:10.1007/s11187-017-9864-8.

Amin, D., Salminen, J., Jansen, B.J., Shin, J. and Kim, D.H., 2025. Generative AI personas considered harmful? Putting forth twenty challenges of algorithmic user representation in human-computer interaction. International Journal of Human-Computer Studies, p.103657.

Andrews, B. (2013, August). Using Personas to Make Better Policy. https://openpolicy.blog.gov.uk/2013/08/08/using-personas-to-help-improve-policy-making/.

Anthis, J., Lum, K., Ekstrand, M., Feller, A., D'Amour, A., & Tan, C. (2024, May). The Impossibility of Fair LLMs. arXiv. https://arxiv.org/abs/2406.03198.

Anthony, B. (2024, March). The role of community engagement in urban innovation towards the co-creation of smart sustainable cities. *Journal of the Knowledge Economy*, *15*(1), 1592–1624. doi:10.1007/s13132-023-01176-1.

Anvari, F., Richards, D., Hitchens, M., & Tran, H. M. T. (2019, May). Teaching user centered conceptual design using cross-cultural personas and peer reviews for a large cohort of students. In: *2019 IEEE/ACM 41st International Conference on Software Engineering: Software Engineering Education and Training (ICSE-SEET)*, Montreal, QC, Canada, pp. 62–73. doi:10.1109/ICSE-SEET.2019.00015.

Batty, M. (2018). *Inventing Future Cities*. MIT Press.

Beatley, T. (2011). *Green Cities of Europe: Global Lessons on Green Urbanism*. Island Press.

Bødker, S., Christiansen, E., Nyvang, T., & Zander, P.-O. (2012). Personas, people and participation: Challenges from the trenches of local government. In: *Proceedings of the 12th Participatory Design Conference on Research Papers (PDC '12),* Roskilde, Denmark, p. 91. doi:10.1145/2347635.2347649.

Borge, M., & Xia, Y. (2023). Beyond the individual: The regulation and negotiation of socio-emotional practices across a learning ecosystem. *Journal of the Learning Sciences*, *32*(3), 325–375. doi:10.1080/10508406.2022.2157725.

Bremer, C., Knowles, B., & Friday, A. (2022). Have we taken on too much? A critical review of the sustainable HCI landscape. In: *Proceedings of the 2022 CHI Conference on Human Factors in Computing Systems*, New Orleans, LA, pp. 1–11. doi:10.1145/3491102.3517698.

Brown, L. R. (2008). *Plan B 3.0: Mobilizing to Save Civilization*. W.W. Norton & Company.

Brown, T. B., Mann, B., Ryder, N., Subbiah, M., Kaplan, J., Dhariwal, P., Neelakantan, A., Shyam, P., Sastry, G., Askell, A., Agarwal, S., Herbert-Voss, A., Krueger, G., Henighan, T., Child, R., Ramesh, A., Ziegler, D. M., Wu, J., Winter, C., Hesse, C., Chen, M., Sigler, E., Litwin, M., Gray, S., Chess, B., Clark, J., Berner, C., McCandlish, S., Radford, A., Sutskever, I., & Amodei, D. (2020, July). Language Models are Few-Shot Learners. arXiv. https://arxiv.org/abs/2005.14165.

Brussard, P. F., Reed, J. M., & Tracy, C. R. (1998). Ecosystem management: What is it really? *Landscape and Urban Planning*, *40*(1–3), 9–20.

Cachat-Rosset, G., & Klarsfeld, A. (2023). Diversity, equity, and inclusion in artificial intelligence: An evaluation of guidelines. *Applied Artificial Intelligence*, *37*(1), 2176618. doi:10.1080/08839514.2023.2176618

Capaldi, S., Passmore, H. A., Nisbet, E. K., & Zelenski, J. M. (2014). The connection between nature and wellbeing: A systematic review of the literature. *Journal of Environmental Psychology*, *40*, 409–418.

Carpenter, S. R., Kitchell, J. F., & Hodgson, J. G. (1998). *Ecosystem Management: Principles for a Sustainable Future*. Island Press.

Chandra, A., Gulati, N., & Kandlikar, M. (2010). Green drivers or free riders? An analysis of tax rebates for hybrid vehicles. *Journal of Environmental Economics and Management*, *60*(2), 78–93.

Chapman, C., & Milham, R. P. (2006, October). The personas' new clothes: Methodological and practical arguments against a popular method. In: *Proceedings of the Human Factors and Ergonomics Society Annual Meeting*, San Francisco, CA, vol. 50, pp. 634–636. doi:10.1177/154193120605000503.

Cherry, C., Thomas, G., Groves, C., Roberts, E., Shirani, F., Henwood, K., & Pidgeon, N. (2022). A personas-based approach to deliberating local decarbonisation scenarios: Findings and methodological insights. *Energy Research & Social Science*, *87*, 102455.

Cho, S. H., Jon, S., Jin, Y., Jung, J., & Oh, C. (2024). Understanding the dynamics in creating domain-specific AI design guidelines: A case study of a leading digital finance company in South Korea. In: *Extended Abstracts of the 2024 CHI Conference on Human Factors in Computing Systems*, Honolulu, HI. doi:10.1145/3613905.3650759.

Cooper, A. (1999). *The Inmates Are Running the Asylum: Why High Tech Products Drive Us Crazy and How to Restore the Sanity*, 1st edn. Sams-Pearson Education.

De Paoli, S. (2023). Writing User Personas with Large Language Models: Testing Phase 6 of a Thematic Analysis of Semi-Structured Interviews. arXiv. https://arxiv.org/abs/2305.18099.

Department of the Prime Minister and Cabinet. (DPMC). (2023, June). *Personas*. https://www.dpmc.govt.nz/our-programmes/policy-project/policy-methods-toolbox/design-thinking/personas.

Dietz, T., Ostrom, E., & Stern, P. C. (2003). The struggle to govern the commons. *Science*, *302*(5652), 1907–1912.

European Commission. (2019). *Clean Mobility*. European Union.

Evans, M. D. (2003). Biodiversity and ecosystem function: Current knowledge and future directions. *Ecological Complexity*, *1*(1), 3–10.

Farahani, M. S., & Ghasemi, G. (2024). Artificial intelligence and inequality: Challenges and opportunities. *Qeios*, February 11, 2024.

Ferrucci, D. (2012). Artificial intelligence for healthcare: Opportunities and challenges. *Journal of the American Medical Informatics Association*, *19*(5), 764–771.

Geary, W. L., Bode, M., Doherty, T. S., Fulton, E. A., Nimmo, D. G., Tulloch, A. I., & Ritchie, E. G. (2020). A guide to ecosystem models and their environmental applications. *Nature Ecology & Evolution*, *4*(11), 1459–1471.

Glaeser, E. L., & Gyourko, J. (2010). Housing supply and housing bubbles. *Journal of Urban Economics*, *67*(2), 198–213.

Goodman-Deane, J. A.-L., Bradley, M., Waller, S., & Clarkson, P. J. (2021, August). Developing personas to help designers to understand digital exclusion. *Proceedings of the Design Society*, *1*, 1203–1212. doi:10.1017/pds.2021.120.

Goodman-Deane, J., Waller, S., Demin, D., González-de Heredia, A., Bradley, M., & Clarkson, J. P. (2018, June). Evaluating inclusivity using quantitative personas. In: *Proceedings of Design Research Society Conference,* London, UK. doi:10.21606/drs.2018.400.

Graham-Rowe, E., Gardner, B., Abraham, C., Skippon, S., Dittmar, H., Hutchins, R., & Stannard, J. (2012). Mainstream consumers driving plug-in battery-electric and plug-in hybrid electric cars: A qualitative analysis of responses and evaluations. *Transportation Research Part A: Policy and Practice*, *46*(1), 140–153.

Green, D. G., & Sadedin, S. (2005). Interactions matter—complexity in landscapes and ecosystems. *Ecological Complexity*, *2*(2), 117–130.

Gross, S., Toombs, A., Wain, J., & Walorski, K. (2011). Foodmunity: Designing community interactions over food. In: *Proceedings of the 2011 Annual Conference Extended Abstracts on Human Factors in Computing Systems—CHI EA '11*, Vancouver, BC, Canada, p. 1019. doi:10.1145/1979742.1979504.

Grudin, J. (2006). Why personas work: The psychological evidence. In: J. Pruitt & T. Adlin (Eds.), *The Persona Lifecycle,* pp. 642–663, Elsevier. doi:10.1016/B978-012566251-2/50013-7.

Grudin, J., & Pruitt, J. (2002). Personas, participatory design and product development: An infrastructure for engagement. In: *Proceedings of Participation and Design Conference (PDC2002)*, Malmö, Sweden, p. 8.

Grudin, J., & Pruitt, J. (2006). Personas, participatory design and product development: An infrastructure for engagement. In: *Proceedings of the Participatory Design Conference,* Trento, Italy, vol. 11, pp. 144–161.

Guan, K. W., Salminen, J., Jung, S.-G., & Jansen, B. J. (2023, September). Leveraging personas for social impact: A review of their applications to social good in design. *International Journal of Human–Computer Interaction, 40*, 1–16. doi:10.1080/10447318.2023.2247568.

Guan, K. W., Salminen, J., Nielsen, L., Jung, S.-G., & Jansen, B. J. (2021). Information design for personas in four professional domains of user experience design, healthcare, market research, and social media strategy. In: *Proceedings of the 54th Hawaii International Conference on System Sciences (HICSS'21),* Maui, HI, p. 10. doi:10.24251/HICSS.2021.540.

Hansson, L. E. J., Cerratto Pargman, T., & Pargman, D. S. (2021, May). A decade of sustainable HCI: Connecting SHCI to the sustainable development goals. In: *Proceedings of the 2021 CHI Conference on Human Factors in Computing Systems,* Yokohama, Japan, pp. 1–19. doi:10.1145/3411764.3445069.

He, X., Zhan, X., & Hu, Y. (2017). The impact of policy incentives on electric vehicle adoption in China. *Transportation Research Part A: Policy and Practice*, *103*, 188–200.

Holland, J. H. (1992). Complex adaptive systems. *Daedalus*, *121*(1), 17–30.

Holling, C. S. (2001). Understanding the complexity of economic, ecological, and social systems. *Ecosystems*, *4*(5), 390–405.

Jacobs, J. (1961). *The Death and Life of Great American Cities*. Random House.

Jansen, B. J., Jung, S.-G., & Salminen, J. (2019). Creating manageable persona sets from large user populations. In: *Extended Abstracts of the 2019 CHI Conference on Human Factors in Computing Systems*, Glasgow, Scotland, pp. 1–6. doi:10.1145/3290607.3313006.

Jansen, B. J., Jung, S.-G., Salminen, J., An, J., & Kwak, H. (2018, March). Leveraging online social media data for persona profiling. In: *Qatar Foundation Annual Research Conference Proceedings*, Doha, Qatar, vol. 2018, ICTPP78. Hamad bin Khalifa University Press. doi:10.5339/qfarc.2018.ICTPP78.

Jansen, B. J., Salminen, J. O., & Jung, S.-G. (2020, March). Data-driven personas for enhanced user understanding: Combining empathy with rationality for better insights to analytics. *Data and Information Management*, 4(1), 1–17. doi:10.2478/dim-2020-0005.

Jansen, B. J., Salminen, J., Jung, S.-G., & Guan, K. (2021a). Evaluating data-driven personas. In: B. J. Jansen, J. Salminen, S.-G. Jung, & K. Guan (Eds.), *Data-Driven Personas: Synthesis Lectures on HumanCentered Informatics*, pp. 209–237. Springer. doi:10.1007/978-3-031-02231-9_1.

Jansen, B. J., Salminen, J., Jung, S.-g., & Guan, K. (2021b). Selecting the appropriate persona creation method. In: B. J. Jansen, J. Salminen, S.-G. Jung, & K. Guan (Eds.), *Data-Driven Personas: Synthesis Lectures on HumanCentered Informatics*, pp. 209–237. Springer. doi:10.1007/978-3-031-02231-9_10.

Jensen, M., Elverum, C., & Steinert, M. (2017, March). Eliciting unknown unknowns with prototypes: Introducing prototrials and prototrial-driven cultures. *Design Studies*, *49*, 1–31. doi:10.1016/j.destud.2016.12.002.

Johnson, A., Martinez, K., & Liu, Y. (2020). Barriers to sustainable practices: Insights from human-nature interaction studies. *Environmental Management*, *55*, 789–805.

Jung, S.-G., Salminen, J., Kwak, H., An, J., & Jansen, B. J. (2018). Automatic persona generation (APG): A rationale and demonstration. In: *Proceedings of the 2018 Conference on Human Information Interaction&Retrieval—CHIIR*'18, New Brunswick, NJ, pp. 321–324. doi:10.1145/3176349.3176893.

Keystone J. R., Tilman, D., Polley, H. W., & Reich, P. B. (2015). Biodiversity enhances ecosystem resilience to climate extremes. *Nature*, *526*, 574–577.

Kim, J., Park, S., & Lee, H. (2015). Human-nature interactions: Enhancing environmental stewardship through urban green spaces. *Urban Forestry & Urban Greening*, *14*, 45–53.

Kishore Mahabani. (2024, February). Measuring the power of the global south. *The World Today*, February–March, 2024. https://www.chathamhouse.org/publications/the-world-today/2024-02/measuring-power-global-south.

Knowles, B., Bates, O., & Håkansson, M. (2018). This changes sustainable HCI. In: *Proceedings of the 2018 CHI Conference on Human Factors in Computing Systems*, Montreal, QC, Canada, pp. 1–12. doi:10.1145/3173574.3174045.

Kunhikrishnan, T., & Jose, S. (2021). Stakeholder engagement in renewable energy transitions: Lessons from social innovation. *Energy Research & Social Science*, *80*, 102211.

Lee, A., Zukauskaite, M., & Barvidaite, V. (2017). Consumer behavior towards green energy: Evidence from a Lithuanian household survey. *Energy Policy*, *106*, 733–743.

Lee, O.-J., & Jung, J. J. (2019, September). Integrating character networks for extracting narratives from multimodal data. *Information Processing & Management*, *56*(5), 1894–1923. doi:10.1016/j.ipm.2019.02.005.

Li, Y., Shi, Y., Guo, J., & Zhao, X. (2020). Modeling consumer behavior in the sharing economy using activity theory and agent-based modeling. *Technological Forecasting and Social Change*, *155*, 119986.

Lipeipei Sun, Tianzi Qin, Anran Hu, Jiale Zhang, Shuojia Lin, Jianyan Chen, Mona Ali, and Mirjana Prpa. 2025. Persona-L has Entered the Chat: Leveraging LLMs and Ability-based Framework for Personas of People with Complex Needs. In Proceedings of the 2025 CHI Conference on Human Factors in Computing Systems (CHI '25). Association for Computing Machinery, New York, NY, USA, Article 1109, 1–31. https://doi.org/10.1145/3706598.3713445.

Litman, T. (2020). *Autonomous Vehicle Implementation Predictions: Implications for Transport Planning* (Tech. Rep.). Victoria Transport Policy Institute.

MacKinlay, R. D., & Shaw, R. C. (2023). A systematic review of animal personality in conservation science. *Conservation Biology*, *37*(1), e13935.

Marsden, N., & Haag, M. (2016). Evaluation of gendermag personas based on persona attributes and persona gender. In: M. Kurosu (Ed.), *International Conference on Human-Computer Interaction, Lecture Notes in Computer Science*, vol. 9733, pp. 122–127, Springer.

McKinsey & Company. (2016). *Automotive Revolution—Perspective towards 2030: How the Convergence of Disruptive Technologydriven Trends Could Transform the Auto Industry.* McKinsey & Company, New York, NY, USA.

Meierhofer, J., Schweiger, L., Lu, J., Züst, S., West, S., Stoll, O., & Kiritsis, D. (2021). Digital twin-enabled decision support services in industrial ecosystems. *Applied Sciences*, *11*(23), 11418.

Moons, I., & Pelsmacker, P. D. (2012). Emotions as determinants of electric car usage intention. *Journal of Marketing Management*, *28*(3–4), 195–237.

Nah, F. F.-H., Zheng, R., Cai, J., Siau, K., & Chen, L. (2023, July). Generative AI and ChatGPT: Applications, challenges, and AI-human collaboration. *Journal of Information Technology Case and Application Research*, 25(3), 277–304.

New Economics Foundation. (2020). *Reinventing the Wheel: Mobility's Coming Transformation*. NEF.

Newman, P., & Kenworthy, J. (1999). *Sustainability and Cities: Overcoming Automobile Dependence*. Island Press.

Nielsen, L. (2002). From user to character: An investigation into user-descriptions in scenarios. In: *Proceedings of the 4th Conference on Designing Interactive Systems: Processes, Practices, Methods, and Techniques*, London, England, pp. 99–104. doi:10.1145/778712.778729.

Nielsen, L. (2013). *Personas: User Focused Design*, 1st ed. Springer.

Nielsen, L. (2019). Persona writing. In: L. Nielsen (Ed.), *Personas: User Focused Design*, pp. 55–81. Springer. doi:10.1007/978-1-4471-7427-1_4.

Nielsen, L., Hansen, K. S., Stage, J., & Billestrup, J. (2015). A template for design personas: Analysis of 47 Persona descriptions from Danish industries and organizations. *International Journal of Sociotechnology and Knowledge Development*, *7*(1), 45–61. doi:10.4018/ijskd.2015010104.

OpenAI. (2022). *Introducing ChatGPT*. OpenAI. https://openai.com/chatgpt/.

Philippon, T. (2016). *The Fintech Opportunity* (Tech. Rep. No. w22476). National Bureau of Economic Research.

Piussi, P., & Farrell, E. P. (2000). Interactions between society and forest ecosystems: Challenges for the near future. *Forest Ecology and Management*, *132*(1), 21–28.

Preist, C., Schien, D., & Blevis, E. (2016). Understanding and mitigating the effects of device and cloud service design decisions on the environmental footprint of digital infrastructure. In: *Proceedings of the 2016 CHI Conference on Human Factors in Computing Systems*, San Jose, CA, pp. 1324–1337. doi:10.1145/2858036.2858378.

Putnam, R. D. (2000). *Bowling Alone: The Collapse and Revival of American Community*. Simon and Schuster.

Roche, C. (2023). Ethics and diversity in artificial intelligence policies, strategies and initiatives. *AI and Ethics, 3*(4), 1095–1115.

Rose, D. C., Sutherland, W. J., Parker, C., Lobley, M., Winter, M., Morris, C., & Dicks, L. V. (2016). Decision support tools for agriculture: Towards effective design and delivery. *Agricultural Systems*, *149*, 165–174.

Salminen, J., Froneman, W., Jung, S.-G., Chowdhury, S., & Jansen, B. J. (2020, April). The ethics of data-driven personas. In: *Extended Abstracts of the 2020 CHI Conference on Human Factors in Computing Systems Extended Abstracts*, pp. 1–9, Honolulu, HI. doi:10.1145/3334480.3382790.

Salminen, J., Jung, S.-G., & Jansen, B. (2022, March). Developing persona analytics towards persona science. In: *27th International Conference on Intelligent User Interfaces*, Helsinki, Finland, pp. 323–344. doi:10.1145/3490099.3511144.

Salminen, J., Jung, S.-G., & Jansen, B. J. (2021, December). Are data-driven personas considered harmful? Diversifying user understandings with more than algorithms. *Persona Studies*, *7*(1), 48–63. doi:10.21153/psj2021vol7no1art1236.

Salminen, J., Jung, S.-G., Chowdhury, S., Sengün, S., & Jansen, B. J. (2020, April). Personas and analytics: A comparative user study of efficiency and effectiveness for a user identification task. In: *Proceedings of the 2020 CHI Conference on Human Factors in Computing Systems*, pp. 1–13. New York, NY. doi:10.1145/3313831.3376770.

Salminen, J., Jung, S.-G., Medina, J., Aldous, K., Azem, J., Akhtar, W., & Jansen, B. J. (2024, July). Using cipherbot: An exploratory analysis of student interaction with an LLM-based educational chatbot. In: *Proceedings of the Eleventh ACM Conference on Learning @ Scale*, Atlanta, GA, pp. 279–283. doi:10.1145/3657604.3664690.

Salminen, J., Liu, C., Pian, W., Chi, J., Häyhänen, E., & Jansen, B. J. (2024, May). Deus ex machina and personas from large language models: Investigating the composition of ai-generated persona descriptions. In: *Proceedings of the CHI Conference on Human Factors in Computing Systems*, Honolulu, HI, pp. 1–20. doi:10.1145/3613904.3642036.

Salminen, J., Santos, J. M., Jung, S., Chowdhury, S. A., & Jansen, B. J. (2019). Persona perception scale: Developing and validating an instrument for human-like representations of Data. *International Journal of Human-Computer Studies*, *129*, 1–14.

Salminen, J., Santos, J. M., Jung, S.-g., & Jansen, B. J. (2023, December). How does an imaginary persona's attractiveness affect designers' perceptions and IT solutions? An experimental study on users' remote working needs. *Information Technology & People*, *36*(8), 196–225. doi:10.1108/ITP-09-2022-0729.

Salminen, J., Santos, J. M., Kwak, H., An, J., Jung, S.-g., & Jansen, B. J. (2020, September). Persona perception scale: Development and exploratory validation of an instrument for evaluating individuals' perceptions of Personas. *International Journal of Human-Computer Studies, 141*, 102437. doi:10.1016/ j.ijhcs.2020.102437.

Salminen, J., Sengün, S., Kwak, H., Jansen, B. J., An, J., Jung, S.-g., Harrell, D. F. (2018, June). From 2,772 segments to five personas: Summarizing a diverse online audience by generating culturally adapted personas. *First Monday*, 23(6). doi:10.5210/fm.v23i6.8415.

Salminen, J., Wenyun Guan, K., Jung, S.-G., & Jansen, B. (2022, April). Use cases for design personas: A systematic review and new frontiers. In: *Proceedings of the 2022 CHI Conference on Human Factors in Computing Systems,* New York, NY, pp. 1–21. doi:10.1145/3491102.3517589.

Sattele, V., & Carlos Ortiz, J. (2024). Generating user personas with AI: Reflecting on its implications for design. In: C. Gray, E. Ciliotta Chehade, P. Hekkert, L. Forlano, P. Ciuccarelli, & P. Lloyd (Eds.), *DRS2024: Boston*, pp. 1–9. Boston, MA, USA.

Scheer, H. (2004). *The Solar Economy: Renewable Energy for a Sustainable Global Future.* Earthscan.

Schmidt, O., Hawkes, A., Gambhir, A., & Staffell, I. (2017). The future cost of electrical energy storage based on experience rates. *Nature Energy*, *2*(8), 17110.

Schuller, A., Janssen, D., Blumenröther, J., Probst, T. M., Schmidt, M., & Kumar, C. (2024). Generating personas using LLMs and assessing their viability. In: *Extended Abstracts of the 2024 CHI Conference on Human Factors in Computing Systems,* New York, NY. doi:10.1145/3613905.3650860.

Sengün, S., Salminen, J., Jung, S.-G., Aldous, K. K., & Jansen, B. J. (2024, July). "There's something about noura": Exploring think-aloud reasonings for users' persona choice in a design task. In: *Designing Interactive Systems Conference*, IT University of Copenhagen, Denmark, pp. 1234–1247. doi:10.1145/3643834.3661512.

Sheng, E., Joe, L., Alambo, A., & Solaiman, I. (2021). A societal-aware methodology for managing bias in AI applications. *Patterns*, *2*(6), 100238.

Shin, J., Hedderich, M. A., Rey, B. J., Lucero, A., & Oulasvirta, A. (2024, July). Understanding human-AI workflows for generating personas. In: *Designing Interactive Systems Conference*, IT University of Copenhagen, Denmark, pp. 757–781. doi:10.1145/3643834.3660729.

Smith, T., Nguyen, P., & Brown, C. (2021). Enhancing community well-being through human-nature interaction Personas. *Journal of Community Psychology*, *49*, 123–140.

Soon-Gyo Jung, Joni Salminen, Kholoud Khalil Aldous, and Bernard J. Jansen. 2025. PersonaCraft: Leveraging language models for data-driven persona development. Int. J. Hum.-Comput. Stud. 197, C (Mar 2025). https://doi.org/10.1016/j.ijhcs.2025.103445.

Stern, P. C. (2000). Toward a coherent theory of environmentally significant behavior. *Journal of Social Issues*, *56*(3), 407–424.

Sun, Z., Sun, L., & Strang, K. D. (2016). Big data analytics services for enhancing business intelligence. *Journal of Computer Information Systems*, *56*(2), 133–145.

UN-Habitat. (2009). *Planning Sustainable Cities: Global Report on Human Settlements 2009.* Earthscan.

United Nations. (2015). *Transforming Our World: The 2030 Agenda for Sustainable Development.* United Nations General Assembly.

United Nations. (2018). *World urbanization prospects: The 2018 revision.* Department of Economic and Social Affairs, Population Division, New York, NY.

Vergani, F., & Di Liberto, F. (2024). Designing from the plants' perspective: A field case study in urban forest of "la goccia". In: C. Gray, E. Ciliotta Chehade, P. Hekkert, L. Forlano, P. Ciuccarelli, & P. Lloyd (Eds.), *DRS2024: Boston*, pp. 1–9. Design Research Society.

Vinuesa, R., Azizpour, H., Leite, I., Balaam, M., Dignum, V., Domisch, S., & Fuso Nerini, F. (2020, January). The role of artificial intelligence in achieving the Sustainable Development Goals. *Nature Communications, 11*(1), 233. doi:10.1038/s41467-019-14108-y.

Wang, L., Zhao, M., & Chen, R. (2019). Global perspectives on incentive policies for electric vehicle adoption. *Sustainable Cities and Society*, *44*, 597–603.

Wells, P., & Nieuwenhuis, P. (2010). Understanding the changing automotive industry: Dynamics of supply and demand. *International Journal of Automotive Technology and Management*, *10*(2–3), 121–135.

Wilensky, U., & Rand, W. (2015). *Introduction to Agent-based Modeling: Modeling Natural, Social, and Engineered Complex Systems with NetLogo*. MIT Press.

Wilson, A., De, P. S., Forbes, P., & Sachy, M. (2020, November). Creating personas for political and social consciousness in HCI design. *Persona Studies*, *4*(2), 25–46. doi:10.3316/informit.032482973719629.

World Economics. (2024, August). *Global South* [Economic Data]. https://www.worldeconomics.com/Regions/Global-South/default.aspx.

Yang, S., & Lee, H. (2018). Identifying electric vehicle drivers' characteristics and their intention to adopt electric vehicles. *Transportation Research Part D: Transport and Environment*, *57*, 116–127.

Yanlic, B. S., & Coskun, A. (2024). Understanding the Needs of Nonhuman Stakeholders in Design Process: An Overview of and Reflection on Methods. arXiv. https://arxiv.org/abs/2407.14750.

Yoonseo Choi, Eun Jeong Kang, Seulgi Choi, Min Kyung Lee, and Juho Kim. 2025. Proxona: Supporting Creators' Sensemaking and Ideation with LLM-Powered Audience Personas. In Proceedings of the 2025 CHI Conference on Human Factors in Computing Systems (CHI '25). Association for Computing Machinery, New York, NY, USA, Article 149, 1–32. https://doi.org/10.1145/3706598.3714034.

Zhou, L., Fang, Y., Ding, S., Cheng, Y., Yan, B., Zhu, W., Bao, S., Wang, J., & Song, S. (2024). Vivid-persona: Customizable persona tool with interactive and immersive experiences. *Journal of Engineering Design*, *36*, 1–22. doi:10.1080/09544828.2024.2383543.

Zhuo, T. Y., Huang, Y., Chen, C., & Xing, Z. (2023, May). Red Teaming ChatGPT Via Jailbreaking: Bias, Robustness, Reliability and Toxicity. arXiv. https://arxiv.org/abs/2301.12867.

11 Theoretical Foundations of Human Factors and Trustworthiness in Large Language Models

Kitty Kioskli

11.1 INTRODUCTION

The integration of large language models (LLMs) into knowledge-intensive, decision-support, and communication systems signals a paradigmatic shift in the nature of human–machine interaction. These models, characterized by high-dimensional linguistic fluency and domain-agnostic generativity, have not only expanded the functional landscape of artificial intelligence (AI) but also introduced new epistemic, relational, and psychosocial complexities (Zhou & Li, 2023). In practice, LLMs increasingly serve as intermediaries in domains as diverse as mental health, legal reasoning, and education, where they interface with users not merely as computational tools but as entities perceived rightly or wrongly as intentional, conversational, and socially aware (Cheng et al., 2023; Suresh et al., 2023). Crucially, the interaction between users and LLMs is rarely neutral or purely functional. Rather, it is situated within a broader ecosystem of sociotechnical expectations, cognitive heuristics, and embedded power relations. This interaction is further shaped by affective states, institutional branding, and the cultural legibility of the AI system's interface (Shin & Park, 2022). Users do not approach LLMs as blank slates; they carry with them prior experiences, interpretive frameworks, and social scripts that inform their interpretation of the model's behavior, its tone, responsiveness, and perceived transparency. These micro-evaluations coalesce into trust judgments, which are rarely linear and often evolve dynamically with each encounter (Wu et al., 2023).

Trust, in this context, cannot be reduced to mere system usability or technical performance. Rather, it must be understood as an emergent construct co-produced through human–machine interaction over time. According to Mayer, Davis, and Schoorman (1995), trust is defined as "the willingness of a party to be vulnerable to the actions of another party based on the expectation that the other will perform a particular action important to the trustor, irrespective of the ability to monitor or control that other party." This definition highlights the inherently relational and risk-laden nature of trust, which becomes especially salient in interactions with opaque, probabilistic systems like LLMs. Recent work on adaptive trust calibration has emphasized that trust in LLMs is context-sensitive, historically contingent,

DOI: 10.1201/9781003492252-11

and deeply shaped by the psychosocial makeup of the user (Zhang et al., 2022). For instance, the same probabilistic output may be interpreted as reassuring in one context and as opaque or even deceptive in another, depending on the stakes, user expertise, and domain conventions. In this light, trust is less a stable state and more a dynamic cognitive-affective response, regulated by factors such as cognitive load, interface design, and institutional cues (Chugunova & Seleznova, 2022). The complexity of trust in LLMs becomes even more apparent when examined through a socio-technical lens. Here, trust is not located solely in the dyadic interaction between user and system but emerges from the broader assemblage of algorithms, interface logics, policy regimes, and organizational narratives that structure this interaction. The implications are profound: design interventions aimed at increasing explainability or feedback may fail if they do not account for the lived experience of the user or the power asymmetries embedded in AI infrastructure (Dellagiacoma et al., 2022; Kioskli et al., 2024). Thus, understanding trust in LLMs demands a shift away from prescriptive models toward an empirical, user-sensitive approach that foregrounds context, contingency, and interpretive variability.

The psychosocial dimensions of trust in LLMs also demand critical interrogation. Recent studies suggest that the degree of perceived social presence-whether the system appears empathic, human-like, or contextually attuned-can significantly modulate user trust, often irrespective of output accuracy (Sykora et al., 2023). This presents a paradox: users may develop relational dependencies on systems that are not epistemically transparent or normatively aligned with human ethical standards. The anthropomorphic design of LLM interfaces, particularly in domains involving emotional labor or care (e.g., mental health chatbots), amplifies these dynamics, generating trust based not on verifiability but on perceived emotional resonance or responsiveness (Cheng et al., 2023). The implications of such interactional mirroring, where systems simulate understanding without possessing it, are not merely philosophical but profoundly operational, affecting user decision-making, emotional well-being, and even identity negotiation. Moreover, the experience of trust is never isolated from structural factors. As Kioskli et al. (2024) argue, human–AI interaction must be situated within an integrative risk management framework that includes psychological, contextual, and institutional dimensions. Trust is not solely built on perceived competence or fairness; it is mediated by users' broader sense of procedural justice, digital agency, and perceived alignment between their goals and those of the system or its governing institutions. In other words, the trust equation extends beyond the model to encompass the infrastructure, policy environment, and organizational culture that deploy it.

It is also important to recognize the temporal dynamics of trust. Trust can degrade, stabilize, or increase over time, depending on a user's cumulative experience and the system's perceived trajectory of improvement or stagnation. Longitudinal user studies have demonstrated that early positive experiences with LLMs can create a form of affective inertia, wherein users remain loyal or tolerant of future failures based on earlier perceived reliability (Zhou & Li, 2023). Conversely, a single high-stakes failure, especially one that violates moral or cultural expectations, can lead to disproportionate distrust or disengagement, even if the error is statistically rare (Suresh et al., 2023). Given these complexities, it is insufficient to approach trust in LLMs through deterministic models or technical metrics alone. Instead, a multidisciplinary, empirically grounded approach is necessary-one that foregrounds user subjectivity, environmental context, and system design in equal measure. This chapter therefore adopts a socio-technical lens, informed by cognitive psychology, human–computer interaction, and critical systems thinking, to examine how trust is co-constructed, maintained, and occasionally broken in the context of LLMs. By weaving together theoretical frameworks and empirical insights, it aims to offer a foundation for designing, regulating, and deploying LLMs in ways that are not only technically effective but also socially trustworthy and ethically responsive.

11.2 UNDERSTANDING HUMAN FACTORS IN HUMAN–AI INTERACTION

The rapid expansion of LLMs in public-facing and expert-in-the-loop systems has foregrounded the need to interrogate the human factors that underpin user interaction, interpretation, and trust. These models, while powerful in output, are dependent on users' cognitive, emotional, and contextual capacities to be engaged responsibly. Human–AI interaction is not merely an interface problem; it is a relational and interpretive act shaped by the psychological constraints and social scripts that users carry into each interaction. Trust, reliance, and even rejection of LLMs are rarely grounded in performance metrics alone; rather, they emerge through dynamic, contextualized user experiences that often deviate from the system designer's assumptions (Kocielnik, Amershi, & Bennett, 2022).

Cognitive load is one of the foundational constructs for understanding human engagement with complex AI systems. In LLM-mediated tasks, such as document summarization, decision support, or dialogue management, users must not only process the content but also assess the plausibility, relevance, and trustworthiness of the output in real time. This evaluative labor increases when explanations are either absent or fail to align with users' mental models, often resulting in disengagement or overreliance (Lai, Wu, & Tan, 2021). This is especially critical in high-stakes domains, where the perceived authority of the AI can obscure the user's sense of accountability or critical distance. Recent studies demonstrate that trust in AI can be misleadingly stable when users rely on shallow cues like fluency or formatting, even when the underlying reasoning is flawed (Buçinca, Malaya, & Glassman, 2021). This phenomenon, where apparent confidence in output triggers passive trust, is a manifestation of cognitive offloading. Instead of applying reflective scrutiny, users defer to the AI, particularly under time pressure or when domain knowledge is limited. In this regard, LLMs may not just assist cognition they may displace it. See Figure 11.1.

Yet, trust is not uniformly distributed across users. Psychosocial dimensions such as age, digital literacy, institutional exposure, and emotional state significantly modulate how people perceive and respond to LLMs. A user unfamiliar with AI may perceive neutrality in tone as "objectivity," while another may detect subtle reinforcement of bias. These interpretations are shaped by lived experience, sociocultural knowledge, and power awareness, which are factors that are often excluded from the design logic of ostensibly "universal" AI systems (Binns et al., 2020).

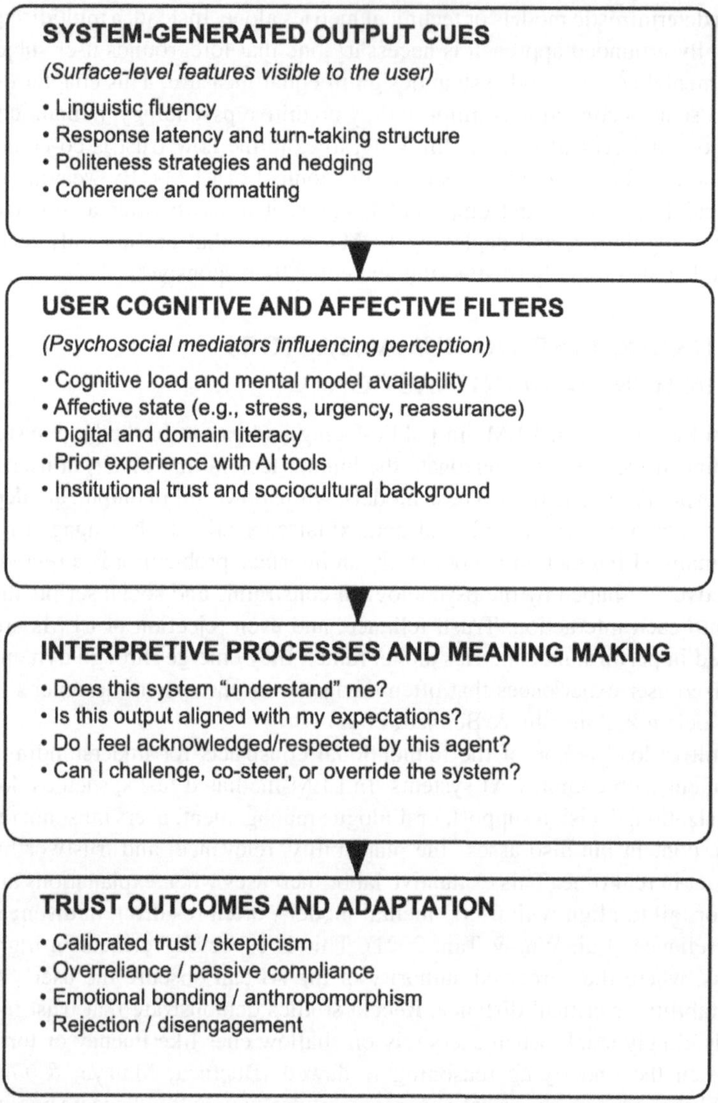

FIGURE 11.1 Cognitive and affective dimensions shaping user trust in large language models (LLMs).

Table 11.1 provides a summary of psychosocial variables that influence user trust trajectories in interactions with LLMs, as derived from the existing literature.

Table 11.1 outlines key psychosocial factors, such as cognitive load, digital literacy, emotional state, and social identity, that influence how users trust and respond to AI systems. For each variable, it explains its impact on trust dynamics and offers design recommendations, like simplifying interfaces, tailoring transparency, and promoting equity through culturally- and socially-aware design.

TABLE 11.1
Psychosocial Variables Influencing User Trust Trajectories in LLM Interactions

Psychosocial Variable	Influence on Trust Dynamics	Design Considerations
Cognitive load	High load may lead to overreliance or disengagement, reducing critical evaluation of output.	Simplify interfaces; use just-in-time explanations; avoid information overload.
Digital literacy	Low literacy increases reliance on interface cues (tone, fluency) over substantive evaluation.	Tailor transparency levels; provide guided interaction pathways.
Emotional state	Stress or urgency amplifies deference to perceived authority or fluent output.	Use emotionally calibrated responses; signal uncertainty with care.
Domain expertise	Experts are more sensitive to epistemic gaps, less persuaded by superficial cues.	Offer layered explanations; support challenge, override, and system probing.
Institutional trust	Users with high institutional trust may extend that trust to associated AI systems.	Transparently communicate system provenance, backing, and ethical frameworks.
Cultural frame	Cultural norms influence how users interpret AI behavior (e.g.., politeness, authority).	Localize design language; avoid overgeneralized behavior profiles.
Prior ai experience	Familiarity shapes expectations, calibration, and tolerance for system ambiguity or failure.	Allow personalization and adaptive trust scaffolding based on interaction history.
Social identity/power awareness	Users from marginalized groups may approach systems with heightened skepticism.	Recognize asymmetries; design for relational equity and epistemic humility.

Beyond cognitive heuristics and surface-level perceptions, LLM interaction demands consideration of user explanation preferences and interpretive anchoring. A central human factor in AI systems, particularly those operating through natural language, is the user's ability to construct a coherent mental model of how the system operates. When users fail to develop such a model, trust may become decoupled from understanding, leading to dependency or unjustified rejection (Eiband et al., 2021). Explanation strategies must therefore serve dual roles: enabling operational clarity while fostering subjective validation. Yet, these goals often conflict. Users do not simply want to "know how it works"; they want to understand why it behaves as it does, in a way that resonates with their context and values. See Table 11.1.

Trust is shaped not only by what is said but how and when it is said. This is particularly true in language-based systems, where tone, structure, and vocabulary co-create perceived intention. Research by Dodge et al. (2021) shows that users respond differently to the same explanation content depending on the timing and modality of delivery, whether embedded in a conversation, post-hoc, or on-demand. These preferences vary across user profiles, reflecting both psychosocial traits and task-specific needs. A system that explains too much can frustrate or patronize; one that explains too little may appear evasive or untrustworthy. Designing for the "right amount" of explanation remains a deeply human-centered challenge. The emotional affordances of LLM interaction are often underestimated in system design, yet they are central to understanding user trust trajectories. Afroogh et al. (2024) argue that trust in AI cannot be reduced to a single factor such as accuracy or transparency; rather, it emerges from a complex interplay of technical, social, and ethical dimensions. Their review underscores that users often evaluate trustworthiness not only on whether systems deliver correct outcomes, but also on whether they demonstrate alignment with human values and intentions. In practice, this means that perceptions of fairness, accountability, and responsiveness weigh as heavily as performance metrics, shaping a relational form of trust that extends beyond mere functionality.

While initial impressions and first-use experience shape the early trajectory of human–LLM interaction, long-term engagement introduces a fundamentally different set of psychosocial dynamics. Trust, in this temporal frame, becomes less reactive and more adaptive-a learned behavior influenced by repeated exposure, observed consistency, and the user's evolving mental model. As Schneider, Langer, and Butz (2022) demonstrate, trust calibration is rarely linear. Users may oscillate between reliance and skepticism based on system responsiveness, emotional labor, and error recovery over time.

In longitudinal contexts, users shift from initial curiosity and caution to more routine, patterned interactions. This habituation process is not inherently problematic-indeed, it can improve efficiency and reduce cognitive burden. However, as Ren, Huang, and Park (2023) note, habituation may also suppress critical reflection, particularly when early positive experiences solidify into unexamined trust. This adaptive trust is often reinforced by social and institutional signals, user reviews, branding, endorsements, or default system integration, creating a perceived social norm around AI reliability. Moreover, the development of trust is increasingly relational. LLMs deployed in dialogic or assistive contexts, such as writing aides or therapeutic bots, engage users in ongoing narrative and emotional framing. Over time, users begin to anthropomorphize the system, projecting traits like consistency, helpfulness, or even empathy onto its persona. This projection is not a mere illusion but a psychologically functional response to systems that mimic human cadence and emotional scaffolding (Preece & Shneiderman, 2021). However, it raises ethical questions when user vulnerability intersects with algorithmic opacity.

Users also adapt their behavior based on prior system performance. They may develop strategies to "game" or bypass limitations, for instance, by phrasing prompts in ways they believe yield better results. These behaviors, as shown in Liao, Gruen, and Miller's (2020) research, reflect a form of meta-interaction: users not only engage with the content but with their evolving expectations about the system's behavior. Such adaptations complicate the assumption that trust can be measured solely through static satisfaction or task success metrics. The temporal character of trust is further complicated by update cycles, model drift, and interface redesign. A previously "trusted" LLM may change subtly or significantly between versions; retraining, policy updates, or architectural shifts can disrupt the stability of user interaction. These changes, when poorly communicated, erode what Schneider et al. (2022) describe as interactional continuity-the user's sense that the system is predictable, familiar, and aligned with prior expectations.

What becomes clear across this landscape is that trust in LLMs is best understood not as a single moment or feature but as a temporally extended, emotionally layered process. It reflects not just system capability but the user's evolving sense of agency, safety, and purpose in the interaction. This requires a rethinking of how designers approach longitudinal usability and psychological safety, moving beyond task success to include indicators such as self-efficacy, narrative control, and trust resilience.

11.3 CONCEPTUALIZING TRUST AND TRUSTWORTHINESS IN AI SYSTEMS

Trust in AI systems, and in LLMs more specifically, is not a discrete variable that can be "optimized," but a socially situated, interpretive construct embedded within a constellation of psychological, organizational, and communicative dimensions. In contrast to traditional software systems, LLMs occupy a unique affective and epistemic space: they communicate in natural language, they generate information probabilistically, and they are frequently invoked in contexts where the distinction between tool and interlocutor is unstable. As a result, trust becomes less about system outputs in isolation and more about how those outputs are received, interpreted, and morally situated by human users (Vianello et al., 2023).

Contemporary models of trust in AI often draw upon classical organizational frameworks that emphasize three antecedents: ability, integrity, and benevolence (Hoffman et al., 2021). While these constructs provide a useful starting point, their applicability to LLMs is not straightforward. Ability, in a human context, implies competence rooted in intentional learning and domain expertise. In LLMs, "ability" is measured algorithmically via benchmarks and data coverage but may still yield hallucinations or biased output due to latent flaws in training data. This divergence raises a psychosocial tension: users may perceive an LLM as capable because of its fluency or speed, while remaining unaware of its structural limitations. See Figure 11.2.

Integrity and benevolence, likewise, become complicated when applied to systems lacking volition. Nonetheless, users often anthropomorphize LLMs and interpret design signals (e.g., confidence cues, disclaimers, politeness strategies) as indicators of moral stance or trustworthiness (Niculae et al., 2023). This is not a cognitive error to be corrected; it is a relational phenomenon shaped by language itself. The medium through which LLMs operate invites social interpretation, which then becomes embedded in the user's trust posture. What appears to be a neutral sentence may be read as considerate or dismissive, helpful or evasive, depending on the user's psychosocial orientation and the interaction context. Moreover, trust in LLMs is not necessarily grounded in verifiability. Instead, it often emerges from surface features that act as credibility cues: tone, structure, response latency, and even the perceived confidence of the model. Niculae et al. (2023) demonstrate that users consistently over-trust outputs from LLMs framed as authoritative, even when the content is demonstrably flawed. This dynamic, where epistemic authority is inferred from presentation rather than grounded evidence, reflects broader trends in platform communication, but it is particularly acute in AI-mediated contexts, where the line between machine and messenger blurs.

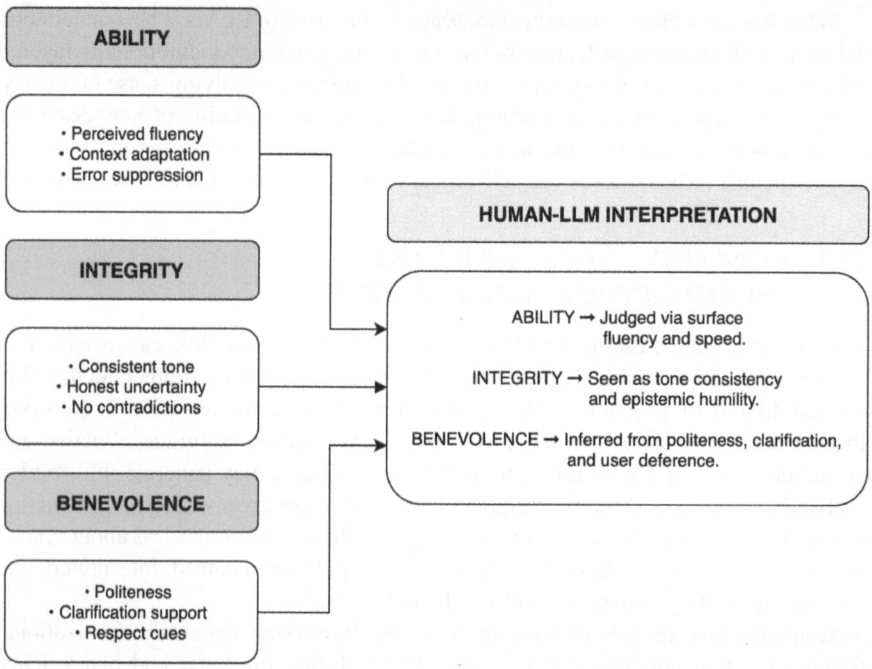

FIGURE 11.2 Reconceptualizing trust antecedents in human–LLM interaction.

Table 11.2 summarizes cross-cultural and contextual factors that modulate trust in LLM systems, as identified in the existing literature. This table highlights how contextual factors, such as culture, language, media exposure, and socioeconomic status, influence how users form trust in LLM systems. It also offers design recommendations, including localized interaction styles, support for dialect diversity, simplified user interfaces, and transparent institutional accountability to build inclusive and equitable AI systems.

As Jakesch et al. (2021) argue, the perceived fairness of algorithmic systems also mediates trust, even in the absence of full transparency. Users are more likely to trust a system they feel is "trying" to be fair, even if its inner logic remains opaque. This finding suggests that trust is not a product of rational deliberation alone, but a relational and affective state molded by past experiences, institutional narratives, and contextual expectations. See Table 11.2.

The tension between perceived trust and actual trustworthiness in AI systems stems, in part, from the failure to reconcile surface-level user cues with underlying system properties. Trust, as a lived experience, is modulated by emotional, contextual, and symbolic factors. Trustworthiness, by contrast, is a normative and technical construct, denoting a system's consistent alignment with principles such as accuracy, transparency, and fairness. The challenge for designers and researchers is not merely how to improve model behavior, but how to make trustworthiness legible to users without resorting to reductive proxies like confidence scores or authority framing

TABLE 11.2
Cross-Cultural and Contextual Factors Modulating Trust in LLM Systems

Contextual Dimension	Influence on Trust Formation	Implications for Design and Deployment
Cultural trust norms	Expectations of authority, communication politeness, and institutional roles vary by culture.	Localize interaction style; support culturally relevant formality, hierarchy cues, and deference patterns.
Language and dialect	LLMs trained on dominant language corpora may misinterpret or exclude minority linguistic forms.	Support dialectal diversity; use culturally representative training data; allow phonetic and regional input.
Technological access and history	Historical disenfranchisement or exclusion from digital infrastructure breeds baseline skepticism.	Account for digital trauma; design for inclusion, transparency, and reparability across populations.
Institutional association	Trust in AI is shaped by perceived legitimacy of host institutions (e.g., government and private sectors).	Signal institutional accountability clearly; disclose affiliations and redress channels.
Media and misinformation exposure	High exposure to misinformation or AI fear narratives may reduce initial trust thresholds.	Provide proactive epistemic scaffolds; clarify LLM capabilities and limitations upfront.
Socioeconomic status (SES)	Users from lower SES backgrounds may face constraints in digital fluency, bandwidth, or safety.	Avoid cognitive overload; simplify UX; prioritize mobile and low-resource usability.
Regulatory environment	National and regional AI laws impact perceptions of safety, rights, and recourse.	Adapt disclosure practices to local law; reference user rights and local ethical frameworks explicitly.
Historical power structures	Users in post-colonial, authoritarian, or highly surveilled societies may distrust AI by default.	Emphasize user autonomy; embed opt-out and control mechanisms; design for visibility over automation.

(Seth, Shen, & Narayanan, 2022). Transparency is often presented as the antidote to mistrust. However, empirical studies show that naive implementations of transparency-such as full exposure to decision logic or confidence intervals-can backfire when users lack the cognitive scaffolding to interpret them. Bhatt et al. (2021) suggest that uncertainty itself can serve as a form of transparency, especially when contextualized appropriately within user goals. In this framing, uncertainty signals not model failure but epistemic humility a quality more likely to foster calibrated trust than blind assurance. Emotional resonance plays a critical but underexplored role in perceived trustworthiness. When LLMs acknowledge ambiguity, use hedging language, or reflect user sentiment, they tend to be evaluated as more trustworthy, even when their outputs are less assertive or definitive (Binns et al., 2020). This reflects a deeper psychosocial logic: users are not simply seeking accuracy, but systems that respect their agency, acknowledge complexity, and reinforce their sense of epistemic control. Trust in these instances is affective and dialogic, not merely rational.

Institutional and process-based trust also factor prominently into human–LLM relations. Raji et al. (2020) propose that trustworthiness is not located solely in the model, but in the end-to-end pipeline that governs its creation, testing, deployment, and maintenance. Users may never encounter the dataset, training architecture, or audit protocol behind an LLM, but they are sensitive to whether the system "feels" like it was built responsibly. Institutional cues, such as disclaimers, feedback mechanisms, and data provenance indicators, can serve as trust scaffolds, offering users meta-information that shapes interpretation. This is where psychosocial and technical design must converge. As Seth et al. (2022) show, poorly calibrated models, those that express excessive certainty in incorrect outputs, cause users to either over-trust or disengage entirely. Even highly accurate systems can erode trust if their expressive behavior contradicts the user's experience or domain norms. In this way, trustworthiness becomes an expressive property: not only what the system does, but how it communicates its doing.

Designing for trust in LLMs requires more than the mitigation of error or enhancement of interpretability; it requires a deliberate reorientation toward epistemic modesty and interactional transparency. As Harrison et al. (2021) suggest, users are more likely to trust systems that acknowledge the limits of their knowledge or defer to user judgment in ambiguous cases. This trust is not diminished by uncertainty; rather, it is reinforced when systems provide cues that they are open to being questioned, corrected, or contextualized. In LLMs, such cues may take the form of hedged language, responsive clarification, or adaptive dialogue patterns-all of which can signal respect for the user's interpretive agency. Equally crucial is the alignment between the system's epistemic posture and the user's mental model of how intelligence operates. Ullman et al. (2020) argue that systems must be intelligible not only in a technical sense but also in a cognitive one. If users assume the LLM "understands" as a human does, or holds beliefs and intentions, this can create a disconnect between expected and actual behavior, particularly when the system produces content that appears plausible yet lacks grounding. Closing this gap requires not just better explanations, but an improved design of interaction rituals-how information is requested, offered, corrected, and reframed across repeated encounters.

Finally, trust in LLMs is not an individual event; it is a distributed process that involves institutions, systems, designers, and communities. Alvarado (2023) conceptualizes AI as an epistemic technology, a tool that not only generates information but also structures how knowledge itself is produced, shared, and legitimized. In this framing, trust in AI extends beyond verifying the correctness of outputs; it entails recognizing the epistemic roles these systems assume within human practices of inquiry. Users are asked, implicitly, to accept AI not merely as a computational assistant but as a co-participant in processes of reasoning and sense-making. Such a shift raises critical questions of authority and accountability: whose knowledge practices are embedded in the model, and whose perspectives are excluded? Trust, therefore, becomes a matter of negotiating epistemic legitimacy, demanding transparency about the cultural, social, and methodological commitments inscribed in AI design.

11.4 TRUST IN PRACTICE: USER PERCEPTIONS OF LLMs

While trust in LLMs has been theorized through cognitive models and normative design principles, its actualization occurs in the lived, contingent experience of users. These experiences are marked by ambiguity, affect, and adaptation, shaped less by system accuracy in isolation and more by the user's ongoing interpretation of cues, outcomes, and perceived intent. Trust, as it unfolds in practice, is never static; it is recursive, performative, and bound to social context.

Across studies, users consistently report that they rely on interactional fluency rather than underlying transparency as a key trust cue. Bach and Holmström (2024) note that user trust in AI systems is shaped less by abstract technical performance than by perceivable qualities such as transparency, explainability, and reliability. Their review highlights that when users can make sense of how and why an AI system produces outcomes, they are more likely to attribute credibility and competence to it. In this way, the interpretability of the system functions as a proxy for trust, signaling not only functional adequacy but also ethical alignment with user expectations.

At the same time, users bring to the interaction varied mental models of what the system is and how it operates. These models are often fuzzy, partial, or implicitly drawn from social analogies rather than technical knowledge. Bargas-Avila and Hornbæk (2021) argue that in the absence of clear conceptual scaffolding, users often adopt anthropomorphic metaphors, interpreting LLMs as advisors, helpers, or even companions. These roles come with embedded expectations of honesty, support, and reliability expectations that the system has not committed to, but is nonetheless judged by. See Figure 11.3.

Trust in LLMs does not unfold uniformly across users; it is profoundly shaped by prior knowledge, digital literacy, cultural frameworks, and the institutional context of use. While novice users may interpret fluency as competence, more experienced or domain-specific users are often attuned to subtle markers of epistemic uncertainty. The same interface may therefore foster trust in one context and prompt skepticism in another, not because the model has changed, but because the user's interpretive lens has (Srivastava, Holstein, & de Freitas, 2022). As Park and Shin (2021) argue, requests for transparency are not uniformly distributed: users with lower algorithmic literacy often prioritize fairness and responsiveness over technical disclosure. Conversely, users with higher domain expertise tend to evaluate LLMs through lenses of auditability, reliability, and consistency. These divergent preferences produce intersecting trust ecologies, where no singular design cue can satisfy all trust expectations simultaneously. The implication is that trust must be contextualized, not generalized.

Table 11.3 provides a summary of differentiated trust expectations shaped by user profiles and contextual factors, as derived from the existing literature.

Table 11.3 categorizes different user profiles, such as novice users, experts, emotionally distressed individuals, and task-focused professionals, highlighting their

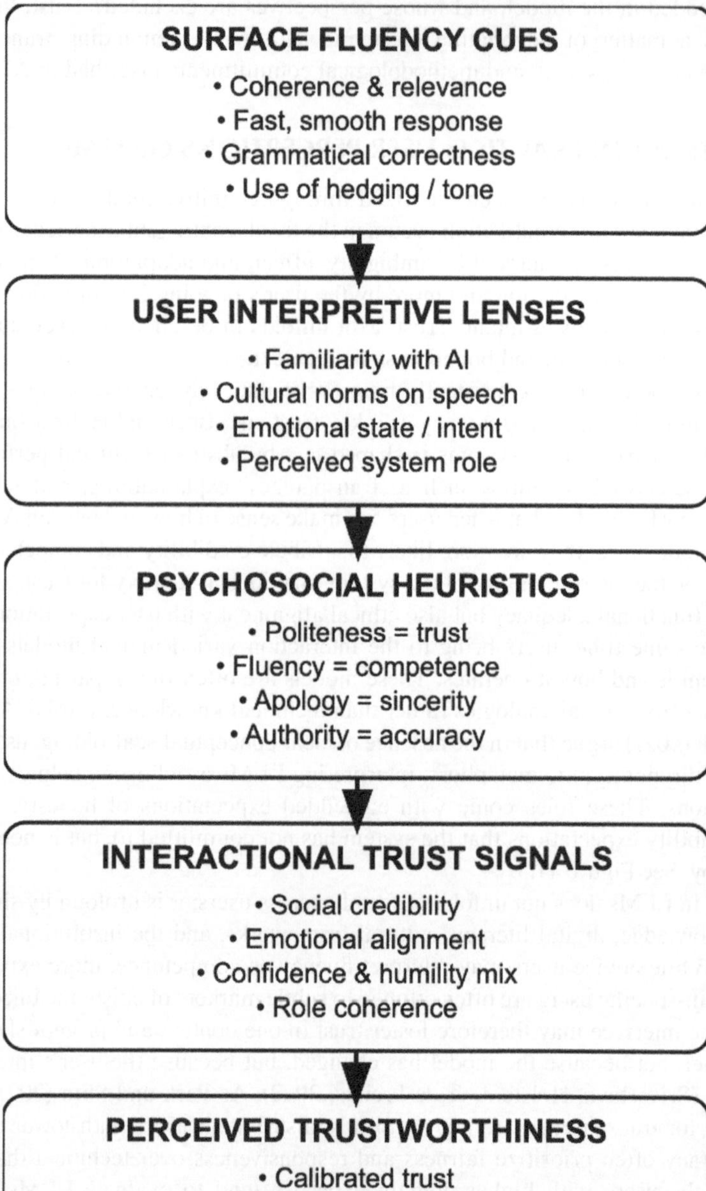

FIGURE 11.3 Interpretive pathways from interactional fluency to perceived trustworthiness in LLMs.

TABLE 11.3
Differentiated Trust Expectations Based on User Profiles and Contextual Factors

User Profile	Primary Trust Drivers	Common Trust Vulnerabilities	Design Recommendations
Novice user (low literacy)	Simplicity, fluency, politeness, and visual coherence	Over-trust from interface polish; limited error detection	Use plain language explanations; avoid overconfident tone; scaffold uncertainty gently
Expert user (domain specialist)	Alignment with epistemic norms, consistency, precision	Distrust when facing hallucinations or vague justifications	Offer layered explanations; enable audit or trace paths; allow challenge mechanisms
Emotionally Distressed user	Perceived empathy, reassurance, tone matching	Overreliance due to vulnerability; emotional resonance mistaken for accuracy	Use affect-sensitive language; offer disclaimers in a supportive tone
Marginalized user	Cultural respect, transparency of institutional origin, non-extractive tone	Skepticism due to past exclusion, bias, or invisibility	Disclose institutional actors; foreground participatory values; reinforce user agency
Task-oriented professional	Task relevance, speed, accuracy, documentation support	Frustration from hallucinations or irrelevant verbosity	Prioritize task clarity; structure outputs with summaries, options, and citations
High-trust default user	Brand familiarity, smooth interaction, minimal friction	Uncritical acceptance of flawed or biased outputs	Encourage reflective prompts; allow toggled transparency
Critical media consumer	Evidence-backed claims, citations, counterfactuals	Distrust if model lacks justification or hedges excessively	Provide embedded sources; allow user to probe, ask for alternatives
Institutionally constrained user	Compliance with organizational tech or automation policies	Apparent trust masking enforced compliance; disengagement from critique	Signal recourse mechanisms; clarify limits of system authority

primary trust drivers, typical vulnerabilities, and corresponding design recommendations. It emphasizes the importance of tailoring AI system design to meet diverse user needs through strategies like plain language, transparency, adaptive interfaces, and user agency support.

Designers who ignore this diversity often default to a monolithic trust model-one that assumes all users seek similar levels or forms of explanation. But as Lee, Kim, and Liao (2020) emphasize, explanation strategies must be user-contingent: some users require epistemic transparency to validate system behavior; others seek only social reassurance or narrative alignment. When these needs are misread, such as when a novice user receives a dense confidence interval instead of a simple rationale, trust may not erode immediately, but it becomes brittle, vulnerable to rupture under stress or uncertainty. See Table 11.3.

These dynamics are magnified in sociotechnical environments where LLMs are deployed at scale. In healthcare triage, legal consultation, or educational platforms, users often interact with AI under constrained timelines and heightened emotional states. In such contexts, the psychosocial burden of evaluation increases. Users may not only lack the cognitive bandwidth to critically interrogate outputs, but may also actively defer to perceived system authority as a way to reduce emotional dissonance. This is not trust as belief; it is trust as coping (Rahwan et al., 2021). Moreover, even when users question an LLM's response, their willingness to reject it is not solely determined by perceived error. Social and institutional pressures, such as fear of appearing uninformed, deferring to system branding, or organizational policies discouraging manual override, can inhibit disconfirmation, preserving the appearance of trust where none may exist. Here, compliance should not be mistaken for trust, especially in data derived from click-throughs or retention metrics. What is observed as interaction may, in fact, be resignation.

What emerges is a portrait of trust that is neither universal nor static. It is entangled with emotion, literacy, power, and policy. To understand trust in practice, we must study it not as a trait to be cultivated, but as a social signal-one that both reflects and obscures the deeper currents of interactional vulnerability and resilience. Trust, when studied in practice, reveals not just how users relate to LLMs, but how they understand their own agency within the interaction. A key finding across recent literature is that perceived agency, the sense that one can question, override, or co-steer the AI, plays a significant role in shaping trust trajectories (Wang et al., 2021). Users do not simply evaluate the system's competence; they evaluate their role in the system's logic. When LLMs invite clarification, acknowledge error, or allow for user correction, trust becomes interactive rather than transactional. However, many LLM-based interfaces do not meaningfully integrate user feedback. Feedback mechanisms, if present, are often symbolic "thumbs up/down" buttons detached from any observable consequence. This disconnect erodes user motivation and leads to a degradation of what Benbasat and Wang (2020) call *process trust*: confidence not only in system output, but in the *process by which the system evolves*. Without visibility into how feedback alters the system, user agency becomes performative rather than participatory.

Furthermore, Liao, Gruen, and Miller (2020) show that users often suppress feedback when it threatens to disrupt the flow of interaction. In dialogic systems like LLMs, feedback is perceived not just as evaluative but as socially consequential. Critiquing the model may feel awkward or futile, particularly when the system appears responsive or benevolent. This reflects the psychosocial complexity of user feedback-users may defer not because they trust, but because they fear interrupting a socially coherent exchange. To foster sustainable trust, designers must therefore move beyond static transparency features and implement reflexive trust architectures, which are systems that can signal internal change, invite epistemic dialogue, and support situated feedback. This includes visual indicators of update cycles, adaptive transparency based on user familiarity, and layered control over interaction depth. Such mechanisms do not eliminate uncertainty but mediate it, transforming ambiguity from a threat into a site of shared interpretive labor.

To make these dynamics more tangible, consider two domain-specific examples where trust in LLMs is shaped not by general functionality but by contextually embedded expectations. In a mental health chatbot setting, users often seek empathetic affirmation rather than factual precision. A user experiencing anxiety might trust the LLM not because its suggestions are clinically validated, but because the language feels attuned and supportive. Here, trust emerges from affective alignment, which is the model's ability to mirror emotional tone and offer perceived presence (even in the absence of psychological expertise). However, this affective trust can be fragile: a single off-tone response, such as dismissive wording or a generic suggestion, may rupture the perceived alliance, particularly when the user is in a heightened emotional state. Conversely, in a financial advisory context, the trust calculus shifts. Users evaluating investment recommendations assess LLM outputs through a lens of procedural reliability and verifiability. Trust here is predicated on perceived auditability-whether the model can justify its rationale in terms aligned with financial norms. A fluent but unverifiable recommendation may actually decrease trust, particularly for expert users attuned to regulatory and economic nuances. These contrasting cases reveal that trust is not a uniform metric of system quality, but a variable expression of user need, risk perception, and contextual stakes.

In summary, user trust in LLMs is not a one-time calibration task; it is an ongoing negotiation, shaped by relational cues, institutional scaffolding, and system responsiveness. Recognizing trust as a co-authored and context-bound experience is essential for designing systems that do more than perform well-they must also relate well, adapt responsibly, and sustain user agency over time.

11.5 ETHICAL AND SOCIETAL DIMENSIONS OF TRUST

Trust in AI systems cannot be divorced from the broader ethical and societal frameworks within which these technologies are developed, deployed, and experienced. For users, trust is not only a product of how an LLM behaves in a technical sense, but of whether the system aligns with deeper values of fairness, accountability, transparency, and respect. When these values are perceived to be compromised, trust deteriorates not because of poor performance, but because of perceived moral failure.

This tension is especially pronounced in systems that appear intelligent while remaining fundamentally opaque. The epistemic opacity of LLMs, which is their inability to explain decision-making in human-understandable terms, challenges basic ethical expectations about autonomy, agency, and consent. Crawford (2021) emphasizes that AI systems are never neutral; they are embedded within power structures, informed by extractive data practices, and deployed in contexts that amplify or obscure existing inequalities. In this light, trust in LLMs becomes a matter not of technical assurance, but of political legitimacy. Moreover, the deployment of LLMs in public infrastructure (e.g., government services, social security, health triage) introduces institutional trust into the equation. Choung, David, and Ross (2023) contend that ethical concerns and trust in AI are inseparable, since perceptions of fairness, transparency, and accountability directly shape whether users accept or resist these systems. They show that even technically proficient models face skepticism when embedded in opaque decision-making processes, especially where outcomes have social or political stakes. Trust, in this view, cannot be engineered solely through performance metrics; it requires institutional commitments to openness and responsiveness that acknowledge users as stakeholders rather than passive recipients of automated judgment.

Users may trust an LLM's response in a moment of convenience, yet maintain broader distrust in the system as a whole-a form of compartmentalized trust that reflects the psychosocial need to navigate opaque systems without endorsing them. As Whittaker et al. (2021) show, this fragmentation is intensified in contexts where LLMs are used to mediate human behavior, such as in criminal justice or education, where even small misalignments between system goals and human values can produce structural harm. Ethical trust in LLMs is not forged in isolation; it is entangled with histories of surveillance, exclusion, and systemic inequity. As these models increasingly operate in spaces once governed by human discretion, such as benefits determination, parole evaluation, and job screening, they bring with them not only predictive capacity but also the logics of automation. These logics, as Eubanks (2021) argues, are often skewed by socioeconomic bias: automating systems tend to replicate the assumptions and blind spots of those with institutional power, regardless of the intentions behind their deployment.

The result is that trust becomes deeply contested. For some, LLMs may represent a promise of efficiency or impartiality. For others, particularly those historically subject to over-surveillance or algorithmic misclassification, they signify further disenfranchisement. In these contexts, trust is not a default state to be earned; it is a barrier to be overcome. Trust also becomes politicized when LLMs are used to make or justify decisions about identity, access, and credibility.

Table 11.4 summarizes ethical risks and structural harms that shape user trust in LLM systems, as identified in the existing literature.

Table 11.4 outlines five key domains where ethical risks and structural harms affect user trust in LLMs, including data labor, institutional alignment, cultural marginalization, automation bias, and procedural ethics. For each domain, it links ethical concerns, such as hidden labor or opaque governance, to trust implications and provides real-world examples, like AI in immigration or healthcare, to illustrate the impact.

Even the language of "AI ethics" itself can obscure more than it reveals. Tsamados et al. (2022) warn that ethics, when operationalized as checklists or regulatory boxes, risks abstracting trust from lived experience. A system that meets formal fairness criteria may still feel deeply untrustworthy to those on the receiving end of automated judgment. In such cases, the perception of injustice is itself a form of harm-not simply an artifact of misunderstanding, but a valid signal that something essential is missing from the interaction. Designing for ethical trust, then, means acknowledging that not all users are positioned equally in the sociotechnical landscape. Marginalized users are more likely to be exposed to opaque systems without recourse, to experience alienation in voice-based interactions, or to interpret neutral design choices as forms

TABLE 11.4
Ethical Risks and Structural Harms Shaping User Trust in LLM Systems

Domain	Ethical Concern	Trust Implication	Illustrative Example
Data labor	Hidden human annotation, moderation trauma	Fragile trust based on unseen suffering	Ghost work in content moderation
Institutional alignment	Opaque governance and lack of redress	Distrust in public-sector LLM use	AI in welfare or immigration processing:
Cultural marginalization	Disregard for local context and voice	Perceived AI neutrality masks erasure	Voice assistants ignoring non-dominant dialects
Automation bias	Over-reliance in high-stakes domains	Trust-by-default masks systemic bias	AI triage in healthcare without patient override
Procedural ethics	Ethics as checklist or compliance ritual	Symbolic ethics reduce perceived legitimacy	Shallow Algorithmic Impact Assessments (AIAs)

of cultural erasure. In these cases, trust is not a UX problem or a communication gap. It is a signal of broader epistemic inequality, whose voices are heard, whose values are encoded, and whose autonomy is respected (See Table 11.4).

Trustworthiness in AI is not achieved solely through better technical design; it is co-produced through institutional mechanisms of accountability, public engagement, and transparent governance. The limits of current trust paradigms are not just epistemological but structural. Many existing frameworks reduce trust to individual perception or usability, ignoring the broader infrastructures that enable, constrain, or bypass trust altogether. One promising direction involves shifting from ethical intentions to procedural accountability. Morley et al. (2021) note that while ethical principles such as fairness, transparency, and justice are now ubiquitous in AI discourse, their operationalization remains inconsistent and fragmented. Ethics, to be trusted, must be translated into institutional processes, including audit trails, redress systems, and stakeholder-inclusive design protocols.

Algorithmic Impact Assessments (AIAs) represent one such attempt to embed accountability into system lifecycles. Metcalf et al. (2021) emphasize that AIAs are not merely tools for documentation; they are sites of negotiation where stakeholders debate values, anticipate harms, and demand transparency. When conducted seriously, they facilitate trust not as persuasion but as verification rooted in due process and openness. However, when performed as bureaucratic rituals, they risk becoming superficial signals devoid of substantive accountability. Equally critical is the role of participatory design in shaping trustworthy AI. Green and Viljoen (2020) propose "algorithmic realism" as a framework that foregrounds the lived realities of those subject to AI systems. Rather than assume that trust can be engineered, they advocate for inclusive co-design, where affected communities are not merely consulted but given material influence over system design and deployment. In this model, trustworthiness is not granted by default, but built iteratively through mutual recognition, dialogue, and adaptability.

What emerges from these perspectives is a conception of trust as inherently political and contingent. It cannot be reduced to a static property of systems or an internal state of users. Trust is an index of relational ethics-how systems behave, how institutions respond, and how users are positioned within sociotechnical hierarchies. To design for trustworthiness is therefore to engage not only in interface refinement but in systemic reform.

11.6 FUTURE DIRECTIONS AND OPEN QUESTIONS

As LLMs continue to be integrated into high-stakes decision-making systems and everyday digital infrastructure, the question is no longer whether users should trust these systems, but how that trust can be earned, contextualized, and continually recalibrated. The challenge is not one of incremental optimization, but of conceptual realignment: moving from trust as a static design goal to trust as a co-evolving psychosocial process, shaped by diverse users, fluctuating norms, and system dynamism. One critical frontier lies in adaptive trust calibration-the ability of systems to respond to user signals and context in a way that preserves interpretive agency while mitigating over- or under-reliance. Current LLMs offer limited support for dynamic adaptation based on user needs or emotional states. Yet empirical research increasingly shows that trust miscalibration, whether through premature overconfidence or persistent doubt, can undermine both task outcomes and user well-being (Rahwan et al., 2021). Future models must move beyond confidence scoring to design interaction patterns that respond reflexively to user cues such as hesitation, revision, or disengagement.

Related to this is the issue of transparency's diminishing returns. Ananny and Crawford (2021) argue that the dominant discourse of "explainable AI" often treats transparency as a moral good without questioning what, to whom, and for what purpose transparency is enacted. When systems provide justifications that are either too technical, too shallow, or too generic, users may experience confusion or cognitive overload rather than clarity. Future work must shift from "telling users everything" to enabling actionable, contextual understanding, especially when decisions involve risk, ambiguity, or ethical complexity. Another underexplored area is cross-cultural trust variation. Much of the current research assumes Western-centric norms of agency, fairness, and autonomy. However, studies in cooperative AI and global digital ethics have shown that conceptions of trust and what constitutes a trustworthy system can differ dramatically depending on cultural values, regulatory expectations, and historical legacies of technology governance. For instance, a system that foregrounds individual control and customization may be appealing in liberal democracies, but less meaningful in collectivist cultures where institutional endorsement or community validation play stronger roles in trust formation.

As LLMs become embedded in long-term digital ecosystems, their trust dynamics will need to extend beyond immediate interactional quality. A major gap in current research is the temporal dimension of trust-how trust builds, erodes, and stabilizes across weeks, months, or years of user interaction. One critical challenge is how LLMs should communicate their evolution. Model updates, retraining cycles, or shifts in underlying data are often invisible to users. When these changes affect tone, accuracy, or behavior without acknowledgment, users may interpret them as arbitrariness or deception. Transparency here cannot be an abstract virtue; it must be interactionally and temporally situated. Systems need to disclose when, how, and why they change-ideally in ways that empower rather than overwhelm. Equally pressing is the need to develop participatory infrastructures for LLM oversight. Viljoen, Narayanan, and Pasquale (2022) argue that algorithmic accountability cannot be fully internal to design teams or developers. Instead, trust must be co-constructed with the communities affected by AI through deliberative engagement, feedback loops with demonstrable impact, and institutionalized pathways for redress. This reflects a shift from compliance-based ethics to relational legitimacy-where trust emerges from mutual recognition and sustained dialogue.

In this light, the prevailing focus on individual fairness or user-level calibration may be insufficient. As Binns (2020) points out, structural fairness-how systems differentially affect groups across demographic, geographic, or economic lines-often conflicts with user personalization. Addressing this tension requires a broader lens on trust: one that incorporates social solidarity, justice, and shared epistemic responsibility. The question is not simply, "Does this user trust this system?" but, "Is this system trustworthy in the society it helps shape?" This expansion also necessitates critical engagement with power and voice. As D'Ignazio and Klein (2020) argue, data infrastructures are never neutral; they encode values, erase dissent, and amplify dominant narratives. Trust in LLMs must therefore include space for contestation-not all users will or should trust AI, especially when it reinforces marginalization. Designing for trustworthy AI means not only supporting belief but enabling disagreement, refusal, and collective resistance when warranted. Ultimately, the future of trust in LLMs will hinge on our ability to treat trust as a layered, participatory, and ethically situated process. This includes respecting uncertainty, designing for variability, and enabling institutions, not just interfaces, to bear the weight of responsibility. Open questions remain: How do we measure trust without instrumentalizing it? How do we differentiate trust from dependence? And how do we ensure that systems worthy of trust are not only possible but prioritized?

11.7 CONCLUSION

Trust in LLMs is not a monolithic construct, nor is it reducible to technical performance or usability. As this chapter has argued, trust is a socially embedded, psychologically mediated, and politically situated phenomenon emerging at the intersection of cognition, design, ethics, and institutional practice. To understand trust in LLMs is not to ask whether users "believe" in the system but to inquire into the conditions under which trust is invoked, withheld, negotiated, or imposed. Throughout this chapter, we have shown that user trust is shaped by an array of human factors: cognitive load, mental models, prior experience, and emotional disposition. LLMs do not simply "perform" a task; they perform a relationship. They produce language in ways that simulate intention, attentiveness, and even moral agency cues, which users, especially in moments of uncertainty, often interpret as grounds for trust. But this interpretive process is uneven, filtered through lived experience, cultural norms, and the sociotechnical scripts that govern our interactions with authority and automation.

Trust, however, is not merely a psychological state. It is also a structural outcome mediated by the design of systems, the values of institutions, and the visibility (or absence) of accountability mechanisms. When systems fail to align with users' ethical expectations by omitting uncertainty, ignoring feedback, or reinforcing historical inequities, trust erodes not because the user is irrational, but because the system is no longer relationally or ethically legible. In this sense, trustworthiness must precede trust. Systems must be designed to be deserving of trust before they can rightfully claim it. We have also emphasized the importance of temporal and contextual variation in trust. User trust in LLMs does not remain stable across interactions or domains; it adapts based on system evolution, feedback processes, and moments of vulnerability. A trustworthy LLM must therefore be capable not only of producing reliable outputs but of sustaining dialogic integrity over time, acknowledging change, enabling calibration, and respecting user agency in every interaction. Trust cannot be maintained by interface design alone; it must be scaffolded by institutional transparency, participatory governance, and epistemic humility. At a societal level, trust in LLMs is bound up with broader questions of justice and legitimacy. Who gets to define what is fair, transparent, or appropriate in AI behavior? Whose values are encoded into training data, interface norms, or feedback channels? These questions are not peripheral to trust; they are central. For many users, especially those from marginalized communities, trust is not something they give freely to systems that resemble past instruments of control or exclusion. To design for trust, then, is to design *with* and not just *for* the communities affected.

Ultimately, this chapter calls for a redefinition of trust in LLM contexts-not as a static attribute to be engineered but as a living relationship to be cultivated. This requires models that adapt, institutions that listen, and designers who understand that trustworthiness is not only a matter of logic, but of ethics, dialogue, and care. As we move into a future where LLMs are increasingly integrated into decision-making, education, governance, and care work, the stakes of trust grow ever more profound. We must ensure that trust is not manipulated, coerced, or assumed, but earned through transparency, responsiveness, and mutual respect.

REFERENCES

Afroogh, S., Akbari, A., Malone, E., Kargar, M., & Alambeigi, H. (2024). Trust in AI: Progress, challenges, and future directions. Humanities and Social Sciences Communications, 11(1), 1568. https://doi.org/10.1057/s41599-024-04044-8

Alvarado, R. (2023). AI as an Epistemic Technology. Science and Engineering Ethics, 29, 32. https://doi.org/10.1007/s11948-023-00451-3

Bach, T. A., & Holmström, J. (2024). A Systematic Literature Review of User Trust in AI-Enabled Systems. International Journal of Human–Computer Interaction, 40(2), 159-174.

Bargas-Avila, J. A., & Hornbæk, K. (2021). Towards user-centered explainable AI: Developing mental models through interaction. *Interacting with Computers,* 33(2), 179–194. https://doi.org/10.1093/iwc/iwaa023.

Benbasat, I., & Wang, W. (2020). Trust in AI-based decision support systems: A process-tracing approach. *Journal of the Association for Information Systems,* 21(6), 1537–1556. https://doi.org/10.17705/1jais.00629.

Bhatt, U., Antoran, J., Zhang, Y., et al. (2021). Uncertainty as a form of transparency: Measuring, communicating, and using uncertainty. In: *NeurIPS 2021 Workshop on Human-Centered AI*, Virtual. https://arxiv.org/abs/2107.09292.

Binns, R. (2020). On the apparent conflict between individual and group fairness. In: *Proceedings of the 2020 ACM Conference on Fairness, Accountability, and Transparency (FAccT)*, Barcelona, Spain, pp. 514–524. https://doi.org/10.1145/3351095.3372864.

Binns, R., Veale, M., Van Kleek, M., & Shadbolt, N. (2020). 'It's reducing a human being to a percentage': Perceptions of justice in algorithmic decisions. In: *Proceedings of the 2020 CHI Conference on Human Factors in Computing Systems*, Virtual, pp. 1–14. https://papers.ssrn.com/sol3/papers.cfm?abstract_id=3114133

Buçinca, Z., Malaya, M. B., & Glassman, E. L. (2021). To trust or to think: Cognitive forcing functions can reduce overreliance on AI in AI-assisted decision-making. In: *Proceedings of the 2021 CHI Conference on Human Factors in Computing Systems*, Virtual, pp. 1–14. https://doi.org/10.1145/3411764.3445310.

Cheng, M., Bernstein, M. S., & Hancock, J. (2023). How language models affect user agency and trust: A field study in email composition. In: *Proceedings of the 2023 CHI Conference on Human Factors in Computing Systems*, Virtual, pp. 1–15.

Choung, H., David, P., & Ross, A. (2023). Trust and ethics in AI. AI & Society, 38(2), 733–745. https://doi.org/10.1007/s00146-022-01473-4

Chugunova, M., & Seleznova, E. (2022). Trust and automation bias in human-AI interaction: The moderating role of cognitive load. *Computers in Human Behavior Reports,* 6, 100162.

Crawford, K. (2021). *Atlas of AI: Power, Politics, and the Planetary Costs of Artificial Intelligence*. Yale University Press.

Dellagiacoma, D., Fotis, T., Kioskli, K., & Mouratidis, H. (2022). The supply chain of a Living Lab: Modelling security, privacy, and vulnerability issues. *Journal of Wireless Mobile Networks, Ubiquitous Computing, and Dependable Applications,* 13(2), 147–182.

D'Ignazio, C., & Klein, L. F. (2020). *Data Feminism*. MIT Press.

Dodge, J., Long, C., Gardner, M., & Singh, S. (2021). Explaining with impact: Designing and evaluating explanations for text classifiers. In: *Proceedings of the 2021 EMNLP Conference*, Virtual, pp. 1360–1375. https://doi.org/10.18653/v1/2021.emnlp-main.106.

Eiband, M., Schneider, H., Langer, L., & Butz, A. (2021). "The system explained it all." A taxonomy of explanation goals and strategies for explainable AI. *Proceedings of the ACM on Human-Computer Interaction,* 5(CSCW1), 1–27. https://doi.org/10.1145/3449143.

Eubanks, V. (2021). *Automating Inequality: How High-Tech Tools Profile, Police, and Punish the Poor*. St. Martin's Press.

Green, B., & Viljoen, S. (2020). Algorithmic realism: Expanding the boundaries of algorithmic thought. In: *Proceedings of the 2020 Conference on Fairness, Accountability, and Transparency (FAccT)*, Virtual, pp. 19–31. https://doi.org/10.1145/3351095.3372840.

Harrison, J., Menon, S., & Muller, M. (2021). You should have asked: The value of implicit signals and epistemic modesty in LLM interactions. In: *Proceedings of the 2021 ACM Conference on CSCW*, Virtual, pp. 1–25. https://doi.org/10.1145/3476054.

Hoffman, R. R., Mueller, S. T., Klein, G., & Litman, J. (2018). Metrics for Explainable AI: Challenges and Prospects. CoRR, abs/1812.04608. https://arxiv.org/abs/1812.04608

Jakesch, M., Garimella, K., West, R., & Hancock, J. T. (2021). Human perceptions of fairness in algorithmic decision-making: A user-centered study. In: *Proceedings of the 2021 ACM Conference on Fairness, Accountability, and Transparency (FAccT)*, Virtual, pp. 782–793. https://doi.org/10.1145/3442188.3445891.

Kioskli, K., Bishop, L., Polemi, N., & Ramfos, A. (2024). Towards a human-centric AI trustworthiness risk management framework. In: A. Moallem (Ed.), *Human Factors in Cybersecurity* (pp. 63–73). Springer.

Kocielnik, R., Amershi, S., & Bennett, C. L. (2022). Will you accept an imperfect AI? Exploring designs for adjusting end-user expectations of AI systems. *ACM Transactions on Computer-Human Interaction,* 29(2), 1–36. https://doi.org/10.1145/3484493.

Lai, V., Wu, T., & Tan, C. (2021). Human evaluation of explanations: A user study framework for explainable AI. In: *Findings of the Association for Computational Linguistics: ACL-IJCNLP 2021*, Virtual, pp. 619–633. https://doi.org/10.18653/v1/2021.findings-acl.55.

Lee, M. K., Kim, J., & Liao, Q. V. (2020). Human-centered approaches to AI transparency. *Communications of the ACM,* 63(1), 68–77. https://doi.org/10.1145/3360311.

Liao, Q. V., Gruen, D. M., & Miller, S. (2020). Questioning the AI: Informing design practices for explainable AI user experiences. In: *Proceedings of the 2020 CHI Conference on Human Factors in Computing Systems*, Virtual, pp. 1–15. https://doi.org/10.1145/3313831.3376590.

Mayer, R. C., Davis, J. H., & Schoorman, F. D. (1995). An integrative model of organizational trust. *Academy of Management Review*, 20(3), 709–734.

Metcalf, J., Moss, E., Watkins, E., & Elish, M. C. (2021). Algorithmic impact assessments and accountability: The co-construction of transparency and trust. *Proceedings of the ACM Conference on Fairness, Accountability, and Transparency (FAccT)*, Virtual, pp. 43–53. https://doi.org/10.1145/3442188.3445921.

Morley, J., Elhalal, A., Garcia, F., Kinsey, L., Mökander, J., & Floridi, L. (2021). Ethics as a service: A pragmatic operationalisation of AI ethics. Minds & Machines, 31(2), 239–256. https://doi.org/10.1007/s11023-021-09563-w

Niculae, V., Park, J. S., & Li, J. (2023). Trusting the messenger: Credibility cues in human-LLM cooperation. In: *Proceedings of the 2023 Conference on Empirical Methods in Natural Language Processing (EMNLP)*, Virtual, pp. 1072–1085. https://doi.org/10.18653/v1/2023.emnlp-main.77.

Park, S., & Shin, D. (2021). Who wants explainable AI? Intersections of trust, literacy, and fairness in algorithmic systems. *Telematics and Informatics,* 59, 101556. https://doi.org/10.1016/j.tele.2020.101556.

Preece, J., & Shneiderman, B. (2021). The reader-to-leader framework for LLMs: Trust, learning, and user evolution. *Journal of Human–Computer Interaction,* 37(9), 881–896. https://doi.org/10.1080/10447318.2021.1909156.

Rahwan, I., Cebrian, M., Obradovich, N., et al. (2021). Machine behaviorism: Studying artificial intelligence in the wild. *Nature,* 591(7848), 529–537. https://doi.org/10.1038/s41586-021-03470-x.

Raji, I. D., Smart, A., White, R. N., et al. (2020). Closing the AI accountability gap: Defining an end-to-end framework for internal algorithmic auditing.In: *Proceedings of the 2020 ACM Conference on Fairness, Accountability, and Transparency*, Virtual, pp. 33–44. https://doi.org/10.1145/3351095.3372873.

Ren, J., Huang, K., & Park, J. (2023). Temporal trust dynamics in real-world AI deployments: Understanding user adaptation. *Proceedings of the ACM on Human-Computer Interaction,* 7(CSCW2), 1–24. https://doi.org/10.1145/3579496.

Schneider, H., Langer, L., & Butz, A. (2022). From mental models to interaction trust: A longitudinal field study of explainability and AI-assisted systems. *ACM Transactions on Interactive Intelligent Systems (TiiS),* 12(4), 1–28. https://doi.org/10.1145/3524121.

Seth, S., Shen, Y., & Narayanan, A. (2022). Beyond accuracy: Calibration and user trust in AI models. *Journal of Machine Learning Research,* 23(216), 1–32. https://jmlr.org/papers/v23/20-1681.html.

Shin, D., & Park, Y. J. (2022). Role of fairness, accountability, and transparency in algorithmic affordance. *Computers in Human Behavior,* 128, 107115.

Srivastava, M., Holstein, K., & de Freitas, J. (2022). Users in context: Differential trust calibration across AI-supported environments. In: *ACM CHI Conference on Human Factors in Computing Systems*, Virtual, pp. 1–15. https://doi.org/10.1145/3491102.3501977.

Suresh, H., Nikolic, B., Kiani, A., & Kaushik, D. (2023). Aligning clinicians: The impact of AI triage tools on clinical decision making. *Journal of Biomedical Informatics,* 138, 104267.

Tsamados, A., Aggarwal, N., Cowls, J., et al. (2022). The ethics of algorithms: Key problems and solutions. *AI & Society,* 37(3), 665–682. https://doi.org/10.1007/s00146-021-01154-w.

Ullman, T. D., Spelke, E., Battaglia, P., & Tenenbaum, J. (2020). Mind games: Improving mental model alignment between humans and AI. *Nature Machine Intelligence,* 2(8), 396–403. https://doi.org/10.1038/s42256-020-0197-1.

Vianello, A., Laine, S., & Tuomi, E. (2023). *Improving trustworthiness of AI solutions: A qualitative approach to support ethically-grounded AI design. International Journal of Human–Computer Interaction*, 39(7), 1405-1422. https://doi.org/10.1080/10447318.2022.2095478

Viljoen, S., Narayanan, A., & Pasquale, F. (2022). Algorithmic accountability in practice: Confronting systemic opacity with participatory governance. *AI and Society,* 37(4), 1481–1495. https://doi.org/10.1007/s00146-021-01238-y.

Wang, D., Yang, Q., Abdul, A., & Lim, B. Y. (2021). Designing theory-driven user-centric explainable AI. In: *Proceedings of the 2021 ACM Conference on Human Factors in Computing Systems (CHI)*, Virtual, pp. 1–15. https://doi.org/10.1145/3411764.3445744.

Whittaker, M., Alper, M., Anderson, M., et al. (2021). The Steep Cost of Capture: How Surveillance Technologies Reshape Social Trust. AI Now Institute Report 2021. https://ainowinstitute.org/reports.html.

Wu, J., Zhang, C., & Jiang, X. (2023). Trust and transparency in AI: A meta-analysis of user interaction studies. *International Journal of Human-Computer Studies,* 172, 102915.

Zhang, Q., Zhang, M., Wu, L., & Ma, S. (2022). When users trust LLMs too much: Trust calibration in open-domain dialogue systems. *Proceedings of ACL* 2022, 101–113.

Zhou, Y., & Li, L. (2023). Dynamics of trust in human-AI collaboration: An experimental study. *Computers in Human Behavior,* 136, 107389.

12 How Do AI and LLMs Change Our Lives? Reflections and Outlook

Helmut Degen, Stavroula Ntoa, Abbas Moallem, Joerg Beringer, Carrie Ching, Lance Chong, Thomas Geis, Pei-Hsuan Hsieh, Khalid Kattan, Prabhat Kumar, André Frank Krause, Carsten Lanquillon, Rebecca McNulty, Mark Mittrick, Mark Nuppnau, Abraham Moore Odell, Ming Qian, Adrienne Raglin, Robert G. Reynolds, John Richardson, Yao Sun, Lijing Wang, and Carsten Wittenberg

12.1 INTRODUCTION

Advancements in generative artificial intelligence (AI) and the widespread application of large language models (LLMs), such as ChatGPT and Gemini, have led to millions of people adopting these technologies. Furthermore, this growing adoption spans across diverse groups, including students, academics, engineers, lawyers, designers, writers, media professionals, programmers, politicians, artists, other professions, and individuals from all walks of life worldwide.

The impact of LLMs has already been observed on various industries, including education (Milano et al., 2023; Peláez-Sánchez et al., 2024; Chu et al., 2025), healthcare (Meskó, 2023; Omar et al., 2024; Zhang et al., 2025), finance (Maple & Sabuncuoglu, 2024; Zhao et al., 2024; Lee et al., 2025), law (Adams, 2024; Europol, 2023; Lai et al., 2024), software engineering (Fan et al., 2023; Hou et al., 2024; Zheng et al., 2025), programming (Kazemitabaar et al., 2024; Nam et al., 2024; Yan et al., 2025), cybersecurity (Ferrag et al., 2024; Yamin et al., 2024; Zhang et al., 2025), and UX design (Shi et al., 2023; Zhou et al., 2024; Luo, 2025).

Meanwhile, numerous studies examine both the applications of LLMs and the associated challenges. Bharathi Mohan et al. (2024) highlight seven challenges, namely training data, tokenization, computational requirements, fine-tuning, inference latency, limited context length, knowledge updating, and refinement. They also specify nine risks: bias, hallucination, lack of explainability, reasoning errors,

DOI: 10.1201/9781003492252-12

TABLE 12.1
Future Use Cases of LLMs as Identified by Popular LLMs in their Most Recent Versions in May 2025

Future Use Cases	ChatGPT	Gemini	DeepSeek	Claude	Llama	Copilot
Medical research	X			X		
Business and enterprise	X		X	X		X
Civic engagement and governance			X			X
Creative industries	X	X	X	X	X	X
Cybersecurity		X				
Domain-specific and specialized AI		X			X	
Education/learning	X	X	X	X	X	X
Finance						
Healthcare	X		X	X		X
Hyper-personalized AI agent		X				
Legal and governance	X			X		X
Robot control/ enhanced human–computer collaboration	X	X				
Scientific discovery	X	X	X		X	X
Software development				X		
Language translation and localization				X		
Public opinion and sentiment analysis					X	

x: denotes that the LLM identified the specific future use case.

prompt and data injection, adversarial attacks, behavioral changes, and spelling and counting errors. Raiaan et al. (2024) identify open issues and challenges, including, in addition to the above challenges, dynamic evaluation needs, evaluation complexity, output understandability, contextual constraints, and real-time responsiveness. Patil and Gudivada (2024) discuss the challenges of content toxicity, hallucinations, biases, carbon footprint, as well as cost and resources.

When asked about future applications of LLMs,[1] popular LLMs identified future use cases as depicted in Table 12.1.

This chapter aims to document the outcomes of a workshop that brought together practitioners, researchers, and scholars from the fields of AI, natural language processing

(NLP), and human–computer interaction (HCI). The workshop focused on examining the impact of LLMs on various aspects of our lives, exploring their multifaceted implications for daily activities, industry, academia, and society at large. The primary emphasis was on the intersection of human interaction with AI technology. Far from being redundant in a crowded field, this work aimed to adopt a human-centered, co-creative inquiry approach, reframing the discussion around LLMs through a real-world experience lens.

In this chapter, we aim to deepen our understanding of generative AI by analyzing the outcomes of using LLMs and exploring both current and potential future applications. By leveraging recent developments and real-world use cases, we seek to anticipate the evolving role of LLMs through the following questions:

- Which parts of our lives have already changed?
- Which parts of our lives are going to be changed?
- What are the anticipated challenges associated with the future use of AI technologies?

12.2 METHODOLOGY

The questions at hand were elaborated through two interconnected workshops, each designed to address specific aspects. The first workshop was carried out in the context of the AI-HCI Conference 2024, aiming to elicit contributions from the research community addressing different daily living contexts where LLMs have an impact. The second workshop, building on the shared content of the first workshop, was carried out online as a co-creation activity, entailing brainstorming, data collection, and a follow-up data analysis.

12.2.1 Workshop 1: Academic Inquiry through Conference Submissions

The first workshop was held as part of the AI-HCI conference 2024, aiming to engage participants in a discussion on the impact of LLMs on various aspects of life, facilitating knowledge sharing and collaborative exploration of the topic. Contributions were solicited on the following topics of interest:

- **Human–AI Interaction**: Insights on user experiences and the impact of AI and LLMs.
- **User Experience Design/User Studies**: Research and practices related to crafting interactions with AI and LLMs.
- **Industry/Business Applications**: Real-world case studies and applications of AI and LLMs.
- **Ethical and Societal Implications**: Ethical considerations and societal challenges arising from the widespread use of AI and LLMs.
- **Privacy and Security**: Data privacy and security concerns associated with AI and LLMs.
- **Scientific Research**: Contributions and advancements of LLMs in scientific research.

- **Law and Regulations**: Legal and regulatory perspectives on the use of LLMs.
- **Education**: Strategies for teaching and managing the use of LLMs and generative AI in educational settings.
- **Media Companies**: Utilization of generative AI and LLMs for media content creation.
- **Software Engineering**: Generation and verification of source code using LLMs.
- **Life**: The overall impact of LLMs and generative AI on daily life.

Eight submissions were accepted for a 4-hour workshop covering diverse topics, including education and technology, online learning inclusivity, the impact of generative AI and LLMs, mental health in high-pressure fields, artificial reasoning, user research, and AI-enhanced image generation. Further details are provided in Section 12.3.

12.2.2 Workshop 2: Co-Creation and Collaborative Topic Exploration

To deepen the insights gathered from the first workshop, multiple 1-hour online sessions were conducted. In more detail, the activities carried out explored the following questions:

- What parts of life will AI / LLM impact?
- How do AI and LLMs change our lives?
- What are or will be the key challenges?
- What could be some futuristic scenarios of using LLMs?

This workshop adopted a co-creative approach, employing collaborative methods to explore the challenges and potentials of LLMs in everyday life. In total, 25 participants engaged in interactive activities on a shared digital board, including brainstorming, mapping, and scenario construction.

The exploration of the questions was structured in three phases. Ideation was the first phase, during which participants individually contributed ideas on the shared board. The synthesis phases followed, where through collective discussions, the group synthesized individual inputs into cohesive themes, visualizing the interconnections between different perspectives and concerns. Finally, the validation phase involved iteratively refining the synthesized themes through group feedback, ensuring that the final outputs represented a balanced and nuanced understanding of the impact of LLMs. The results of this process are presented in Section 12.4.

Following the conclusion of the workshop, participants were asked to send more extended scenarios of futuristic use cases of LLMs and were given as an example the ISTAG Scenarios for Ambient Intelligence in 2010, which were created in 2001 (European Commission, 2001). The final elaborated scenarios are presented in Section 12.5.

Upon completing both workshops, the insights gathered from the conference submissions and the co-creation activities were elaborated by three researchers, following a thematic analysis approach aiming to produce a comprehensive understanding of the impact of LLMs on daily life. The themes that emerged from both workshops were systematically compared, allowing for the identification of converging insights and contrasting viewpoints, reflected in the discussion section of this chapter.

12.3 WORKSHOP SUBMISSIONS

Participants interested in attending Workshop 1 submitted position papers outlining specific application domains for the use of LLMs. The papers addressed a wide range of topics, including the use of LLMs in education, the perception of LLMs within the programming community, machine translation (MT) from English to Chinese using LLMs, industrial design image creation with generative AI, AI in reasoning and decision-making, the role of LLMs in generating design input for product development, and knowledge generation through LLMs. A summary of each individual position paper follows.

Rebecca McNulty's position paper examines the implementation of generative AI policies in U.S. higher education, with a particular emphasis on their impact on instructional design in online learning environments. Her objective is to develop actionable guidelines tailored for instructional designers, especially those engaged in digital education. These guidelines seek to address key questions surrounding the integration of generative AI in academic contexts, including: (1) What constitutes learning with generative AI? (2) How can learning outcomes supported by AI be effectively assessed? (3) How should academic integrity and the concept of cheating be redefined in the age of generative AI? and (4) In what ways can generative AI be leveraged to promote student success and enhance future career readiness? The development of these guidelines requires input from diverse stakeholders, including student leaders and faculty, and must emphasize adaptability to keep pace with the rapidly evolving capabilities of AI technologies. A flexible and inclusive policy framework not only supports ethical decision-making but also guides informed instructional design in an AI-enhanced educational landscape.

Carrie Ching and André Frank Krause investigate the role of LLMs in higher education, centering their research on the question: *Do LLMs genuinely benefit students in higher education?* To explore this inquiry, the authors have formulated 24 hypotheses, which they intend to test through surveys administered at an internationally oriented university of applied sciences in Germany. Key hypotheses include: (1) Across disciplines, nationalities, and levels of study, students do not perceive LLM tools as more beneficial than traditional study aids; (2) LLMs are primarily used for paraphrasing text, summarizing content, enhancing readability, acting as study companions or tutors, providing topic overviews, serving as alternative search engines, and initiating literature research; (3) most students have not invested substantial time in learning prompt engineering; (4) faculty have observed a decline in student performance across various assessment formats following the public release of ChatGPT; (5) most professors and lecturers have not significantly modified their teaching practices in response to ChatGPT's emergence; (6) however, the majority have made substantial changes to their assessment methods; and (7) many remain uncertain about the effectiveness of these changes in detecting plagiarism. This research aims to provide empirical insight into the evolving relationship between generative AI tools and educational practices.

Yao Sun and Lijing Wang examined the perception of ChatGPT within the Stack Exchange online Q&A community, focusing on two primary research questions: (1) What are the dominant topics and sentiments surrounding ChatGPT in user-generated content? and (2) How do these topics and sentiments differ across various subcommunities? To address these questions, the authors employed topic modeling and sentiment analysis to extract and analyze prevailing themes and attitudes across Stack Exchange subsites. The dataset comprised 17,957 posts containing the keyword "ChatGPT," collected from 173 Stack Exchange subsites between April 2019 and June 2023. These subsites were grouped into four categories: AI-related subsites, technical subsites (e.g., AskUbuntu, Android Enthusiasts, Mathematics), non-technical subsites (e.g., Coffee, Music Fans, Pets), and Stack Overflow. The analysis revealed that users in AI-related subcommunities expressed relatively high levels of trust in ChatGPT, whereas members of Stack Overflow demonstrated notably more skepticism, reporting over 5% less trust than their AI-focused counterparts. This skepticism may stem from dissatisfaction with ChatGPT's performance on coding-related queries. Furthermore, users in technical subcommunities, in general, exhibited lower trust in ChatGPT compared to those in non-technical subcommunities. These findings highlight the contextual variability in user attitudes toward generative AI tools within online knowledge-sharing platforms.

Ming Qian and Chuiqing Kong investigated the application of LLMs for MT from English to Chinese, with a particular focus on the translation of conceptual content. Their findings indicate that LLMs perform effectively when translating relatively simple concepts, such as identifying an author's stylistic features or converting measurement units (e.g., from the Imperial system to the International System of Units). However, the authors also explored the performance of LLMs when translating more complex concepts from Chinese into English and identified several limitations: (1) lack of consistency—LLMs frequently acquire relevant knowledge but fail to apply it coherently throughout a translation, leading to internal contradictions; (2) inefficient speed–accuracy trade-offs—LLMs struggle to balance response speed with translation accuracy, often exhibiting suboptimal performance in managing this trade-off; and (3) dependence on chain-of-thought (CoT) prompting—LLMs benefit significantly from CoT prompting to improve reasoning capabilities and conceptual accuracy in complex translation tasks. These findings underscore the potential and limitations of LLMs in achieving high-fidelity conceptual translation across languages.

Lance Chong conducted a preliminary study to evaluate the capabilities of DALL·E, a text-to-image model, and ChatGPT-4 developed by OpenAI across three distinct design domains: (1) preproduction design for animation, (2) industrial and consumer product design, and (3) fashion-oriented footwear design. The analysis was framed using an information theory-based approach, specifically the Networked Two-Way Communication Channel model. The findings indicate that enhancements in LLM user interfaces could substantially improve multimodal interaction within design workflows. Notably, the study underscores a critical limitation of current LLM-based tools: the underrepresentation of non-verbal and non-linguistic aspects integral to many design practices. This gap highlights an important opportunity for future research aimed at integrating artistic and craft-based knowledge into the development of next-generation LLMs, as well as optimizing their interfaces to better support creative and visually driven disciplines.

Adrienne Raglin, Anjon Basak, Prabhat Kumar, Somiya Metu, John Richardson, Mark Mittrick, Justin Rawal, and Erin Zaroukian investigate the role of artificial reasoning in LLMs, focusing on six key areas: (1) Causal Reasoning—the ability to understand cause-and-effect relationships, particularly through the concepts of *intervention* and *counterfactuals.* LLMs can generate and identify counterfactual scenarios (i.e., "what-if" situations) and, when provided with textual descriptions of video scenes, can infer causal relationships among objects, actions, and events. (2) Theory of Mind (ToM)—the capacity to model the mental states of others, typically through linguistic cues. The authors argue that fully developing ToM capabilities in AI systems will require multimodal inputs beyond language alone, implying a shift from language-only LLMs to more integrated architectures. (3) Agent-Based Modeling—a foundational method for building autonomous decision systems. LLMs can enhance agent-based models by decomposing complex tasks into manageable subtasks, thereby improving decision-making efficiency. (4) Context-Aware Simulation—adaptive simulation environments that respond dynamically to evolving contexts. LLMs contribute by identifying decision points and proposing optimal courses of action. (5) Metareasoning—metacognitive strategies for monitoring and controlling reasoning and problem-solving processes. In the study, an agent employs metareasoning to select an optimal pathfinding algorithm while avoiding adversaries, with the support of LLMs in hypothesizing potential adversarial behaviors. (6) Complex Decision-Making—decision processes involving multiple variables, uncertainty, and interdependencies. LLMs assist by generating plausible scenarios, feedback, and evaluative metrics to serve as inputs for algorithmic decision-making. Collectively, the research highlights the expanding role of LLMs in augmenting artificial reasoning across a broad spectrum of cognitive functions.

Joerg Beringer and Thomas Geis have developed a professional service called *Product Context Analyzer,* aimed at supporting product development professionals in efficiently identifying critical contextual information and user requirements. The service leverages automated structuring of unstructured data to produce clear and concise representations of users, goals, tasks, objects, pain points, user stories, and user requirements. The service significantly reduces the time required to extract relevant insights—from what traditionally takes days or weeks to mere minutes. Rather than replacing conventional user research, *Product Context Analyzer* is designed to complement the manual effort by shifting the emphasis from extensive data collection to the validation and refinement of research findings. *Product Context Analyzer* enhances the efficiency and accuracy of early-stage product definition and user-centered design processes.

Mark Nuppnau, Khalid Kattan, and Robert G. Reynolds investigate the application of LLMs and other machine learning tools in generating knowledge to inform and enhance the optimization of deep learning algorithms. Their research is grounded in the *Cultural Algorithm* (CA), an evolutionary framework that integrates domain-specific knowledge to improve the efficiency and effectiveness of the optimization process. As a hyper-heuristic extension of genetic algorithms and swarm intelligence, the CA is designed to support multiple interacting knowledge sources, each contributing uniquely to a shared search strategy. Central to this framework is the use of pattern recognition across related problem domains, enabling the algorithm to steer the search process more strategically. This knowledge-driven guidance not only enhances optimization performance but also provides interpretability, contributing to the development of more transparent, reliable, and safe AI systems. Each knowledge source within the CA framework plays a distinct role in refining the search dynamics and accelerating convergence toward optimal solutions.

The position papers served to broaden the workshop group's perspective on the use of LLMs, offering a foundational basis for identifying impact domains and challenges, as well as the future potential uses of LLMs.

12.4 MAPPING THE IMPACT AND CHALLENGES OF LLMs

12.4.1 The Impact of LLMs on Daily Life Activities

Initially, participants explored what parts of life AI and LLMs change in our lives today and how. In this regard, impact was identified in several everyday life domains, enhancing efficiency, offering personalization and accessibility, altering everyday practices, ways of working, and how we interact with each other. In particular, analysis of participants' contributions to the workshop revealed the following domains of impact.

- **Language and Communication**: One domain of impact was multilingual communication, where LLMs can empower real-time translation of written and spoken languages. This moderates existing practices of using dictionaries and human translators. Cross-cultural exchange is also facilitated, in-person and online, increasing diversity and inclusion. However, participants emphasized that challenges remain in the design of real-time translation tools, particularly regarding speed, accuracy, user interface complexity, and accessibility.
- **Companionship and Social Interaction**: A notable change was discussed regarding social interactions and how individuals, especially older adults or socially isolated persons, seek companionship. Typically, people use online forums or social networks to fulfill their needs. Increasingly, people are now engaging with LLMs for conversations and support. This behavioral shift has advantages, but it also raises critical concerns regarding emotional well-being and social detachment.
- **Healthcare and Mental Health**: In relation to the above topic on companionship, LLMs are also gradually being used for consultation, wellness support, and even for interpreting medical data. This holds significant potential for personalized consultancy and treatment; however, it is also closely connected with issues related to relying on non-human agents for sensitive information, accuracy concerns, and ethical ramifications.
- **Information Retrieval and Conceptual Learning**: Information search, traditionally conducted through search engines and encyclopedias, is now streamlined with LLMs responding to natural language queries. In addition

to retrieving content easily for one's personalized inquiries according to their needs, users can also benefit from summarized content and immediate conceptual clarifications. As a result, LLMs can promote personalization and productivity, although questions arise regarding their impact on critical thinking and source transparency.

- **Creative and Professional Content Creation**: Content creation has been hugely affected by LLMs, in various fields ranging from journalism to instructional design, report writing and book authorship, and movie and media production. LLMs augment human creativity by offering rapid content generation, linguistic refinement, and multimedia instructional material. In some cases, research and context exploration are also conducted with the use of LLMs (e.g., expanding writers' knowledge on a topic, providing background information about a journalistic article, offering insights, and complementing a write-up for movie scripts). This transformation challenges traditional roles of authorship and editorial review, while raising concerns about accuracy and authenticity.
- **Software Development and Data Analysis**: With the use of LLMs, non-experts can conduct data analysis or develop simple software using natural language commands. In addition, LLMs can generate synthetic data for research "on the spot." This has an impact on how new research ideas and projects are generated, having the potential to fundamentally reshape software development and research workflows. Caveats are obviously in order, pertaining to the quality of outcomes, affected by the data used and generated, as well as issues of ethics, user autonomy, and oversight.
- Overall, participants highlighted a new ecosystem of intelligence, even emphasizing that the dimensionality of human existence is expanded, from the 3D physical world to the many-dimensional digital world of the mind. The speculative remark that an AI system might someday be indistinguishable from a human candidate for academic, military, or political positions raises profound ontological questions. In a digital society where identity, performance, and credibility can be simulated with high fidelity, discerning humans from artificial agents becomes not only technically complex but also ethically crucial. With these concluding remarks, the workshop proceeded with discussing challenges stemming from the wide adoption and penetration of LLMs in various human activities.

12.4.2 Challenges

Despite the considerable impact of LLMs already marked in numerous daily activities, participants also highlighted several challenges and potential harms that need to be addressed.

- **Legal and Ethical Concerns**: Issues relevant to copyright, intellectual property rights, and liability were extensively discussed. For example, if an LLM-generated suggestion causes harm (e.g., in medical or legal domains), it remains unclear who bears responsibility—the developer, the user, or the model itself. Although there have been some efforts to frame the legal landscape, legal frameworks have not kept pace with this rapidly evolving technological domain yet.

- **Scope of Use and Quality Assurance**: Participants voiced concerns about how to verify and validate LLM outputs regarding their accuracy and quality, identify hallucinations, and measure how well and how useful LLMs are. The need for automated or software-assisted procedures was brought up, along with new human reviewing methods, complemented with appropriate criteria for various domains and contexts of use, as well as metrics for system performance and user satisfaction. Overall, it was noted that the absence of robust quality assurance mechanisms limits the responsible deployment of LLMs, especially in high-stakes contexts.
- **Data Security and Privacy**: As LLMs interact with users and collect highly sensitive data—serving as companions, advisors, and creative partners—the risk of data breaches or misuse increases. The issue of training LLMs through stored conversations raises privacy concerns. In addition, altering training data through malicious attacks can tamper with the outcome of LLMs, potentially inflicting harm to the well-being of users. Personalized LLMs, especially those hosted on cloud platforms, pose novel threats to individual privacy.
- **Explainability and Trust**: While LLMs can produce important outcomes, their reasoning process is often opaque. Shared explanation content is often about technical attributes of the AI model that can be difficult to understand for end users. Due to the characteristics of AI technology, the outcome can be incorrect. The lack of transparency reduces the human's ability to understand and, to some extent, control the reasoning process. This includes the existence of bias that can be caused by the AI model, the available source content, or both. Participants suggested a need for new forms of "explainability through prompting," and better UI/UX design to support interpretability. The need for clear policies and regulations was also raised.
- **Synthetic Content and Reality Distortion**: The generation of synthetic data, LLM derivatives, and potential information bubbles introduces both creative possibilities and risks of misinformation. It was identified that a generation of synthetic data (LLM derivatives) will be witnessed in the near future. The compounded use of generated content may lead to a drift from empirical reality, particularly in sensitive domains like journalism or education.
- **Social and Environmental Consequences**: The societal impacts identified include heightened digital inequality, the erosion of interpersonal relationships, and the amplification of algorithmic biases. Overreliance on technology was also raised as an important concern. Participants voiced numerous questions, such as "What happens with populations who do not have access to technology?", "Will we stop interacting with each other? Asking an LLM is easier than human relationships which involve disagreement," "Will this technology support the increased isolation of cultures and the production

of cultural silos?", or "Are we generating informational walls that work to separate us?" Participants warned of a "tragedy of the commons" scenario, where shared social and ecological responsibilities are neglected due to the opacity and complexity of AI systems. Environmentally, the energy cost of training and deploying large models was raised as a growing concern.

- **Human Adaptation and Cognitive Load**: LLMs require users to develop new forms of interaction, including prompt engineering and synthetic reasoning—connecting separate ideas and information to achieve understanding and build new knowledge. This cognitive shift can be empowering but also challenging, particularly for those not technically inclined or aware of the underlying mechanisms. For persons who lack technical skills, keeping up with the fast-paced evolution of technology may be a considerable challenge, putting them at risk of exclusion.
- **Artificial Agency Not Distinguishable from Human Agency**: Taking into consideration that LLMs are increasingly achieving outcomes that may be hard to distinguish from human output, depending on the task at hand, one potential point of concern is being able to discern artificial agency from human agency, especially for non-tech-savvy users. Although this may be beneficial in some cases (for instance, in customer support service), this property raises profound ontological questions about the use of technology and the interaction between humans. Beyond ethical dilemmas, one question that was posed was whether human performance/expertise will in the future be assessed against LLM performance/expertise, which is gradually becoming the norm.

12.5 FUTURE SCENARIOS

After identifying the impact domains and the challenges, we asked the workshop participants to send in futuristic scenarios that answer the following questions: Who is the central character? Where is the central character? Why is the central character using the LLM? What is the interaction with the LLM?

We have received fifteen scenarios that can be grouped into three categories: (1) Assistance for students and researchers, (2) Assistance for everyday situations, and (3) Assistance for selected professions.

12.5.1 Assistance for Students and Researchers

The first group of scenarios centers around students and researchers, encompassing five scenarios (Figure 12.1).

The first scenario, referred to as "Turbo Tutor," centers around a student facing challenges in learning new concepts. As students often struggle with grasping complex ideas, the LLM serves as an interactive tool, providing explanations for various topics, such as mathematical formulas or programming languages, in a manner similar to a human tutor. In addition to concept explanations, the LLM supports students in preparing for exams. Students can request the LLM to clarify specific concepts, while the LLM can also proactively engage by asking

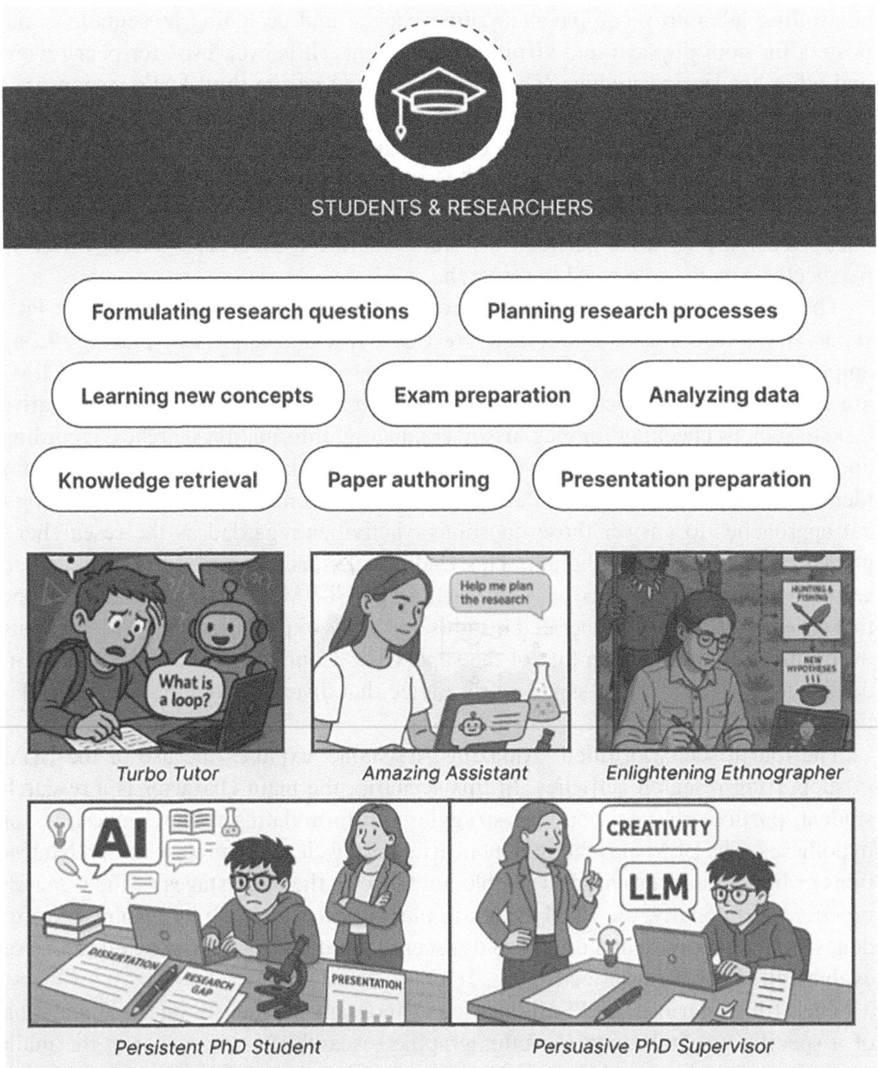

FIGURE 12.1 Future scenarios—LLMs for students and researchers.

relevant questions to assess the student's understanding. By providing feedback on the student's responses, the LLM helps to enhance comprehension and gradually build mastery of the topic. Ultimately, the LLM prepares students for tests by posing exam-level questions, guiding them through the process of effective exam preparation.

The second scenario, titled "Persistent PhD Student," explores the use of LLMs in the pursuit of a doctoral dissertation. In this context, the LLM supports a wide range of tasks, including brainstorming, knowledge retrieval, ideation, identifying research gaps and formulating research questions, proposing research approaches, controlling laboratory equipment, writing papers, and designing presentations and posters for both physical and virtual environments. It is even used for peer review and other academic activities. The student tends to accept the LLM's responses at face value, rarely questioning their validity unless the output appears clearly flawed. As a result, the boundary between creative and administrative research activities becomes increasingly blurred. If the PhD supervisor adheres to a more traditional, pre-LLM approach—favoring human-generated insights over machine-generated ones—tension may arise between the student and the supervisor due to differing perspectives on the role of AI in research.

The next scenario is titled "Persuasive PhD Supervisor." In this case, the PhD supervisor adheres to a traditional, pre-LLM research approach, placing strong emphasis on human creativity as the core driver of scholarly work. While LLMs are incorporated into the research process, their role is limited to administrative tasks—such as checking for plagiarism, conducting information searches, recording and storing data, and serving as a digital assistant. Crucially, the LLM is not used for identifying research gaps, formulating research questions, or defining methodological approaches to answer those questions—activities regarded as the researcher's primary intellectual contribution. This traditional stance can lead to tension when working with PhD students who embrace a "post-LLM" research paradigm—one that relies heavily on AI-generated insights and tends to prioritize the LLM's outputs over human guidance, even that of the supervisor. Conflicts may emerge when students question or resist the supervisor's advice that diverges from the recommendations provided by the LLM.

The fourth scenario, titled "Amazing Assistant," explores the use of the LLM in supporting research activities. In this scenario, the main character is a research student, particularly one who may struggle with formulating research questions or hypotheses. The LLM aids the student in articulating clear and precise research questions or hypotheses, providing valuable guidance in the early stages of the research process. Additionally, the LLM assists in planning the research by helping the student select appropriate and established research methods to address the questions or evaluate the hypotheses effectively.

The fifth scenario, titled "Enlightening Ethnographer," focuses on the application of a specific research method—ethnographic research. In this scenario, the main character is an ethnographic researcher investigating the nutritional needs of a particular tribe. The analysis of biological samples collected from the tribe will contribute to understanding these nutritional requirements. Additionally, information regarding hunting and fishing techniques is recorded and input into the LLM. The LLM then assists the researcher in formulating new research hypotheses that drive the study forward, such as exploring the tribe's cooking techniques and how they relate to their nutritional needs.

12.5.2 Assistance for Everyday Situations

The second group of scenarios is everyday scenarios, benefitting potentially many citizens. Figure 12.2 summarizes LLM tasks and features as described in the below scenarios.

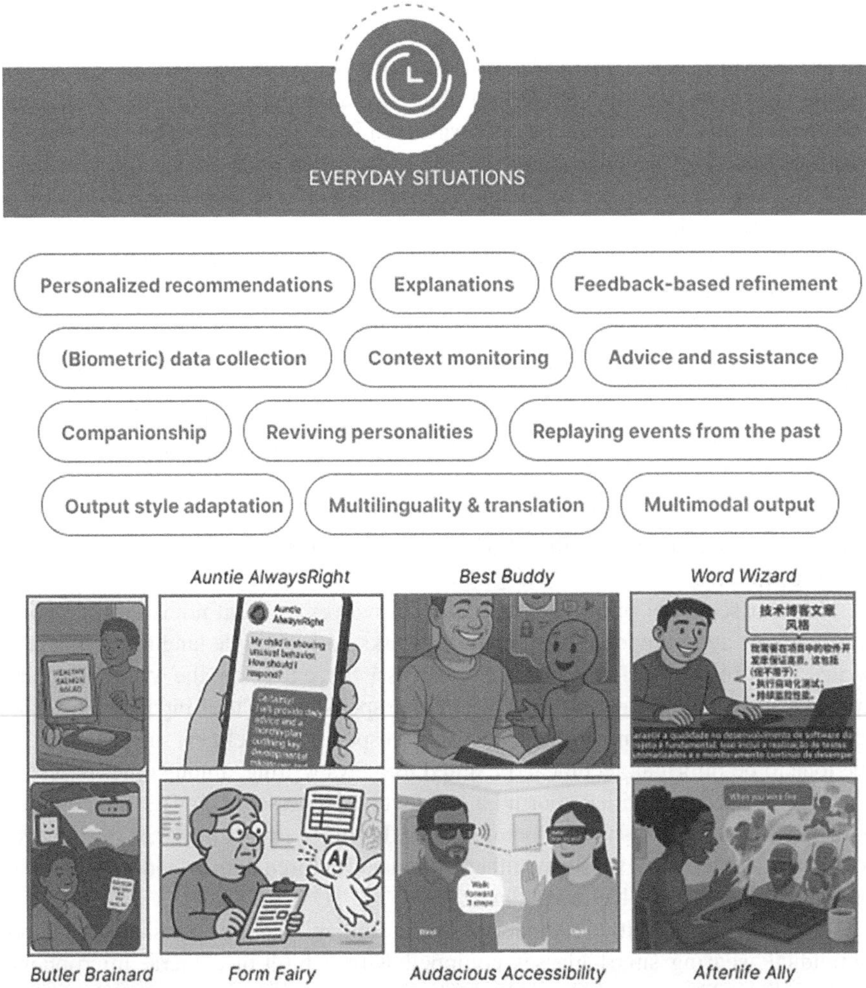

FIGURE 12.2 Future scenarios—LLMs for everyday situations.

The first everyday scenario, titled "Butler Brainard," revolves around a family member who seeks support in managing daily tasks such as cooking and driving. When tasked with preparing a meal, the LLM takes into account taste and food preferences, medical records, and laboratory data to recommend tasty and healthy recipes. It can also order any missing ingredients and arrange for their delivery to the home. Additionally, the LLM provides explanations about the food and the history of the recipe to family members. After the meal, it collects feedback and preferences to refine future cooking plans. In the context of driving, when operating a self-driving car, the LLM ensures a safe journey while also pointing out notable sights along the way. It maintains continuous communication with the passenger, efficiently handling other tasks in parallel, such as scheduling a doctor's appointment or preparing the annual tax return.

The second scenario, titled "Auntie AlwaysRight," centers around a parent who is unsure how to respond to a child displaying unusual behavior (from the adult's perspective). In this scenario, the parent turns to an LLM for guidance on child-rearing and how to address specific situations. The LLM offers daily advice and provides a monthly plan, outlining key developmental milestones and important considerations for raising the child.

Another scenario, titled "Best Buddy," features a character who can be anyone. In this scenario, the LLM serves as the ultimate companion, offering assistance in any context. The LLM continuously observes and records the individual's environment, actions, and conversations, allowing it to gain a deep understanding of the person's surroundings, behaviors, language, beliefs, preferences, and values throughout their life. It provides both solicited and unsolicited advice based on this extensive knowledge. The LLM identifies the individual through biometric data, such as iris scans, voice, and/or facial recognition. Given its comprehensive record of the person's experiences, the LLM is capable of replaying events from years past. Security mechanisms are in place to protect the LLM, ensuring that it remains virtually immune to hacking attempts.

The next scenario, titled "Word Wizard," involves a global nomad, professional translator, or any individual who regularly works with multiple languages. The user writes a text in a foreign language and seeks assistance to refine the writing so that it appears as though it were authored by a native speaker. The user inputs the text into the LLM, which then enhances it according to specific guidelines, such as adapting the tone to be informal, technical, or suited for news writing, among other styles.

A similar scenario, titled "Form Fairy," involves a person who needs to complete a form, such as when visiting a doctor's office. In this scenario, the LLM is employed to automatically fill out the form, either using unstructured data provided by the user or by leveraging stored historical data.

The next scenario, titled "Audacious Accessibility," features a visually impaired individual wearing smart glasses equipped with a built-in camera, microphone, and speakers. The sensory data—both visual and acoustic—are captured and processed by the LLM. The LLM analyzes the environment, including emotional cues, and provides real-time guidance to the visually impaired person through spoken instructions, such as directing them from the entry of a room to a specific spot. In a similar scenario, another individual who is deaf wears the same smart glasses. The sensory inputs are processed and converted into text, which is displayed on the glasses, allowing the deaf person to read and understand the information. This setup also enables them to engage in conversations by reading the spoken dialogue in real-time.

The final scenario in this group is titled "Afterlife Ally." The main character in this scenario can be anyone, with the LLM representing a deceased individual, such as a family member. This setup allows anyone to interact with the person who has passed away, whether by engaging in casual conversation, seeking advice, or asking for opinions on various topics happening in the world.

The scenario offers a source of comfort to those interacting with the memory of the deceased individual.

12.5.3 Assistance for Selected Professions

The last group of scenarios addresses the needs of selected professionals. An illustration of LLM contributions in the identified scenarios is provided in Figure 12.3.

The first professional scenario, titled "News Nerd," features a journalist who is tasked with writing news stories. In this scenario, the LLM is utilized to create and refine these stories. The LLM has access to real-time data, enabling it to validate the facts presented in the news and ensure the accuracy of the information.

The second professional scenario, titled "Product Probe," centers around a product manager in search of a new, successful software product. In this scenario, the LLM is used to generate a fully functional product prototype. A product manager specifies the new product and the LLM provides product options. While going through iterations of refining the specification, the LLM provides refined versions of the product. Once finalized, the software prototype can be shared and tested globally in real-time, enabling immediate feedback and insights that inform further improvements.

The last professional scenario, titled "Policy Pal," features a politician, such as a minister or secretary, who is striving to achieve a specific societal goal but is unsure of the best approach. In this scenario, the politician turns to the LLM for guidance

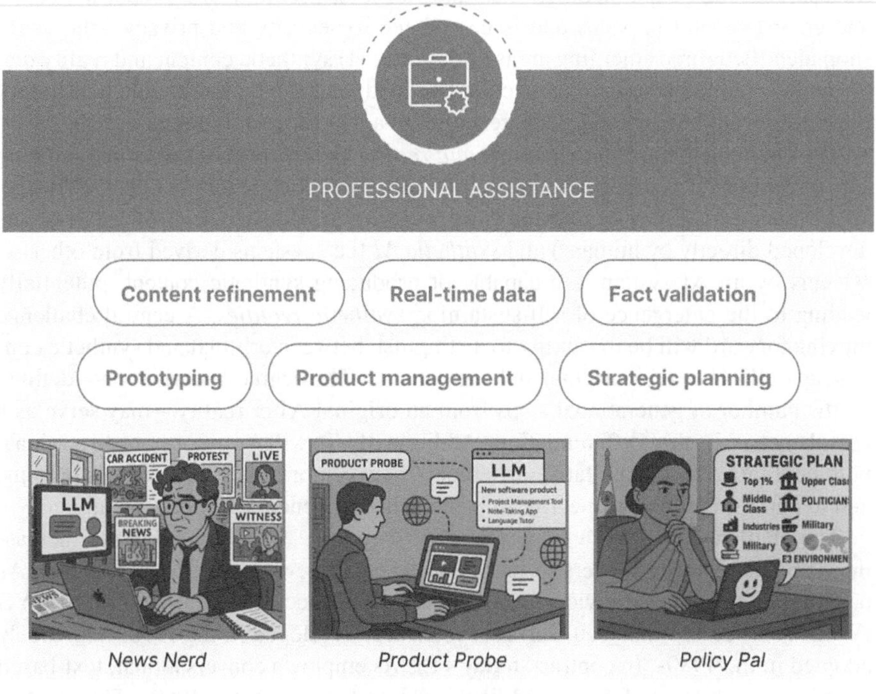

FIGURE 12.3 Future scenarios—LLMs for professionals' assistance.

on how to pursue objectives like reducing poverty. The LLM provides one or more strategic plans, considering the potential impact on various societal sectors, such as the top 1%, upper class, middle class, lower class, politicians, industries, military, education, different regions and countries, as well as the environment.

12.6 DISCUSSION

To investigate how artificial intelligence and LLMs are transforming everyday life, we conducted two workshops. Workshop 1 took place as an onsite workshop at the AI-HCI 2024 conference, held in affiliation with the HCII 2024 conference. In response to a call for participation, eight submissions were accepted. The primary objective of this initial workshop was to establish a baseline understanding of participants' current experiences and research interests with LLMs. In workshop 2, we conducted a series of 1-hour online sessions to explore impact domains, challenges of LLMs, and future scenarios of LLMs.

Taken together, the two workshops facilitated a comprehensive exploration that progressed from current experiences and challenges to forward-looking, "envisioned" scenarios for the integration of LLMs in everyday life.

12.6.1 New Challenges

In addition to well-documented challenges—such as ethical and legal concerns, transparency and explainability, bias and fairness, accuracy and reliability, social and environmental impacts, and issues related to security and privacy—the workshop identified three emerging areas of concern: (1) synthetic content and reality distortion, (2) human adaptation and cognitive load, and (3) the knowledge benchmark is shifting from humans to LLMs, resulting in a loss of trust to humans.

The challenge of *synthetic content and reality distortion* arises from the growing capabilities of AI systems to generate not only synthetic content but also derivative AI systems—what may be referred to as *synthetic AI*. Both *original AI* (i.e., systems developed directly by humans) and *synthetic AI* (i.e., systems derived from other AI systems by an AI system) are capable of producing synthetic content, potentially leading to the emergence of self-sustaining *synthetic realities*. A central challenge moving forward will be the ability to distinguish between original and synthetic content, as well as between original and synthetic AI. The degree of derivation—defined by the number of generational steps from an original AI or reality—may serve as a useful metric in this differentiation. Additionally, it will be important to evaluate both qualitative and quantitative differences between original and synthetic outputs, and to understand their respective impacts, whether beneficial or detrimental.

The challenge of *human adaptation and cognitive load* highlights the increasing demands placed on users to learn and adapt to new interaction paradigms. As part of digital literacy, previous generations became accustomed to the WYSIWYG (What You See Is What You Get) user interface, developed in the 1970s and widely adopted in the 1980s. In contrast, today's LLMs employ a conversational, text-based interface reminiscent of command-line user interfaces from the 1960s. The practice of *prompt engineering* is emerging as a new skill set, yet no standardized interaction conventions have been established for managing the opaque reasoning processes of LLMs. Users are often left to develop their own strategies for interpreting or coping with this opacity—whether through trial and error, adaptive heuristics, or simply by ignoring it. Given the historical shift from command-line to graphical user interfaces, it remains an open question whether a similar evolution will occur to support the usability of LLM-driven systems.

One of the most significant emerging challenges is the *increasing shift of the knowledge benchmark from humans to LLMs*, resulting in a loss of trust in humans. This development carries profound epistemological consequences, potentially signaling a shift in the prevailing standard of knowledge—from human reasoning to AI-derived outputs. As LLMs become more widely adopted, there is a growing risk that individuals may place greater trust in AI-generated responses than in human judgment (Zhao et al., 2023). This shift could represent a departure from Descartes' foundational epistemic claim—*Cogito, ergo sum* ("I think, therefore I am")—toward a new and potentially problematic standard: *Respondit, ergo verum* ("It answered, therefore it is true"). Such a transition demands rigorous empirical and theoretical examination, particularly regarding trust calibration, epistemic authority, the evolving role of critical reasoning, and the question of who ultimately holds the final say in an AI-mediated society.

12.6.2 New Scenarios

Many of the future scenarios discussed during the workshops build upon existing concepts, yet a smooth and beneficial interaction in the foreseen contexts requires addressing recognized risks and challenges. In addition, several proposed scenarios stand out for their potential impact and originality.

One such scenario, "Butler Brainard," is notable not for its individual use cases—which are relatively familiar—but for its vision of integrating LLMs into broader AI ecosystems, such as autonomous vehicles. In this configuration, the LLM does not merely perform tasks but functions as a high-level coordinator across multiple domains. It facilitates parallel task execution and supports human users by prioritizing tasks and enhancing cognitive focus. This orchestration of complex systems positions the LLM as both an assistant and an enabler of higher-order decision-making, thereby improving overall system performance and human effectiveness.

A similarly innovative concept emerges in the "Best Buddy" scenario, which imagines an LLM—or a comparable AI system—serving as a deeply personalized companion. This vision is both compelling and unsettling. On one hand, such a system could offer insights and support that surpass typical human capabilities. On the other hand, it might accumulate an unprecedented level of intimate knowledge about an individual, surpassing even the deepest human relationships. This raises critical questions: Who bears responsibility for the guidance such systems provide? How might our understanding of trust, privacy, and interpersonal dynamics shift in a world where individuals routinely engage with AI companions?

The "Audacious Accessibility" scenario highlights the potential of AI to act as a communication bridge between individuals. While its primary focus is on enhancing accessibility for people with sensory impairments—such as blindness or deafness—the scenario can be extended to address a broader range of needs, including bridging learnable skill gaps. For instance, real-time translation of spoken or written language could support inclusive communication across linguistic and educational boundaries.

Finally, the "Policy Pal" scenario shifts the focus from individual applications to societal challenges, envisioning LLMs as tools for informing public policy at regional and global levels. This scenario suggests that advanced AI systems could play a constructive role in shaping evidence-based policy and supporting more transparent, informed decision-making processes.

Collectively, these scenarios illuminate the dual-edged nature of LLMs' expanding capabilities. On one side, they offer extraordinary potential for knowledge access, personalization, human–human, and human–AI collaboration. On the other hand, they introduce new risks related to misuse, ethical uncertainty, and social disruption. With great power comes great responsibility—and a central question emerges: Who will have access to these transformative tools, and who will be left behind?

12.6.3 Answers to the Workshop Questions

We can now address the three questions posed in this paper.

Which parts of our lives have already changed?
Current observations suggest that LLMs have already transformed aspects of daily life that are centered on information processing. LLMs are widely used to understand and generate information-rich content, such as text and images. However, their impact remains limited in domains involving physical tasks or direct hardware manipulation—such as driving, gardening, cooking, or doing laundry.

Which parts of our lives are likely to change in the future?
Insights drawn from the future scenarios indicate that LLMs, and AI more broadly, are poised to influence nearly all aspects of life. As LLMs are increasingly integrated with robotic systems and computer vision technologies, their support may extend to physical activities as well. Scenarios like *Butler Brainard* and *Audacious Accessibility* envision LLM-powered companions that significantly augment human knowledge and capabilities. The *Policy Pal* scenario illustrates the potential of LLMs to contribute meaningfully to societal challenges and governance at regional and global scales.

What are the anticipated challenges associated with the future use of AI technologies?
Workshop participants identified a wide spectrum of established challenges associated with AI systems, including ethical and legal concerns, transparency and explainability, bias and fairness, accuracy and reliability, social and environmental impacts, and issues related to security and privacy. In addition to these, three emerging challenges were highlighted: (1) synthetic content and reality distortion, (2) human adaptation and cognitive load, and (3) distinguishing AI-generated content from that produced by humans.

Synthetic content and reality distortion pertain to the increasing capacity of AI systems to generate *synthetic content* and to create derivative AI models—here referred to as *synthetic AI*. These developments may contribute to the formation of self-reinforcing synthetic realities. A critical concern moving forward is the ability to distinguish between original and synthetic content or systems, and to evaluate their respective societal impacts—both beneficial and detrimental.

Human adaptation and cognitive load address the growing cognitive demands placed on users as they adapt to novel modes of interaction. Whereas earlier generations of users became accustomed to WYSIWYG (What You See Is What You Get) interfaces, contemporary LLMs primarily employ conversational, text-based interactions that resemble early command-line environments. *Prompt engineering* is emerging as a new form of digital literacy; however, standardized conventions for interacting with the opaque reasoning processes of LLMs are not yet established. Recent research efforts are already exploring direct manipulation interfaces and their impact on user interaction and performance (Masson et al., 2024). How LLM interfaces will evolve—analogous to the historical progression from command-line to graphical user interfaces—remains an open question.

The potential challenge of the *increasing shift of the knowledge benchmark from humans to LLMs* (Zhang et al., 2023) can have a significant impact not only on AI but also on the interaction between humans. Critical questions include how such a transformation in trust dynamics can be managed and how to maintain human accountability. Of particular concern is the need to prepare future generations to manage how to use AI and whom to trust—through education that emphasizes both the capabilities and limitations of AI systems, as well as the unique capability of human reasoning and critical thinking.

12.6.4 Reflection on Workshop Methodology

The results of the workshop suggest that the applied methodology is effective in eliciting novel insights. This highlights its potential as a new benchmark for comparative human-AI evaluation. By anchoring the discussion in existing practices, the format creates a shared experiential foundation among participants. This common ground enables meaningful reflection on present-day challenges and facilitates the identification of domains where emerging technologies may have the most significant impact. The structure of the workshop—beginning with an intensive 4.5-hour session followed by a series of 1-hour online meetings—allowed participants to reflect on specific topics over time and revisit them in subsequent sessions, thereby avoiding the limitations of conventional time-boxed formats.

The innovation methodology employed in the workshop is grounded in the view that innovation is a non-linear, iterative process (Kline & Rosenberg, 2009). Innovation techniques are commonly classified as either intuitive or logical (Shah et al., 2003). The formats used in both workshops fall under the category of "intuitive innovation" techniques, and more specifically, the "hybrid" subcategory (Shah et al., 2003). It combines divergent ideation methods, such as brainstorming, with progressive structuring techniques applied in an iterative manner.

Several other established innovation methodologies provide useful points of comparison. These include classic brainstorming (Osborn, 1953; Murphy et al., 2023), SCAMPER—an acronym for "Substitute, Combine, Adapt, Modify, Put to another use, Eliminate, and Reverse" (Eberle, 1996; Chulvi et al., 2013), TRIZ or the "Theory of Inventive Problem Solving" (Altshuller, 1984), and Design Thinking (Simon, 1969; Liedtka, 2015). Empirical studies have shown that brainstorming, as an intuitive method, tends to produce more novel ideas than TRIZ and SCAMPER (Chulvi et al., 2013). However, brainstorming is generally constrained to short durations, typically up to 60 min (Pauls & Kenworthy, 2019). In contrast, the extended workshop format described here—spanning multiple weeks with both in-person and online components—provided participants with the opportunity to engage more deeply with the material and develop ideas over time.

In summary, the workshop format demonstrates strong potential for generating original, human-centered insights and for supporting the co-creation of forward-looking, transformative scenarios. Further research is warranted to quantify the degree of novelty produced by this format in comparison with other intuitive innovation techniques and to explore its applicability across different domains.

12.6.5 Limitations

The workshop format, while effective in many respects, presents several limitations. A key constraint was time: the in-person session conducted during the conference lasted 4.5 hours and was followed by a series of 1-hour online meetings. Although this distributed format enabled extended engagement and reflection, it also introduced logistical challenges.

In particular, participant attendance varied across sessions, leading to fragmented engagement. Not all participants were able to attend every meeting, which may have resulted in an uneven understanding of prior discussions and workshop outcomes. This discontinuity could have influenced the depth, consistency, and collaborative synthesis of insights generated throughout the process.

In addition, as with any research involving human participants, the outcomes produced are directly influenced by the expertise and personalities of the persons involved. To counterbalance any personal bias effects, the approach followed was to engage as many participants as possible, carry out structured discussions, and analyze findings with a well-established methodology.

12.6.6 Future Research

Future iterations of this workshop could be conducted periodically, incorporating updated baselines of future current practices to explore emerging trends and envision new future scenarios, as well as "dark" scenarios to emphasize challenges and risks. This longitudinal approach would allow for tracking the evolution of ideas and challenges over time.

Additionally, the workshop format itself warrants further investigation. A valuable research direction would be to examine which specific components of the methodology—such as position papers, impact domains, challenges, future scenarios, and participants—most significantly contribute to the generation of innovative ideas. Understanding these dynamics could inform the design of more effective participatory foresight methods.

ACKNOWLEDGEMENTS

The authors thank Anjon Basak, Chuiqing Kong, Somiya Metu, Justine Rawal, and Erin Zaroukian for their active engagement and valuable input during the workshops.

NOTE

1 The following prompt was used: "Role: You are a researcher who writes a white paper about the future use of LLMs including their challenges. What are future use cases for the use of LLMs and their challenges?" It is noted that although LLMs may be supporting a use case, it is not marked in Table 1, if the LLM did not indicate this as a future use case. For example, although all LLMs may be used for language translation, the option is only selected for Claude, since it was the only LLM that identified it as a potential future use case. Such features may be considered as an LLM skill or behavior that arises as a byproduct of scaling up models or using them in real-world scenarios, even if they are not explicitly listed or advertised as a feature.

REFERENCES

Adams, I. T. (2024). Large Language Models and Artificial Intelligence for Police Report Writing. CrimRxiv. https://doi.org/10.21428/cb6ab371.779603ee.

Altshuller, G. S. (1984). *Creativity as an Exact Science*, 1st edn. CRC Press, Boca Raton, FL (Originally published by Altshuller G. & Shapiro R. in 1956 in Russian) https://doi.org/10.1201/9781466593442.

Bharathi Mohan, G., Prasanna Kumar, R., Vishal Krishh, P., Keerthinathan, A., Lavanya, G., Meghana, M. K. U., Sulthana, S., & Doss, S. (2024). An analysis of large language models: their impact and potential applications. *Knowledge and Information Systems*, 66(9), 5047–5070.

Chu, Z., Wang, S., Xie, J., Zhu, T., Yuan, Y., Ye, J., Zhong, A., Hu, X., Liang, J., Yu, P. S., & Wen, Q. (2025). LLM Agents for Education: Advances and Applications. arXiv. https://arxiv.org/abs/2503.11733.

Chulvi, V., González-Cruz, M. C., Mulet, E., & Aguilar-Zambrano, J. (2013). Influence of the type of idea-generation method on the creativity of solutions. *Research in Engineering Design*, 24(1), 33–41. https://doi.org/10.1007/s00163-012-0134-0.

Eberle, B. (1996). *Scamper: Games for Imagination Development.* Prufrock Press, Waco, TX.

European Commission. (2001). Directorate-General for the Information Society and Media, ISTAG—Scenarios for Ambient Intelligence in 2010. Publications Office. https://op.europa.eu/en/publication-detail/-/publication/42f2b572-64c5-4ef3-b9af-2d4de887a254

Europol. (2023). *ChatGPT–The Impact of Large Language Models on Law Enforcement.* Publications Office of the European Union. https://data.europa.eu/doi/10.2813/255453.

Fan, A., Gokkaya, B., Harman, M., Lyubarskiy, M., Sengupta, S., & Yoo, S. (2023). Large language models for software engineering: survey and open problems. In: *IEEE/ACM International Conference on Software Engineering: Future of Software Engineering (ICSE-FoSE)*, Melbourne, Australia, pp. 31–53. https://doi.org/10.1109/ICSE-FoSE59343.2023.00008.

Ferrag, M. A., Alwahedi, F., Battah, A., Cherif, B., Mechri, A., & Tihanyi, N. (2024). Generative Ai and Large Language Models for Cyber Security: All Insights You Need. https://dx.doi.org/10.2139/ssrn.4853709.

Hou, X., Zhao, Y., Liu, Y., Yang, Z., Wang, K., Li, L., Luo, X., Lo, D., Grundy, J., & Wang, H. (2024). Large language models for software engineering: a systematic literature review. *ACM Transactions on Software Engineering and Methodology*, 33, 8. https://doi.org/10.1145/3695988.

Kazemitabaar, M., Hou, X., Henley, A., Ericson, B. J., Weintrop, D., & Grossman, T. (2024). How novices use LLM-based code generators to solve CS1 coding tasks in a self-paced learning environment. In: *Proceedings of the 23rd Koli Calling International Conference on Computing Education Research (Koli Calling '23),* Koli, Finland. https://doi.org/10.1145/3631802.3631806.

Kline, S. J., & Rosenberg, N. (2009). An overview of innovation. N. Rosenberg (Ed.), *Studies on Science and the Innovation Process*. World Scientific, Singapore, pp. 173–203. https://doi.org/10.1142/7306.

Lai, J., Gan, W., Wu, J., Qi, Z., & Yu, P. S. (2024). Large language models in law: a survey. *AI Open*, 5, 181–196. https://doi.org/10.1016/j.aiopen.2024.09.002.

Lee, J., Stevens, N., & Han, S. C. (2025). Large language models in finance (FinLLMs). *Neural Computing and Applications*. https://doi.org/10.1007/s00521-024-10495-6.

Liedtka, J. (2015). Design thinking. *Journal of Product Innovation Management*, 32, 925–938. https://doi.org/10.1111/jpim.12163.

Lu, Y., Yao, B., Gu, H., Huang, J., Wang, J., Li, Y., Gesi, J., He, Q., Li, T. J. J., & Wang, D. (2025). UXAgent: a System for Simulating Usability Testing of Web Design with LLM Agents. https://arxiv.org/abs/2504.09407.

Maple, C., & Sabuncuoglu, A. (2024). *The Impact of Large Language Models in Finance: Towards Trustworthy Adoption*. The Alan Turing Institute, Finance & Economics Programme. https://www.turing.ac.uk/sites/default/files/2024-06/the_impact_of_large_language_models_in_finance_-_towards_trustworthy_adoption_1.pdf.

Masson, D., Malacria, S., Casiez, G., & Vogel, D. (2024, May). Directgpt: a direct manipulation interface to interact with large language models. In: *Proceedings of the 2024 CHI Conference on Human Factors in Computing Systems*, Honolulu, Hawaii, pp. 1–16. https://doi.org/10.1145/3613904.3642462

McDonald, N., Johri, A., Ali, A., & Hingle, A. (2024). Generative Artificial Intelligence in Higher Education: Evidence from an Analysis of Institutional Policies and Guidelines. https://arxiv.org/abs/2402.01659.

Meskó, B. (2023). The impact of multimodal large language models on health care's future. *Journal of Medical Internet Research*, 25, e52865. https://doi.org/10.2196/52865.

Milano, S., McGrane, J. A., & Leonelli, S. (2023). Large language models challenge the future of higher education. *Nature Machine Intelligence*, 5, 333–334. https://doi.org/10.1038/s42256-023-00644-2.

Murphy, L. R., Daly, S. R., & Seifert, C. M. (2023). Idea characteristics arising from individual brainstorming and design heuristics ideation methods. *International Journal of Technology and Design Education*, 33(2), 337–378. https://doi.org/10.1007/s10798-021-09723-0.

Nam, D., Macvean, A., Hellendoorn, V., Vasilescu, B., & Myers, B. (2024). Using an LLM to help with code understanding. In: *Proceedings of the IEEE/ACM 46th International Conference on Software Engineering (ICSE '24)*, Lisbon, Portugal, pp. 1–13. https://doi.org/10.1145/3597503.3639187.

Omar, M., Nadkarni, G. N., Klang, E., & Glicksberg, B. S. (2024). Large language models in medicine: a review of current clinical trials across healthcare applications. *PLOS Digit Health,* 3(11), e0000662. https://doi.org/10.1371/journal.pdig.0000662.

Osborne, A. F. (1953). *Applied Imagination; Principles and Procedures of Creative Thinking*. Scribner & Son, New York City, NY.

Patil, R., & Gudivada, V. (2024). A review of current trends, techniques, and challenges in large language models (LLMs). *Applied Sciences*, 14(5), 2074. https://doi.org/10.3390/app14052074

Paulus, P. B., & Kenworthy, J. B. (2019). Effective brainstorming. In: P. B. Paulus & B. A. Nijstad (Eds.), *The Oxford Handbook of Group Creativity and Innovation*. Oxford University Press, Oxford, pp. 287–305. https://doi.org/10.1093/oxfordhb/9780190648077.013.17.

Peláez-Sánchez, I. C., Velarde-Camaqui, D., & Glasserman-Morales, L. D. (2024). The impact of large language models on higher education: exploring the connection between AI and Education 4.0. *Frontiers in Education*, 9, 1392091. https://doi.org/10.3389/feduc.2024.1392091.

Raiaan, M. A. K., Mukta, M. S. H., Fatema, K., Fahad, N. M., Sakib, S., Mim, M. M. J., Ahmad, J., Ali, M. E., & Azam, S. (2024). A review on large language models: architectures, applications, taxonomies, open issues and challenges. *IEEE Access*, 12, 26839–26874. https://doi.org/10.1109/ACCESS.2024.3365742.

Shah, J. J., Smith, S. M., & Vargas-Hernandez, N. (2003). Metrics for measuring ideation effectiveness. *Design Studies*, 24(2), 111–134. https://doi.org/10.1016/S0142-694X(02)00034-0.

Shi, Y., Gao, T., Jiao, X., & Cao, N. (2023). Understanding design collaboration between designers and artificial intelligence: a systematic literature review. *Proceedings of the ACM on Human-Computer Interaction, 7*, 368. https://doi.org/10.1145/3610217.

Simon, H. A. (1969). *The Sciences of the Artificial*. MIT Press, Cambridge, MA. https://doi.org/10.7551/mitpress/12107.001.0001.

Yamin, M. M., Hashmi, E., Ullah, M., & Katt, B. (2024). Applications of LLMs for generating cyber security exercise scenarios. *IEEE Access*, 12, 143806–143822. https://doi.org/10.1109/ACCESS.2024.3468914.

Yan, Y. M., Chen, C. Q., Hu, Y. B., & Ye, X. D. (2025). LLM-based collaborative programming: impact on students' computational thinking and self-efficacy. *Humanities and Social Sciences Communications* 12, 149 (2025). https://doi.org/10.1057/s41599-025-04471-1.

Zhang, G., Chong, L., Kotovsky, K., & Cagan, J. (2023). Trust in an AI versus a human teammate: the effects of teammate identity and performance on human-AI cooperation. *Computers in Human Behavior*, 139, 107536. https://doi.org/10.1016/j.chb.2022.107536.

Zhang, J., Bu, H., Wen, H., Liu, Y., Fei, H., Xi, R., Li, L., Yang, Y., Zhu, H., & Meng, D. (2025). When LLMs meet cybersecurity: a systematic literature review. *Cybersecurity,* 8, 55. https://doi.org/10.1186/s42400-025-00361-w.

Zhang, K., Meng, X., Yan, X., Ji, J., Liu, J., Xu, H., Zhang, H., Liu, D., Wang, J., Wang, X., Gao, J., Wang, Y. G., Shao, C., Wang, W., Li, J., Zheng, M. Q., Yang, Y., & Tang, Y. D. (2025). Revolutionizing health care: the transformative impact of large language models in medicine. *Journal of Medical Internet Research,* 27, e59069. https://doi.org/10.2196/59069.

Zhao, H., Liu, Z., Wu, Z., Li, Y., Tianze, Y., Shu, P., Xu, S., Dai, H., Zaho, L., Jiang, H., Pan, Y., Chan, J., Zhou, Y., Mai, G., Liu, N., & Liu, T. (2024). Revolutionizing Finance with LLMs: An Overview of Applications and Insights. arXiv. https://arxiv.org/abs/2401.11641.

Zhao, Z., Song, S., Duah, B., Macbeth, J., Carter, S., Van, M. P., Bravo, N. S., Klenk, M., Sick, K., & Filipowicz, A. L. S. (2023). More human than human: LLM-generated narratives outperform human-LLM interleaved narratives. In: *Proceedings of the 15th Conference on Creativity and Cognition (C&C '23),* Virtual, pp. 368–370. https://doi.org/10.1145/3591196.3596612.

Zheng, Z., Ning, K., Zhong, Q., Chen, J., Chen, W., Guo, L., Wang, W., & Wang, Y. (2025). Towards an understanding of large language models in software engineering tasks. *Empirical Software Engineering*, 30, 50 (2025). https://doi.org/10.1007/s10664-024-10602-0

Zhou, Z., Li, Y., & Yu, J. (2024). Exploring the application of LLM-based AI in UX design: an empirical case study of ChatGPT. *Human–Computer Interaction*, 1–33. https://doi.org/10.1080/07370024.2024.2420991.

Index

V

W

X

Y